Teleconnecti
understanding
over relative
contributions Books in the field, which together provide the first comprehensive
review of this important subject.

The main focus of the book's attention is on the event referred to as ENSO (El
Niño–Southern Oscillation). Both the scientific basis for teleconnections and the
environmental (ecological and societal) impacts of such linkages are examined.
Experiments using coupled atmospheric–oceanic general circulation models, as well as
empirical evidence in support of 'observed' connections, are discussed. The book also
reviews some of the important societal impacts of previous climate anomalies
allegedly connected to ENSO. The implications for science and society of forecasts of
future climate anomalies, potentially made possible on the basis of teleconnections,
are also addressed. The support of the United Nations Environment Programme,
through the Working Group on the Societal Impacts of ENSO Events, has been
invaluable in the creation of this volume.

This book will be of importance to all professional scientists and researchers in
climatology and meteorology, particularly those concerned with air–sea interactions
and their environmental impacts and the physical basis for and societal responses to
forecasting. Graduate students in environmental science, meteorology, and climate-
related impact assessments will also find the book useful.

Teleconnections linking worldwide climate anomalies

scientific basis and societal impact

Teleconnections linking worldwide climate anomalies

scientific basis and societal impact

Edited by
MICHAEL H. GLANTZ
National Center for Atmospheric Research

RICHARD W. KATZ
National Center for Atmospheric Research

NEVILLE NICHOLLS
Bureau of Meteorology, BMRC

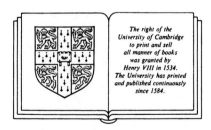

The right of the
University of Cambridge
to print and sell
all manner of books
was granted by
Henry VIII in 1534.
The University has printed
and published continuously
since 1584.

CAMBRIDGE UNIVERSITY PRESS

Cambridge
New York   Port Chester
Melbourne   Sydney

CAMBRIDGE UNIVERSITY PRESS
Cambridge, New York, Melbourne, Madrid, Cape Town, Singapore, São Paulo, Delhi

Cambridge University Press
The Edinburgh Building, Cambridge CB2 8RU, UK

Published in the United States of America by Cambridge University Press, New York

www.cambridge.org
Information on this title: www.cambridge.org/9780521106849

First published 1991
This digitally printed version 2009

A catalogue record for this publication is available from the British Library

Library of Congress Cataloguing in Publication data

Teleconnections linking worldwide climate anomalies /
    edited by M. Glantz, R. Katz, N. Nicholls.
        p.   cm.
    Includes index.
    ISBN 0-521-36475-2
    1. Climatic changes.   2. Southern oscillation.   3. El Niño
Current.   I. Glantz, Michael H.   II. Katz, Richard W.
III. Nicholls, N. (Neville)
QC981.8.C5T45 1991
551.69–dc20   90-47885   CIP

ISBN 978-0-521-36475-1 hardback
ISBN 978-0-521-10684-9 paperback

Dedication

to

Jerome Namias

*a pioneer in long-range weather and climate forecasting,
whose efforts helped to spark a renewed surge of interest in
air–sea interaction and climate fluctuations.*

Contents

Part IV: **Regional impacts of climate anomalies: environmental and societal impacts**

Part V: **Implications for ENSO forecasts**

1

Introduction

MICHAEL H. GLANTZ

Environmental and Societal Impacts Group
National Center for Atmospheric Research*
Boulder, CO 80307

Teleconnections, the linkages over great distance of seemingly disconnected weather anomalies, have been identified through the appearance of geophysical processes, through statistical correlations (in space and in time) and through recognition that many atmospheric processes are manifested as waves.

The earliest known use of this term was by Ångström (1935), who wrote an article of the teleconnections related to an atmospheric "seesaw" mechanism, since referred to as the North Atlantic Oscillation (van Loon and Rogers, 1978), although the notion of teleconnection had been recognized decades earlier. The term became very popular in articles about the societal and environmental impacts of worldwide weather anomalies that have been associated with the 1972–73 El Niño. An El Niño event can generally be identified with the invasion of warm water from the western equatorial Pacific into the central and/or eastern equatorial Pacific Ocean, in conjunction with a cessation of upwelling of cold water along the equator.

A hundred years ago one of the major concerns about El Niño events was their impacts on Peruvian guano production. Guano, produced by sea birds on the rock islands along the coast of Peru, is a highly valued fertilizer. During El Niño events biological productivity in the upwelling region along the coast of Peru is greatly reduced. As a direct result, there was a sharp increase in mortality within the bird colonies. Birds in search of their staple food, anchoveta, could not find enough to stay alive and perished in the search, only to wash up along the shore by the million. Thus, with an El Niño, Peruvian guano production and guano exports declined. Other local impacts were also noted during El

* The National Center for Atmospheric Research is sponsored by the National Science Foundation.

Niño events: heavy rains in normally arid parts of western South America, mudslides, damage to infrastructure, the appearance of fish species more attuned to warmer sea surface temperatures, a dislocation within the artisanal fishing communities with the disappearance of traditionally fished species. In sum, El Niño was in general seen as a local problem with local impacts.

In the 1950s and 1960s, fishmeal, a highly valued and highly sought after animal feed supplement, became an important industry within the Peruvian economy, generating about one-third of its foreign exchange. With the rapid development of this industry, interest in El Niño events became heightened. The major El Niño of 1972–73 and a recruitment failure, together with overfishing, devastated the anchoveta fishery and the Peruvian fishmeal industry, and caused disruptions in the lucrative fishmeal export market. In addition to these local impacts, droughts, floods, and food shortages around the world were also blamed on the 1972–73 El Niño event. Since then the scientific community has deepened its interest in understanding the El Niño phenomenon.

Today, El Niño has been overshadowed by the broader concept of El Niño–Southern Oscillation (ENSO), the linking together of oceanic and atmospheric changes in the Pacific Ocean region. In January 1985, an international scientific research program called TOGA (Tropical Ocean–Global Atmosphere) was established to improve the understanding of the interannual variability in climate that results from the interaction between the tropical ocean and the atmosphere (NAS/NRC, 1983). The Southern Oscillation is basically, as Trenberth notes (Chapter 2) "a see-saw in atmospheric mass involving exchanges of air between eastern and western hemispheres ... with centers of action located over Indonesia and the tropical South Pacific Ocean." ENSO events are now acknowledged to have global implications. The study of the ENSO process has led to the identification of "warm" events and "cold" events. These are often identified by monitoring the Southern Oscillation Index (the difference between sea level pressures at Darwin, Australia and Tahiti), although other indices are sometimes used. High negative values of the SOI indicate a warm event and high positive values indicate a cold event (also referred to as La Niña). However, it is important to note that there is not a one-to-one correspondence between the occurrence of Southern Oscillation events and El Niño events.

Observers have noted that there have been weather anomalies, both within and outside the tropics, that appear to have occurred in association with the warm events. It has been suggested that these anomalies may be correlated, or linked (that is, "teleconnected") to ENSO events. ENSO events have been associated with droughts in Indonesia, India, the Philippines, Australia, Northeast Brazil, Ethiopia, and southern Africa, as well as other regions. They have been also been associated with floods in southern Brazil, Peru, and Ecuador, warm winters in the U.S.S.R., mild winters in eastern Canada, and cool summers in northeast China. ENSO events have been blamed for contributing to the collapse of the Peruvian anchoveta fishery, the rise of the Chilean sardine fishery, and the collapse of the salmon fishery off the coast of the U.S. Pacific Northwest. They have also been blamed for encephalitis outbreaks in the U.S. Northeast, the increase in the incidence of plague in the U.S. Southwest, various disease outbreaks in South America and in Australia, and so forth. This is not to say that each of these claims has strong scientific support. One cannot blame every societal ill on the ENSO phenomenon. Clearly, its impacts in the high latitudes do not occur as reliably as in the tropical Pacific. Some of these "associated" temperature and rainfall anomalies are depicted in Figs. 1.1 and 1.2 (adapted from WMO, 1984).

After the 1982–83 ENSO event, interest in teleconnections associated with ENSO took on heightened levels. This interest broadened in scope to include extratropical teleconnections. It is now widely appreciated that the ENSO phenomenon is a global-scale phenomenon that now holds out to societies in many parts of the world the prospect of the forecasting of its occurrence and concomitant climate anomalies some months in advance.

It does not take much effort to show how information about ENSO can benefit societies in many countries around the world. Some of those countries are only just beginning to realize the importance of this phenomenon to the management of their national, regional and local economies. The highest level of government in Ethiopia, for example, has instructed its national meteorological service to inform it about ENSO events so that it might have some forewarning about the possible development of drought and, therefore, famine conditions.

Even those countries for which it is less clear that their economic activites might be directly affected by ENSO events are showing an

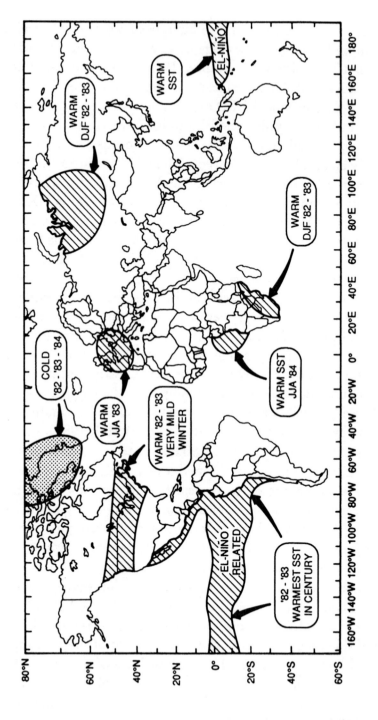

Fig. 1.1 Selected extreme temperature events that persisted for a season or longer in the 1982–84 period.

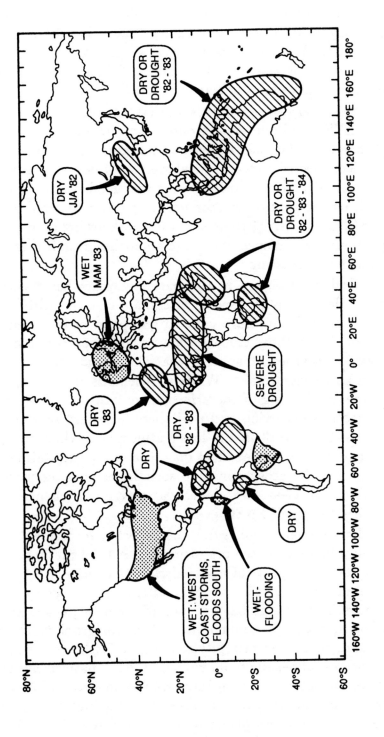

Fig. 1.2 Selected extreme continental precipitation (wet and dry areas) that persisted for a season or longer in the 1982–84 period.

increased interest in the phenomenon. While Kenya, for example, may not be directly affected by ENSO, it does grow crops that compete for sale in the international marketplace. It is, therefore, not so difficult to show why it is in Kenya's interest to know how its competitors in the coffee trade – Brazil, Ethiopia, Southeast Asian countries – have been affected by ENSO.

Of importance to decisionmakers is that an improved understanding of ENSO and related teleconnections can provide a forecast tool that could be used at least to mitigate the impacts on society of teleconnected meteorological events. At best such a forecast tool can aid decisionmakers in coping with food shortages, adverse health effects, and other economic impacts associated with recurrent ENSO events. There is also value in knowing about the contemporaneous occurrence of climate anomalies even if there is no lead time for forecasting.

A great deal of attention in the past few years has focused on the scientific and policy aspects of climate change. Although tens of millions of dollars have gone to groups involved in monitoring or predicting those climate changes that may result from anthropogenic activities, it appears that the global warming issue still remains controversial. While all scientists seem to agree that an increased loading of the atmosphere with radiatively active trace gases is occurring and that these gases are producing a "greenhouse-like effect" in the lower atmosphere, they do not agree on what the greenhouse effect may mean for precipitation in specific regions or for ecosystems and societies around the world. Nor is it yet at all clear how global warming might change the frequency or intensity of ENSO events.

Global climate change will surely be reflected in regional climates. Today there are attempts, using a variety of methods, to ascertain what the impacts of a global climate change will be at the regional and local levels. Many countries are engaged in research to identify how a global warming of a few degrees Celsius might affect their coastal communities, their vulnerability to severe storms, to changes in precipitation, and so forth. Yet, we are still in the process of seeking to understand today's natural climate variability on the inter- as well as intra-annual time scales. We are searching for such information in order to better understand the interactions between atmospheric processes and societies. Thus, in a sense, attention to these climate and global change issues has served to underscore how much more we need to know about our present day global and regional climate regimes. Thus, efforts by physical and social scientists to understand such a global phenomenon as ENSO should be strengthened because ENSO affects so many physical and societal processes and it holds out promise for forecasting climate anomalies on a worldwide basis. The editors and contributors to this volume have, therefore, chosen to focus on the concept of teleconnections.

For the most part the concept of teleconnections relates to climate anomalies around the globe that have been associated with ENSO events in the Indian Ocean–Pacific Ocean field of action. However, because of the recent interest in forecasting droughts in the countries bordering the Atlantic Ocean (especially in Africa), a chapter has been included on possible teleconnections to changes in the difference between sea surface temperatures in the northern and southern Atlantic Oceans. This chapter, in addition to shedding light on this particular teleconnection, serves to remind the readers that other regional teleconnections, not particularly linked to ENSO events, have been suggested to exist.

The contributions in this book have been divided into five parts: Introduction to Teleconnections, Regional Case Studies of Teleconnections: Physical Aspects, A Scientific Basis for Teleconnections, Regional Impacts of Climate Anomalies: Environmental and Societal Aspects, and Implications for ENSO Forecasting.

In the introductory part Trenberth discusses the general characteristics of the ENSO phenomenon providing an outline of "the main physical mechanisms leading to the coupled atmosphere–ocean interannual variability ...." More specifically he examines the mean annual cycle and observed interannual fluctuations involving the sea surface temperatures, atmospheric winds, surface convergence, and links with convection and rainfall, along with how they may contribute to the different character of each El Niño event.

The regional case studies part (physical aspects) contains six case studies from around the globe. Chu's chapter focuses on assessing the level of strength of the linkages between rainfall in various parts of Brazil (the Nordeste, the Amazon Basin and southern Brazil) and ENSO. The teleconnection to the drought-plagued Nordeste, as Chu points out, was noted by Walker in the mid-1920s. Chu also reviews similar linkages (or teleconnections) with circulation anomalies over the Atlantic Ocean and addresses questions relating to the use of diagnostic relationships in long-range forecasting.

Australian social scientists as well as policymakers have become increasingly aware of the impact of ENSO events on their environment and society, especially following the severe 1982–83 event. In his chapter, Allan presents an historical account of scientific interest in teleconnections between ENSO as well as anti-ENSO (i.e., cold) events and climate anomalies in the Australasia region. He describes several of the significant correlation patterns that have been investigated between ENSO and rainfall, wind patterns, mean sea level pressure, sea level, sea surface temperatures, air temperatures and cloudiness. Allan also addresses the prospects for using regional teleconnections in long-range forecasting.

Barnett and colleagues focus their chapter on the connections between Asia snow cover, monsoons and ENSO. As the authors note, the connection between Asia snow cover and the monsoons in Asia was suggested as early as 1884 by Blanford. This chapter discusses using sophisticated numerical models to show "... the hypothesis that snow-induced changes in the monsoon are significant ..." and that "changes in the monsoon can affect the SST in the tropical Pacific and hence induce ENSO events ... ." Note that this is an instance in which the climate anomaly leads ENSO rather than vice versa. They also include a brief section on some of the problems associated with their assessment.

In their chapter, Lau and Sheu assess teleconnections in global rainfall anomalies in order to determine whether major precipitation anomalies in geographically separated regions occur independently (i.e., simultaneously only by chance) or whether "... they are the manifestation of a global scale coherent shift in rainfall patterns that is likely to recur in the future." Such a study, they assert, is important for scientific as well as for societal reasons.

Gray and Sheaffer provide geophysical and statistical evidence for teleconnections between El Niño events and tropical cyclone activity in the Atlantic–Caribbean basin. They conclude that "seasonal Atlantic hurricane activity during El Niño years has been much suppressed in comparison with hurricane activity occurring during non-El Niño years." They provide observational evidence for mechanisms to explain the linkages between El Niño and seasonal hurricane activity in the western parts of the Atlantic Ocean.

event evolved differently (Rasmusson and Wallace, 1983). Unusually warm water first developed in the western and central Pacific during 1982 and subsequently spread eastward. Also the timing relative to the annual cycle of the 1986–87 event was rather unusual (Trenberth, 1989). Possible reasons for these differences in evolution are examined in the following sections.

The opposite phase of the SO and EN has attracted less attention, presumably because it appears more like an extreme case of the normal pattern with a cold tongue of water along the equator, but the "Cold Event" in 1988, which has been linked to the extensive, severe, and persistent 1988 North American summer-time drought (Trenberth et al., 1988), has been widely publicized and has led to an acceptance of the term "La Niña"* for the colder than normal tropical Pacific SSTs. Van Loon and Shea (1985) have documented the earlier historical Cold Events.

The SO has effects globally even though its main centers of action are in the tropics. The links to higher latitudes are less clear in Fig. 2.2 because of seasonal changes in the teleconnections. The teleconnections tend to be strongest in the winter of each hemisphere and feature sequences (wavetrains) of high and low pressures accompanied by distinct waves in the jet stream (Trenberth and Paolino, 1981; (see Chapters 9 and 10 by Tribbia and by Rasmusson, this volume). Ropelewski and Halpert (1987, 1989) have updated the consistent rainfall relationships associated with ENSO and La Niña found by Walker and Bliss, as shown schematically in Fig. 2.4. Thus, during ENSO events, islands in the central tropical Pacific dry zone and Peru and Ecuador experience excessive rains while Indonesia, Australia, southern Africa and Northeast Brazil experience dry conditions. At the same time, changes in oceanic conditions have disastrous consequences for the fishing and guano industries along the South American coast.

In the following sections we introduce the main physical features of importance to the ENSO phenomenon and outline the mechanisms for the coupled variations of the atmosphere–ocean. The mean annual cycle sets the stage for considering the interannual variations. We begin with the broad view and discuss the

---

* La Niña ("the girl" in Spanish, opposite of El Niño, the boy) was introduced by Philander (1985) to replace a previously used term for Cold Events, "anti-El Niño."

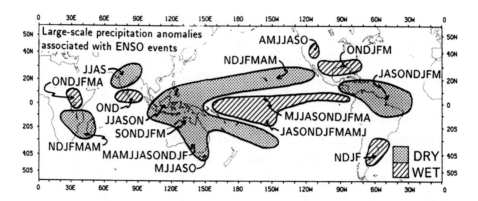

Fig. 2.4 Schematic of areas with a consistent ENSO precipitation signal. For
each region a listing is given of the months during which it is wetter
or dryer than normal, in each case beginning in the ENSO year.
Adapted from Ropelewski and Halpert (1987) and Rasmusson (1985).

overall general mechanisms playing a role, and to go to increased
complexity as we consider how the system evolves.

The mean annual cycle

The existence of the ENSO phenomenon is dependent on the east–
west variations in SSTs (Fig. 2.5) in the Pacific, the close links
between SSTs and surface pressures and thus surface winds in the
tropics, and the connections with the major areas of rainfall and
the direct thermal circulations that dominate the tropical atmosphere on time scales beyond a few weeks.

The very distinctive tropical SST patterns (Figs. 2.5 and 2.6)
in the Pacific feature an extensive warm pool in the west, with
the warmest surface water anywhere on the globe extending to
depths of more than 100 m, and a cold tongue in the east near the
equator that is strongest about October and weakest in March–
April. These SST patterns are brought about by the pattern of
winds and strong solar radiation in the tropics. In the west, the
relatively light winds allow solar heating to warm the ocean and
minimize heat loss from evaporation while water tends to pile up
from the surface westward flowing ocean currents driven by the
strong trade winds further east. At the same time the southeast
trade winds from the Southern Hemisphere cause upwelling of cold
water from below along the coast of South America and along the
equator in the eastern Pacific. The result is an upward slope in

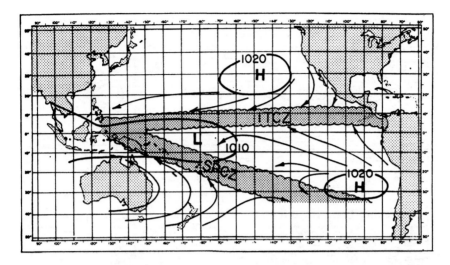

Fig. 2.9 Schematic view of the main convergence zones, the ITCZ and SPCZ, along with the annual mean sea level pressure contours and surface wind streamlines.

ITCZ is weakest in January in the NH when the SPCZ is strongest in the SH.

As the trade winds blow westward, down the pressure gradient, the air is warmed and moistened, thereby providing fuel for the thermally driven east–west circulation; note the "dry zone" along the equator in the eastern Pacific in Figs. 2.5 and 2.10, coinciding with the cold tongue evident in SSTs. The warm moist air rises and latent heat is released aloft by the precipitation in the extensive convection in the west. A return eastward flow in the upper atmosphere provides the link to the dry cool air sinking in the east. Bjerknes (1969) hypothesized that this Walker circulation was driven by the SST gradient across the Pacific.

However, as can perhaps be inferred from Figs. 2.5 and 2.10, the Walker circulation is but part of a fairly complex thermally direct circulation in the tropics involving overturnings in all directions; see Fig. 2.11. The latter clearly shows the main outflows coming from the tropical western Pacific and the seasonal migration. Generally the flow is from heat source to heat sink. The east–west cells are not just confined to the equator but are much broader in extent and are intertwined with the north–south component known as the Hadley cell. The strong meridional component northwards in DJF and southwards in JJA depicts the upper branch of the

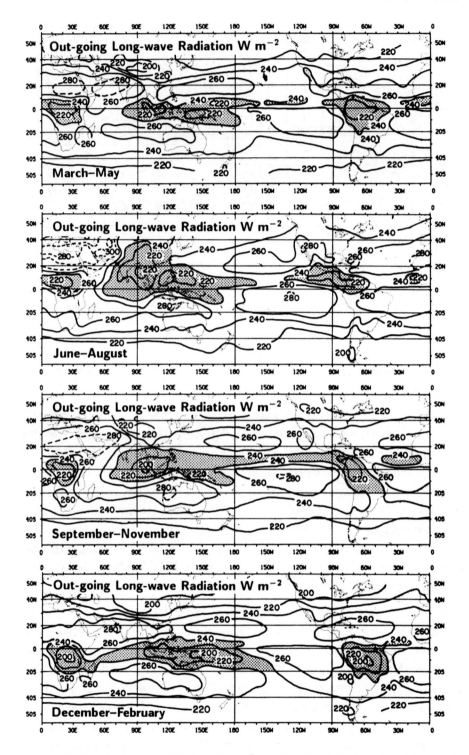

Fig. 2.10 Seasonal mean OLR in W/m², as in Fig. 2.5; but for (a) MAM, (b) JJA, and (c) SON, (d) DJF.

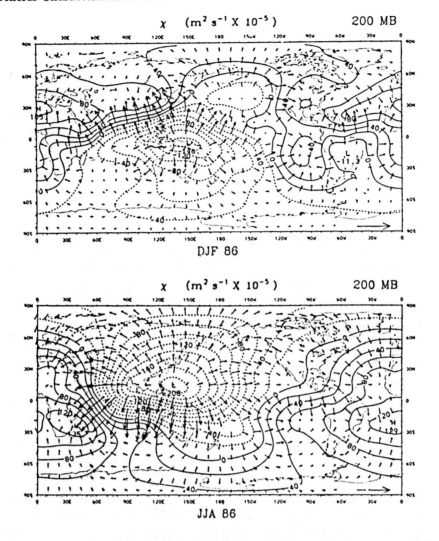

Fig. 2.11 Seasonal mean velocity potential and divergent component of the
wind vectors at 200 mb for (a) December–February 1985–86; and (b)
June–August 1986 which depict fairly normal conditions. The vector
plotted at lower right represents 10 m/s.

Hadley cell, which is strongest in the winter hemisphere. The
upward branch of the Hadley cell is closely linked to the ITCZ.

Nevertheless, it can be seen that we have described a situation
in which the atmospheric winds are largely responsible for the
tropical SST distribution which, in turn, is very much involved in
determining the tropical atmospheric circulation. That is to say,
the atmosphere and ocean are very strongly coupled in the tropics.

Interannual variations

Atmospheric links to sea surface temperatures in the tropics

The local atmospheric response to changes in SSTs depends greatly on the relationship between SST and convection, which is not as simple as some studies imply. Early suggestions of a simple linear relationship clearly do not hold up since the largest SST anomalies form in the eastern Pacific, in the cold tongue area, while the largest rainfall anomalies typically occur near the warmest water, which is near the date line in most ENSO events. Other suggestions that heavy rainfall occurs over SSTs greater than about 28°C (Gadgil et al., 1984; Graham and Barnett, 1987) do not explain all the observations because, although deep convection certainly occurs in such regions, there are times when and places where it does not. Modeling studies (e.g., Shukla and Wallace, 1983) reveal that precipitation anomalies are primarily associated with low level moisture convergence rather than local evaporation anomalies.

Accordingly, the question reduces to why the low level convergence occurs where it does and whether this relates to SSTs. One factor is the nonlinearity in the Clausius–Clapeyron equation which expresses the saturation mixing ratio r_s as a function of temperature. At 1000 mb for SSTs of 20° to 30°C, r_s changes from 15 to 28 g/kg. Moreover, the increase in r_s per degree increase in SST at 30°C is 1.7 times that at 20°C. Because the low troposphere is well mixed, the time mean moisture and moist static energy are closely tied to the absolute SST values and, therefore, low level convergence and convection are favored over regions of warmer water (Neelin and Held, 1987). In addition, SST gradients cause a direct, essentially hydrostatic, change in surface pressures which, in turn, drive the low level winds and convergence and thus the distribution of convection and rainfall. Evidently the latent heat release in the mid and upper troposphere is less important in this respect (Lindzen and Nigam, 1987).

Therefore on time scales beyond a few weeks, the atmospheric boundary layer convergence, which leads to the organized convective areas, preferentially tends to occur over the warmest region, which may be over land on some occasions (especially in the Indian Ocean–Asian monsoon region). Also, because convection cannot occur everywhere at once, the result is effectively a competition within the atmosphere for where the convergence will occur. Because the warmest observed water is ~29°C (Fig. 2.6) only regions where SSTs exceed 28°C can compete. But the whole pattern of SST, including the gradients and absolute values, as well as the proximity of major land masses, are all factors as important as the actual SST anomaly in determining the rainfall anomalies (Trenberth, 1989).

One reason why each ENSO event has its own character and develops in a somewhat different way relates to the above factors. Because SST gradients are weak in the tropical western Pacific (Fig. 2.5), fairly small changes in SST can have profound implications, as the region of warmest water shifts from one region to another perhaps thousands of kilometers away. This changes the favored location of the convergence zones, the ITCZ and the SPCZ. Observations show that during ENSO events, the ITCZ tends to move south and the SPCZ moves northeast (Trenberth, 1976; Heddinghaus and Krueger, 1981; Liebmann and Hartmann, 1982), with the result that the convergence zones come together in the western Pacific and widespread convection often extends east of the date line in the vicinity of the equator, eroding the dry zone but, to a large extent, following and lying just west of the region of warmest water. The Hadley circulation is enhanced at the expense of a weakened Walker circulation.

In the extreme 1982–83 event, warm water in excess of 28°C spread all the way across the Pacific near the equator, forming anomalies of over 4°C. It was followed by the convection which formed what might be interpreted as one massive convergence zone incorporating both the SPCZ and ITCZ. During the 1986–87 event the warmest water was displaced eastward to near 170°W and even though the SST anomalies there were a modest 1°C or so, it was sufficient to produce ensuing strong convection in that area (Trenberth, 1989). Figure 2.12 shows the 200 mb outflow patterns in DJF 1986–87 and, contrasting this with Fig. 2.11, reveals the change in divergent winds. Note the distinct eastward shift in the center of divergent outflow from north of Australia to about the dateline compared with the year earlier. The changes in 1986–87 are more typical of most ENSO events.

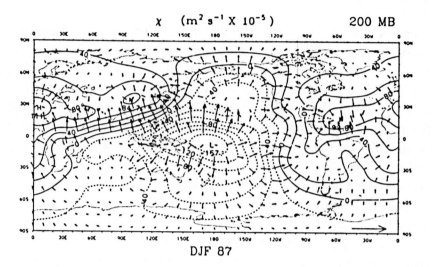

χ $(m^2 s^{-1} \times 10^{-5})$ 200 MB

DJF 87

Fig. 2.12 Seasonal mean velocity potential and divergent component of the wind vectors at 200 mb for December–February 1986–87. As the 1986–87 ENSO developed after August 1986 the contrast of this with Fig. 11a shows the change in upper level divergence accompanying the 1986–87 ENSO. The vector plotted at lower right represents 10 m/s.

Atmosphere–ocean coupling

Lau (1985) and Kang and Lau (1986) have specified the observed SSTs from 1962 to 1976 in the tropical Pacific as lower boundary conditions in a General Circulation Model (GCM) simulation and were able to reproduce aspects of the observed SO variations, provided fluctuations on time scales of less than about three months were ignored. This provides strong support for the view that most of the atmospheric pressure patterns of the SO and associated surface winds are the result of the SST changes. Seager (1989) has been able to reproduce many of the gross aspects of the tropical SST variations from 1970 to 1987 by specifying the observed surface wind forcing on the model ocean. Therefore, it is also established that the SST changes are primarily caused by changes in the atmospheric winds.

Bjerknes recognized that the coupling between atmosphere and ocean provided a positive feedback system and that the interactions between the atmosphere and the ocean were the key to ENSO variations. Philander et al. (1984) emphasized the potentially unstable nature of air–sea interactions in the tropics so that even modest anomalies could amplify temporally and spatially to affect the entire tropical Pacific Ocean and global atmosphere. Zebiak and Cane (1987) encapsulate the basic ideas of Bjerknes into a simple atmosphere–ocean coupled model that successfully reproduces an oscillation which has some features of those observed (see Chapter 11 by Cane).

A key to whether the tropical ocean–atmosphere system can develop important instabilities is the extent to which the two can develop in concert. In low latitudes, the Coriolis effects of rotation are small and thus the oceanic adjustment time to a change in surface winds is a few weeks or months. This is in contrast to higher latitudes where the oceanic response is much more sluggish, effecting a decoupling between the atmosphere and the ocean.

As an example, if the SSTs were to increase in the east, for whatever reason, thereby reducing the SST gradient, then there should follow a corresponding decrease in the atmospheric Walker circulation as the warmer, moister atmosphere leads to falls in the surface pressures over the warmer water. The resulting decrease in surface trade winds leads directly to a decrease in oceanic upwelling along the equator and a change in the ocean currents and thus in advection of warm water, further increasing the SSTs in the east. The associated changes in the atmospheric pressure gradient, although most intense locally, are manifest globally as a swing in the Southern Oscillation. Even though the weakened trade winds would reduce evaporation into the atmosphere, the higher SSTs offset this and the atmosphere warms while the tropical ocean cools. In very broad terms, this is what happens when El Niño conditions develop.

As an alternative to the above, changes in the winds may set the loop off, much as first described by Wyrtki (1975). In particular, changes in the winds in the western Pacific appear to be capable of generating Kelvin waves* in the equatorial ocean (Knox and Halpern, 1982; Eriksen et al., 1983; Cane, 1983) that propagate eastwards to reach the South American coast in just a few months and are associated with a deepening of the thermocline and a surface warming in the eastern Pacific as warmer water is

---

* These are not surface waves normally thought of in conjunction with the ocean but internal modes. See Chapter 11 for a discussion of Kelvin wave theory.

upwelled (see also Barnett, 1977; Luther et al., 1983; Harrison and Schopf, 1984). The reduced surface wind stress results in a drop in sea level over the western Pacific and a rise in sea level by several tens of centimeters in the east and along the coastal Americas causing coastal erosion problems. At the same time, changes in winds affect the ocean currents and thermal advection, and thus the SSTs. But the details also presumably depend on the changes in the curl of the wind stress and the effects on the off-equatorial currents. During most ENSO events, the wind changes in such a way that the SEC weakens while the eastward flowing countercurrent intensifies (Philander, 1983) and thus the anomalous component advects warm water to the east and contributes to the El Niño warming. The loop is completed as the changing SSTs affect precipitation, surface atmospheric pressures and surface winds.

This positive feedback loop also works in the opposite direction. Thus if the trade winds become strong in the eastern tropical Pacific (as they did early in 1988), upwelling of cold water increases, westward surface ocean currents strengthen, and the thermocline becomes shallower, leading to colder SSTs along the equator in the eastern Pacific, increasing the SST gradient and thereby further enhancing the Walker circulation and sustaining the stronger trade winds.

Because higher pressures form over the colder upwelled water, a pressure gradient is set up to the west of the original cooling, thereby favoring development of stronger easterlies just to the west of the original perturbation. The enhanced surface wind stress results in further upwelling to the west, so that the whole coupled disturbance develops toward the west. Or, in the ENSO case, the same westward progression occurs but with essentially the opposite sign. This appears to be one possible mechanism, at least in the tropical eastern Pacific, in the canonical ENSO of Rasmusson and Carpenter (1982) accounting for the evolution of the patterns seen in Fig. 2.3. Horel (1982) suggested this mechanism as an important part of the mean annual cycle evolution of SSTs in the eastern tropical Pacific. In this case changes in SSTs are dominated by upwelling rather than advection.

On the other hand, changes in advection were apparently more prominent in causing the development of SST anomalies and the subsequent eastward spread during the 1982–83 event (Gill and Rasmusson, 1983). The anomalous currents, in turn, were caused by changes in the zonal component of the surface winds in the western Pacific. Harrison (1984) documented the changes in winds in the west in early 1982. Displacement of the warm water and thus the atmospheric convergence eastward brings with it a westerly component to the anomalous winds to the west of the convergence, increasing any anomalous eastward oceanic currents and furthering the anomalous advection, all the while sustaining the warm pool in the ocean in the west by reducing upwelling or even inducing downwelling. Suppressed evaporation in the weaker trade winds in the west also contributes to warming there (Seager, 1989). More generally, the advective processes are probably also important in other ENSO events in the western and central Pacific (Wyrtki, 1975, 1977), but their importance may well change from event to event.

Understanding the relative importance of the two competing processes of upwelling versus advection appears to be central to determining whether the El Niño progresses eastward or westward and thus in understanding the differences in evolution of different ENSO events; see Hirst (1986) for a theoretical discussion. Philander (1983, 1985) suggests that a local influence of SSTs on the atmosphere is necessary for these feedback mechanisms to work. Because the surface winds are weakest in March–May, atmospheric–ocean coupling is weak, but, as the ITCZ is at its farthest south over the Pacific at that time, it also apparently affords an opportunity for coupled anomalies to become organized and thus grow over the next six months. Apparently, coupled growth most readily occurs from July to September as the SSTs decrease seasonally in the Zebiak and Cane (1987) model. The need for fairly local interactions may explain why the ENSO depends more on the SSTs in the central and western Pacific than in the east and why there is a tendency for the timing of ENSO events to be related to the annual cycle.

Evolution of ENSO

Clearly the ocean, as a source of moisture and with enormous heat capacity, acts as the flywheel that drives the whole system. Bjerknes was not able to determine how the SO reverses or why the ENSO changes from warm to cold conditions, and this is still a subject of intense research. The previous section dealt with positive feedback the effect of which is that after an ENSO event begins, the atmosphere and ocean act in harmony. Some intriguing questions remaining are what sets the sequence off and what are the negative feedbacks and the role of other influences.

It is clear that the most intense phase of each event tends to last about a year. Although there are considerable variations in details among the different ENSO events, the change in phase preferentially occurs about March–April, which is the time of year when the trade winds and the SST cold tongue are at their weakest. The average time between subsequent ENSO events averages about four years, but varies from two to about 10 years (Trenberth, 1976).

Because Darwin is at one center of the SO, changes in Darwin sea level pressure can be used as an index of the SO and presumably the major ENSO events of the past. Figure 2.13 shows the Darwin pressure anomalies, smoothed to eliminate noise (fluctuations with periods less than about a year), beginning in 1882, when the Darwin record began. Positive values are times of ENSO events, and the 1982–83 event stands out as the largest on record.

The prominence of the SO has varied throughout this century. Spectral, cross spectral and cross correlation analyses indicate that very strong ENSO events occurred in the first 25 years of this century (see Fig. 2.13), leading Walker to document and name the phenomenon. Further strong coherent oscillations occurred after about 1950 but there were few events of note from 1925 to 1950, with the major exception of the 1939–41 event (Trenberth, 1976; Trenberth and Shea, 1987). A majority of recent studies into the ENSO phenomenon have dealt mostly with the post-1950 era (e.g., Rasmusson and Carpenter, 1983; Barnett, 1985; van Loon and Shea, 1985, 1987; Wright et al., 1988). However, the behavior of the SO during this interval is probably somewhat atypical, especially insofar as it was dominated by about a four-year period over most of the Pacific (Trenberth and Shea, 1987).

Similar kinds of analytic methods also reveal that not everything happens at the same time. As discussed earlier, SSTs often increase first along the coast of South America and the anomalies subsequently spread westwards into the central Pacific (Rasmusson and Carpenter, 1982) but in the 1982–83 ENSO the main progression appeared to be eastwards (Rasmusson and Wallace, 1983). In the atmosphere, changes in the South Pacific Ocean are

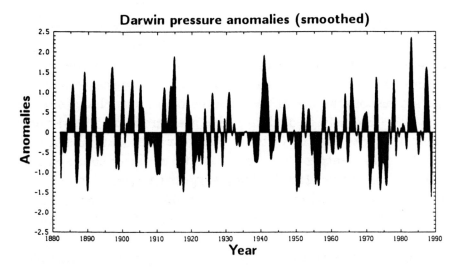

Fig. 2.13 Darwin pressure anomalies since 1882 smoothed with the 11-point low-pass filter.

the most consistent in systematically leading changes in the SO by one to three seasons, pointing to the possible role of extratropical changes in initiating changes in phase of the SO (Trenberth, 1976; Trenberth and Shea, 1987). van Loon and Shea (1985, 1987), in particular, have described southern precursors to ENSO events and have focused attention especially on the likely role of the SPCZ (von Storch et al., 1988).

Barnett (1985) offers an alternative interpretation, and for 1950–80 finds a large-scale eastward propagating feature in sea level pressure that seems to link the SO to the Southeast Asian monsoon. Meehl (1987) linked the eastward progression to a modulation of the mean seasonal cycle of convection in the Indian and west Pacific sectors. Barnett et al. (1989) showed with climate model experiments how anomalous snowfall over Asia could influence the subsequent summer monsoon, with heavier snow causing a delay in continental heating (see Chapter 6 by Barnett et al.). Moreover, they found indications that the subsequent changes in the surface winds over the Pacific could possibly trigger an El Niño event. Although most likely a factor, snow amount variations are not well observed and the overall importance of this mechanism remains to be determined.

Overall, the evidence for eastward progression is less clear (Trenberth and Shea, 1987). Figure 2.14 (from Wright et al., 1988) shows the evolution in time of pressures over broad latitudinal bands. Since these are regressions or correlations with an annual mean Darwin pressure, the patterns can take on either sign, but are depicted with the sign of ENSO events. In all three bands, pressures change first in the eastern Pacific with the eastern center of action responding about five months before the changes occur in the west. From 10 to 38°S negative pressure anomalies develop west of the date line in October the year preceding the ENSO and it was this feature that was found to be fairly consistent by Trenberth and Shea (1987). They suggest that, most often, disturbances from the SH out of the extratropics may play a key role, perhaps as they appeared to do in the 1982–83 event (Harrison, 1984).

But because the tropical Pacific coupled system is fundamentally unstable and is especially sensitive to conditions around March–April, the exact trigger may not be important. Rather, multiple possible triggers appear to exist; other possibilities include the 40–50 day Madden–Julian wave disturbances in the tropics, and cold surges from the NH (Lau, 1982). In all of these cases it is essentially the random weather "noise" that initiates events in the coupled system.

One possible negative feedback mechanism limiting ENSO events involves the notion that there is an inverse variation in the strengths of the Walker and Hadley circulations in the atmosphere, with the Hadley cell becoming relatively stronger when there is more zonal symmetry during an ENSO event and the Walker cell is weakened. Simple models implementing idealized wind changes associated with this (McWilliams and Gent, 1978; McCreary, 1983) show that oscillations can occur in the coupled atmosphere–ocean system.

Another probably important factor in the whole ENSO cycle is the amount of heat storage contained within the tropical Pacific Ocean. Wyrtki (1975, 1985) showed that the amount of warm water in the tropics, as measured by the upper layer volume between the sea level and the thermocline, builds up prior to and is then depleted during ENSO. Apparently, during the cold phase, solar radiation in the dry zone acts to heat up the ocean (Weare, 1983) and the heat is redistributed by currents, with most of it being stored in the deep and warm pool in the west. During El Niño, either heat is transported out of the tropics within the ocean toward higher latitudes in response to the changing currents

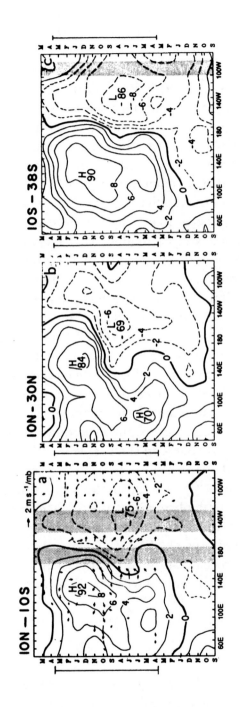

Fig. 2.14 Time-longititude plots of (a) regression coefficients, and (b) and (c) correlations (× 10) of monthly mean pressure with an annual mean Darwin pressure index for April–May (see vertical bar at side of each panel) at various leads and lags. Time runs upward beginning September prior to and ending May following the ENSO. In (a) corresponding winds for which correlations exceed 0.5 are plotted and contours are in tenths of mb anomaly per mb of Darwin pressure. Stippling shows data void regions where values are interpolated. (a) 10N–10S, (b) 10N–30N, (c) 10S–38S; from Wright et al. (1988).

and/or increased heat is released into the atmosphere mainly in the form of increased evaporation, thereby cooling the ocean (Philander and Hurlin, 1988). Strong evidence exists for a general warming of the global atmosphere a few months after a strong ENSO event (Newell and Weare, 1976; Horel and Wallace, 1981). Wyrtki (1985) therefore suggests that the time scale of ENSO is determined by the time required for an accumulation of warm water in the tropics to essentially recharge the system, plus the time for the El Niño itself to evolve as the heat is discharged toward higher latitudes and into the atmosphere. Zebiak and Cane (1987) indicate that the time to refill the tropical Pacific Ocean heat reservoir is the main determining factor in setting the time scale for the ENSO-like variations in their simple model. However, they also note that this result depends on the strength of the atmosphere–ocean coupling in the tropics and this varies seasonally.

Other theories suggest that the tropical atmosphere–ocean system may act as a natural oscillator by itself, without the need for extratropical influences. Graham and White (1988) summarize the results from many simplified atmosphere–ocean models in which the previous ENSO event sows the seeds for the development of the next one. Off equatorial Rossby modes in the ocean along with reflections off the western boundary of the tropical Pacific provide a delayed feedback necessary to produce the quasiperiodicity (Schopf and Suarez, 1988). Of course in the real world the western boundary is not solid and its capacity for reflecting Rossby waves efficiently enough to produce the results in the simple models is not certain. Nevertheless, some observational evidence does show precursive evidence for Rossby wave activity in the western Pacific associated with ENSO (White et al., 1985; Kousky and Leetmaa, 1989). Further analysis of recent and future events will be needed to resolve the roles of internal oceanic wave reflections and weather activity.

Other tropical oceans

Mean conditions in the tropical Indian Ocean are quite different, owing to the merging of the eastern Indian Ocean with the western Pacific in the vicinity of the warm pool, as can be seen in Figs. 2.5 and 2.6. The configuration of surrounding land mass also means that changes in the Indian Ocean are dominated by

the monsoons and the associated large reversals in winds seasonally. Consequently, unlike in the Pacific, there is an absence of
the east–west contrast across the Indian Ocean and El Niño-like
events do not occur. Instead, the interannual changes in the Indian
Ocean are linked to those in the Pacific since the annual migration
of monsoons is influenced by the changes in convection associated
with ENSO. Connections between the Indian monsoon and conditions in the tropical Pacific were noted in the seminal studies by
Walker early this century, and formalized by more recent studies
(e.g., Rasmusson and Carpenter, 1983).

The Atlantic tends to be dominated by the annual cycle in general. However, there was a substantial El Niño-like event in the
tropical Atlantic in 1984 with SST anomalies of 4°C, perhaps as
an aftermath of the 1982–83 ENSO event in the Pacific (see Philander, 1986, and a number of other articles in the same issue of
Nature). The changes in Atlantic SSTs clearly have a major influence on rainfall over Africa and Brazil as the position of the
ITCZ is shifted south. However, the east–west contrast along the
equator is not as large in the Atlantic as in the Pacific, and missing in the Atlantic is a feature corresponding to the SPCZ and
the eastward migration of the convection along with the changes
in SST. Aside from this, the main difference between the Atlantic
and the Pacific relates to their size, so that the warm reservoir
of the western tropical Atlantic and the SSTs there are modest
compared with the Pacific (see Fig. 2.5), and the Pacific dominates with the Southern Oscillation. In addition, adjustments to
changes in the Atlantic Ocean occur more rapidly simply because
of its smaller size.

Perhaps another important difference between the oceans lies
in the nature of their tropical western boundary. Africa lies to
the west of the Indian Ocean and South America is to the west
of the Atlantic; both feature strong migrations in convection back
and forth across the equator with the annual cycle (Fig. 2.10). In
contrast, the main convection in the tropical western Pacific lies
mostly over the ocean, thereby permitting the atmosphere and
ocean to interact more freely and the convection to respond to
changes in SSTs.

Concluding remarks

In this chapter we have described the El Niño and Southern Oscillation phenomena and the main physical mechanisms leading to
the coupled atmosphere–ocean interannual variability that has led
to the international Tropical Oceans Global Atmosphere (TOGA)
Program. The changes in tropical atmospheric circulation and,
especially, the changes in the divergent wind outflow patterns provide the origins of the teleconnections to higher latitudes as the
source of stationary Rossby waves. These aspects are dealt with
further in Chapter 9, from a theoretical and modeling viewpoint,
and Chapter 10, from an observational viewpoint. The changes in
both the tropical and extratropical circulation are responsible for
the more regional effects and impacts dealt with in the rest of the
book.

References

Barnett, T.P. (1977). An attempt to verify some theories of El Niño. *Journal of Physical Oceanography*, **7**, 633–47.

Barnett, T.P. (1985). Variations in near global sea level pressure. *Journal of the Atmospheric Sciences*, **42**, 478–501.

Barnett, T.P., Dümenil, L., Schlese, U., Roeckner, E. & Latif, M. (1989). The effect of Eurasian snow cover on regional and global climate variations. *Journal of the Atmospheric Sciences*, **46**, 661–85.

Bjerknes, J. (1969). Atmospheric teleconnections from the tropical Pacific. *Monthly Weaather Review*, **97**, 103–72.

Cane, M.A. (1983). Oceanographic events during El Niño. *Science*, **222**, 1189–95.

Cane, M.A. & Zebiak, S.E. (1985). A theory for El Niño and the Southern Oscillation. *Science*, **228**, 1085–7.

Colin, C., Hanon, C., Hisard, P. & Oudot, C. (1971). Le Courant de Cromwell dans Le Pacifique central en février, *Cahier ORSTOM Serie Oceanographie*, **9**, 167–86.

Deser, C. & Wallace, J.M. (1987). El Niño events and their relation to the Southern Oscillation. *Journal of Geophysical Research*, **92**, 14189–96.

Eriksen, C.C., Blumenthal, M.B., Hayes, S.P. & Ripa, P. (1983). Wind-generated equatorial Kelvin waves observed across the Pacific Ocean. *Journal of Physical Oceanography*, **13**, 1622–40.

Gadgil, S., Joseph, P.V. & Joshi, N.V. (1984). Ocean–atmosphere coupling over monsoon regions. *Nature*, **312**, 141–3.

Gill, A.E. & Rasmusson, E.M. (1983). The 1982–83 climate anomaly in the equatorial Pacific. *Nature*, **306**, 229–34.

Graham, N.E. & Barnett, T.P. (1987). Sea surface temperature, surface wind divergence, and convection over tropical oceans. *Science*, **238**, 657–9.

Graham, N.E. & White, W.B. (1988). The El Niño cycle: A natural oscillator of the Pacific Ocean–atmosphere system. *Science*, **240**, 1293–1302.

Harrison, D.E. (1984). The appearance of sustained equatorial surface westerlies during the 1982 Pacific Warm Event. *Science*, **222**, 1099–1102.

Harrison, D.E. & Schopf, P.S. (1984). Kelvin-wave induced anomalous advection and the onset of surface warming in El Niño events. *Monthly Weather Review*, **112**, 923–33.

Heddinghaus, T.R. & Krueger, A.F. (1981). Annual and interannual variations in outgoing longwave radiation over the tropics. *Monthly Weather Review*, **109**, 1208–18.

Hirst, A.C. (1986). Unstable and damped equatorial modes in simple coupled ocean–atmosphere models. *Journal of the Atmospheric Sciences*, **43**, 606–30.

Horel, J.D. (1982). On the annual cycle of the Tropical Pacific atmosphere and ocean. *Monthly Weather Review*, **110**, 1863–78.

Horel, J.D. & Wallace, J.M. (1981). Planetary scale atmospheric phenomena associated with the Southern Oscillation. *Monthly Weather Review*, **109**, 813–29.

Janowiak, J.E., Krueger, A.F., Arkin, P.A. & Gruber, A. (1985). Atlas of outgoing long wave radiation derived from NOAA satellite data. NOAA Atlas 6. Silver Springs, MD: U.S. Department of Commerce.

Julian, P.R. & Chervin, R.M. (1978). A study of the Southern Oscillation and Walker circulation phenomena. *Monthly Weather Review*, **106**, 1433–51.

Kang, I.S. & Lau, N.-C. (1986). Principal modes of atmospheric variability in model atmospheres with and without anomalous sea surface temperature forcing in the tropical Pacific. *Journal of the Atmospheric Sciences*, **43**, 2719–35.

Knox, R.A. & Halpern, D. (1982). Long range Kelvin wave propagation of transport variations in Pacific Ocean equatorial currents. *Journal of Marine Research*, **40** (suppl.), 329–39.

Kousky, V.E. & Leetmaa, A. (1989). The 1986–87 Pacific Warm Episode: Evolution of oceanic and atmospheric anomaly fields. *Journal of Climate*, **2**, 254–67.

Lau, K-M. (1982). Equatorial response to northeasterly cold surges as inferred from satellite cloud imagery. *Monthly Weather Review*, **110**, 1306–13.

Lau, N-C. (1985). Modeling the seasonal dependence of the atmospheric response to observed El Niños in 1962–76. *Monthly Weather Review*, **113**, 1970–96.

Liebmann, B. & Hartmann, D.L. (1982). Interannual variations of outgoing IR associated with tropical circulation changes during 1974–78. *Journal of the Atmospheric Sciences*, **39**, 1153–62.

Lindzen, R.S. & Nigam, S. (1987). On the role of sea surface temperature gradients in forcing low-level winds and convergence in the tropics. *Journal of the Atmospheric Sciences*, **44**, 2418–36.

Luther, D.S., Harrison, D.E. & Knox, R.A. (1983). Zonal winds in the central equatorial Pacific and El Niño. *Science*, **222**, 327–30.

McCreary, J.P. Jr. (1983). A model of tropical ocean–atmosphere interaction. *Monthly Weather Review*, **111**, 370–87.

McWilliams, J.C. & Gent, P.R. (1978). A coupled air and sea model for the tropical Pacific. *Journal of the Atmospheric Sciences*, **35**, 962–89.

Meehl, G.A. (1987). The annual cycle and interannual variability in the tropical Pacific and Indian Ocean Regions. *Monthly Weather Review*, **115**, 27–50.

Morrissey, M.L. (1986). A statistical analysis of the relationships among rainfall, outgoing longwave radiation and the moisture budget during January–March 1979. *Monthly Weather Review*, **114**, 931–42.

Neelin, J.D. & Held, I.M. (1987). Modeling tropical convergence based on the moist static energy budget. *Monthly Weather Review*, **115**, 3–12.

Newell, R.E. & Weare, B.C. (1976). Factors governing tropospheric mean temperatures. *Science*, **194**, 1413–4.

Philander, S.G.H. (1983). El Niño Southern Oscillation phenomena. *Nature*, **302**, 295–301.

Philander, S.G.H. (1985). El Niño and La Niña. *Journal of the Atmospheric Sciences*, **42**, 2652–62.

Philander, S.G.H. & Hurlin, W.G. (1988). The heat budget of the tropical Pacific Ocean in a simulation of the 1982–83 El Niño. *Journal of Physical Oceanography*, **18**, 926–31.

Philander, S.G.H., Yamagata, P., & Pacanowski, R.C. (1984). Unstable air–sea interactions in the tropics. *Journal of the Atmospheric Sciences*, **41**, 604–13.

Rasmusson, E.M. (1985). El Niño and variations in climate. *American Scientist*, **73**, 168–77.

Rasmusson, E.M. & Wallace, J.M. (1983). Meteorological aspects of the El Niño/Southern Oscillation. *Science*, **222**, 1195–1202.

Rasmusson, E.M. & Carpenter, T. H. (1982). Variations in tropical sea surface temperature and surface wind fields associated with the Southern Oscillation/El Niño. *Monthly Weather Review*, **110**, 354–84.

Rasmusson, E.M. & Carpenter, T. H. (1983). The relationship between eastern equatorial Pacific sea surface temperatures and rainfall over India and Sri Lanka. *Monthly Weather Review*, **111**, 517–28.

Ropelewski, C.F. & Halpert, M.S. (1987). Global and regional scale precipitation patterns associated with the El Niño/Southern Oscillation. *Monthly Weather Review*, **115**, 1606–26.

Ropelewski, C.F. & Halpert, M.S. (1989). Precipitation patterns associated with the high index phase of the Southern Oscillation. *Journal of Climate*, **2**, 268–84.

Schopf, P.S. & Suarez, M.J. (1988). Vacillations in a coupled ocean–atmosphere model. *Journal of the Atmospheric Sciences*, **45**, 549–66.

Seager, R. (1989). Modeling tropical Pacific sea surface temperature: 1970–87. *Journal of Physical Oceanography*, **19**, 419–34.

Shea, D.J. (1986). Climatological atlas 1950–79: Surface air temperature, precipitation, sea level pressure, and sea-surface temperature (45°S–90°N). NCAR Tech. Note NCAR/TN-269+STR. Boulder, CO: National Center for Atmospheric Research.

Shukla, J. & Wallace, J.M. (1983). Numerical simulation of the atmospheric response to equatorial Pacific sea surface temperature anomalies. *Journal of the Atmospheric Sciences*, **40**, 1613–30.

Trenberth, K.E. (1976). Spatial and temporal variations of the Southern Oscillation. *Quarterly Journal of the Royal Meteorological Society*, **102**, 639–53.

Trenberth, K.E. (1984). Signal versus noise in the Southern Oscillation. *Monthly Weather Review*, **112**, 326–32.

Trenberth, K.E. (1989). TOGA and atmospheric processes. *Understanding Climate Change. Geophysical Monograph 52, IUGG Vol. 7*, ed. A. Berger, R. E. Dickinson and J. W. Kidson. Washington, DC: American Geophysical Union, 117–25.

Trenberth, K.E. & Shea, D.J. (1987). On the evolution of the Southern Oscillation. *Monthly Weather Review*, **115**, 3078–96.

Trenberth, K.E. & Paolino, D.A. Jr. (1981). Characteristic patterns of variability of sea level pressure in the Northern Hemisphere. *Monthly Weather Review*, **109**, 1169–89.

Trenberth, K.E., Branstator, G.W. & Arkin, P.A. (1988). Origins of the 1988 North American drought. *Science*, **242**, 1640–5.

Troup, A. J. (1965). The Southern Oscillation. *Quarterly Journal of the Royal Meteorological Society*, **91**, 490–506.

van Loon, H. & Shea, D.J. (1985). The Southern Oscillation. Pt IV: The precursors south of 15°S to the extremes of the oscillation. *Monthly Weather Review*, **113**, 2063–74.

van Loon, H. & Shea, D.J. (1987). The Southern Oscillation. Part VI: Anomalies of sea level pressure on the Southern Hemisphere and of Pacific sea surface temperature during the development of a Warm Event. *Monthly Weather Review*, **115**, 370–9.

von Storch, H., van Loon, H. & Kiladis, G.N. (1988). The Southern Oscillation Pt. VIII: Model sensitivity to SST anomalies in the tropical and subtropical regions of the South Pacific Convergence Zone. *Journal of Climate*, **1**, 325–31.

Walker, G.T. & Bliss, E.W. (1932). World Weather V. Mem., *Royal Meteorological Society*, **4** (36), 53–84.

Walker, G.T. & Bliss, E.W. (1937). World Weather VI. Mem., *Royal Meteorological Society*, **4** (39), 119–39.

Weare, B.C. (1983). Interannual variation in net heating at the surface of the tropical Pacific Ocean. *Journal of Physical Oceanography*, **13**, 873–85.

White, W.B., Meyers, G.A., Donguy, J.R. & Pazan, S.E. (1985). Short-term climatic variability in the thermal structure of the Pacific Ocean during 1979–1982. *Journal of Physical Oceanography*, **15**, 917–35.

Wright, P.B., Wallace, J.M., Mitchell, T.P. & Deser, C. (1988). Correlation structure of the El Niño/Southern Oscillation phenomenon. *Journal of Climate*, **1**, 609–25.

Wyrtki, K. (1975). El Niño. The dynamic response of the equatorial Pacific Ocean to atmospheric forcing. *Journal of Physical Oceanography*, **5**, 572–84.

Wyrtki, K. (1977). Sea level during the 1972 El Niño. *Journal of Physical Oceanography*, **7**, 779–87.

Wyrtki, K. (1981). An estimate of equatorial upwelling in the Pacific. *Journal of Physical Oceanography*, **11**, 1205–14.

Wyrtki, K. (1985). Water displacements in the Pacific and the genesis of El Niño cycles. *Journal of Geophysical Research*, **90**, 7129–32.

Zebiak, S.E. & Cane, M.A. (1987). A model El Niño–Southern Oscillation. *Monthly Weather Review*, **115**, 2262–78.

3

Brazil's Climate Anomalies and ENSO

PAO-SHIN CHU

Department of Meteorology
University of Hawaii
Honolulu, Hawaii 96822

Introduction

Interest in the relationship between regional climate anomalies in the world and ENSO is growing rapidly. Northeast Brazil (Fig. 3.1) stands out prominently as a semiarid, tropical region where rainfall variability is generally related to ENSO phenomena. Indeed, this was noted more than 60 years ago by Walker (1924, 1928), in particular his latter pioneering work searching for physical causes of dry years in Northeast Brazil (hereafter referred to as Nordeste).

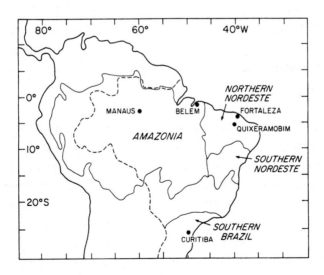

Fig. 3.1 Map showing northeast Brazil, Amazon Basin, and southern Brazil. Broken line is the national boundary of Brazil; dots denote key rainfall stations in each region.

The fickleness of droughts disrupting the normal social and economic structure in Nordeste has been lamented by Euclides da Cunha (1979) and has long stimulated the curiosity of researchers in Brazil (Sampaio Ferraz, 1950; Serra, 1956). More recently, an awareness of the role of short-term climate variations in the national economy has been gained as well by the Brazilian government, which has given priority to the task of contending with Nordeste's droughts, during which famine and mass exodus have been common. The central objectives of this task are to further the understanding of dynamic causes of drought and develop a basis for its prediction.

Since the 1970s, thanks to the availability of the large amount of meteorological and oceanographic data as well as renewed interest in ENSO and global teleconnections in general, studies relating Nordeste's rainfall anomalies to large-scale circulation have received more attention (Namias, 1972; Hastenrath and Heller, 1977; Markham and McLain, 1977; Moura and Shukla, 1981; Chu, 1983; Hastenrath et al., 1984; Kousky et al., 1984; Rao et al., 1986). However, little is known about the meteorology of the vast tropical rainforests of the Amazon basin (Amazonia; Fig. 3.1), although some recent evidence indicates that ENSO teleconnections may also be operative in this region (Kousky et al., 1984; Aceituno, 1988; Rogers, 1988). Rainfall variations in southern Brazil also have been shown to be sensitive to ENSO (Fig. 3.1; Nobre and Oliveira, 1986; Ropelewski and Halpert, 1987). Since agriculture is its major economic base, droughts and floods can affect soil moisture and reduce crop yields, causing socioeconomic repercussions for the region.

This chapter describes the climatic hazards in Nordeste, the Amazon basin, and southern Brazil in relation to ENSO and circulation anomalies over the tropical Atlantic. As will be described later, however, the overall statistical relation between Nordeste's rainfall and ENSO is not particularly strong. Background circulation and rainfall characteristics are given, as is the interannual variability of rainfall in various selected regions. Regional studies of ENSO teleconnections and of circulation anomaly patterns in the Atlantic for composites of extreme years are presented, along with time series of composite monthly rainfall anomalies during ENSO phenomena. Examples of long-range forecasts based on the diagnostic efforts are demonstrated.

Background circulation and rainfall regimes

Climatology of surface circulation over the tropical Atlantic and of
upper-air circulation over South America is described only briefly,
since details are available in numerous recent atlases (e.g., Hastenrath and Lamb, 1977; Chu and Hastenrath, 1982; Sadler et
al., 1987). The understanding of background circulation is instrumental for the subsequent description of the circulation anomaly
mechanisms. The maps present data for only January, April, July,
and October, which represent the approximate midpoints of the
four seasons. Unless otherwise specified, the seasons here refer to
the austral seasons. For instance, the winter season is June to
August, whereas the summer season is December to February.

Circulation over the tropical Atlantic is dominated by two
subtropical highs enclosing an equatorial belt of low pressure
(Fig. 3.2). In the course of a year, the near-equatorial trough
zone is subjected to the influence of the anticyclones of the two
hemispheres, resulting in a southward shift of the trough zone during the summer semester (November–April) and in a northward
shift during the winter semester (May–October).

The zonal wind pattern is characterized by a year-round, easterly flow over the Atlantic, except near the Gulf of Guinea where
westerly flow is observed (Fig. 3.3). The patterns of meridional
wind are depicted in Fig. 3.4. Northerly components dominate
over the North Atlantic, with the largest values in the southeastern sector of the subtropical high. Southerly components occupy a
large part of the South Atlantic, whereas northerlies prevail along
the east coast of Brazil (except in April and May). The zero isopleth in the equatorial Atlantic represents a line where the meridional components of the trades from the two hemispheres meet,
and is embedded in the relatively broad convergence band (Hastenrath and Lamb, 1977). This line of surface wind discontinuity
also lies within the near-equatorial trough zone and reaches its
southernmost position in March/April. Along with the intensification of the cross-equatorial flow from the South Atlantic at the
outset of the southern winter, the wind discontinuity line moves
northward until about August.

The sea surface temperature (SST) pattern is characterized by
relatively warm water ($>26°C$) in the equatorial Atlantic throughout the year (Fig. 3.5). This feature, however, varies seasonally. In

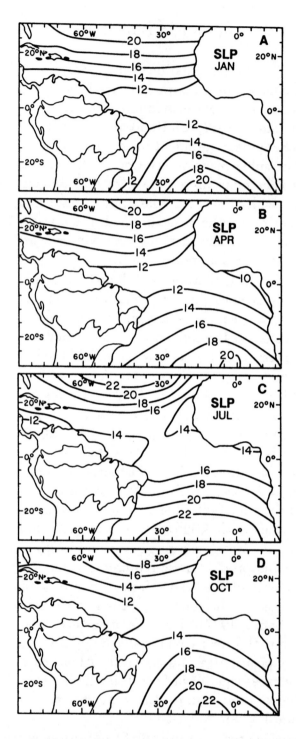

Fig. 3.2 Mean sea level pressure over the Atlantic during January, April, July, and October 1911–70. Values are given as 1000+ mb. Isobar spacing is 2 mb.

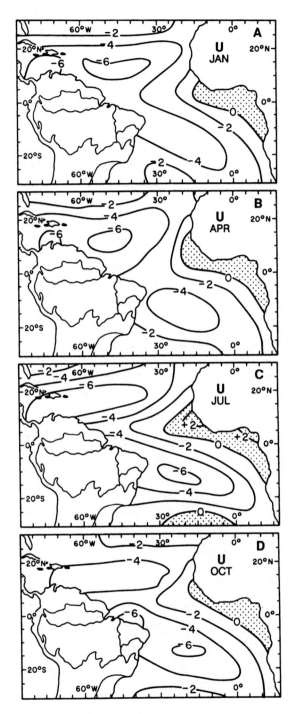

Fig. 3.3 Mean surface zonal wind patterns over the Atlantic during January, April, July, and October 1911–70. Negative values denote easterly component, whereas dot raster denotes westerly component (positive value). Isopleth interval is 2 m/s.

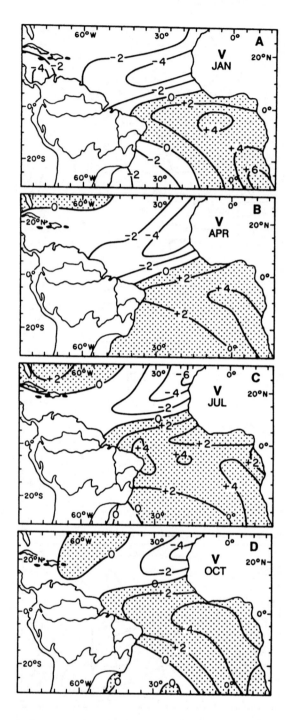

Fig. 3.4 Mean surface meridional wind patterns over the Atlantic during January, April, July, and October 1911–70. Negative values denote northerly component, whereas dot raster denotes southerly wind component (positive value). Isopleth interval is 2 m/s.

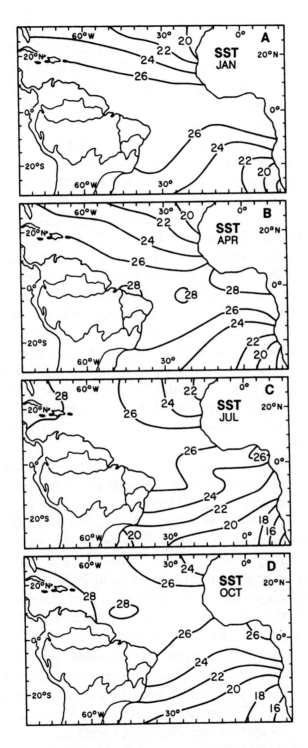

Fig. 3.5 Mean sea surface temperature in the Atlantic during January, April, July, and October 1911–70. Isopleth interval is 2°C.

the winter semester, the warmest water is found further northward. In the summer, the warm pool of surface water shifts equatorward, with the areas of highest SST (>28°C) closest to the equator in March/April. Interestingly, this is also the time of the year when the near-equatorial trough zone and the surface wind discontinuity axis are displaced further equatorward.

The analyses of upper-air circulation use an objective scheme to interpolate values for grids, spaced five degrees latitude and longitude apart, in a rectangular coordinate system. The grid spacing is denser than the upper-air network in most of the map area, except over Nordeste and southern Brazil (Fig. 3.6). Details on the algorithm used are in Chu and Hastenrath (1982) and Chu (1985). Flow in the lower troposphere (Fig. 3.6) in all 12 months is marked by the anticyclonic turning of easterlies from the band of 5–10°S to a more northerly direction in southern Brazil. Consistently under the influence of this counterclockwise flow is Nordeste, interior Brazil, and easternmost Amazonia. A weak easterly flow prevails over Nordeste and to the south of the Amazon River (5°S) in summer when the South Atlantic high is farther away from South America (Figs. 3.2 and 3.6). Concomitant with the enhanced South Atlantic high in winter, the strengthened southeast trades dominate Nordeste, and as a result, easterlies penetrate farther inland.

The intense anticyclone over the Peruvian–Bolivian Altiplano and the well-developed trough downstream over Nordeste are integral parts of upper tropospheric flow patterns in midsummer (Fig. 3.7). This trough tilts from about 10°S along the Nordeste coast, northwestward to the far extreme of Nordeste. The position of the anticyclone is centered around 14°S, 65°W. A band of relatively strong westerlies (>10 m/s) is to the south of the upper-level anticyclone. A comparison of Figs. 3.6 and 3.7 reveals that the direction of the upper tropospheric flow is approximately opposite that of the lower troposphere during summer.

As autumn nears, the upper troposphere over the Altiplano cools. The anticyclone trough system gradually weakens, giving way to westerlies, which prevail over the entire continent until October (Fig. 3.7). The southern subtropical jet, near 30°S, more than doubles its speed from summer to winter.

Mean annual rainfall varies greatly throughout Brazil, from less than 400 mm in interior Nordeste to more than 2800 mm at the

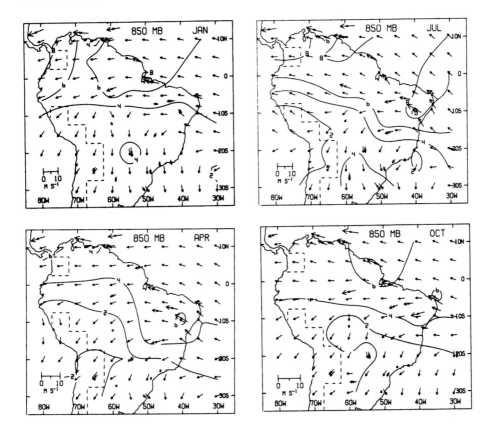

Fig. 3.6 Lower tropospheric (850 mb) mean monthly resultant wind over South America during January, April, July, and October 1970–74. Direction at interpolated grid points is shown by arrows of uniform length, and speed by isotachs labelled in m/s. Upper air stations are marked by crosses; observed resultant wind vectors are indicated by arrows, whose lengths represent wind speed in m/s. Broken line indicates 5°squares with average heights above 1500 m. Sources: Chu and Hastenrath (1982) and Chu (1985).

mouth of the Amazon and in the upper Amazon basin (Fig. 3.8). A large part of Nordeste is characterized by annual rainfall of 300–800 mm, and rainfall increases sharply from the interior eastward to the coast. Moderate rainfall is found in the southern Brazil, where annual values exceed 1600 mm.

The annual cycle of rainfall is not homogeneous across Nordeste. The rainy season in the Northern Nordeste extends from February to May, with a peak in March/April. These two months signal the further equatorward migration of the surface wind discon-

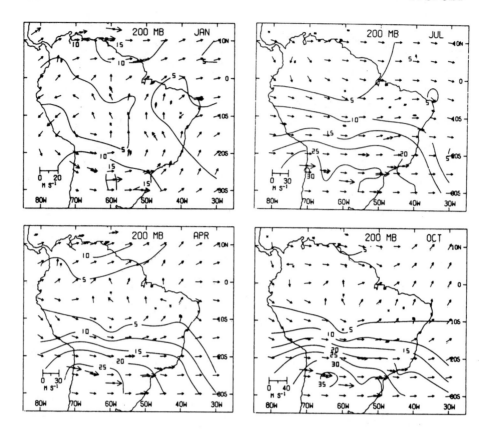

Fig. 3.7 Same as Fig.3 6, but for troposhere (200 mb) and with interval spacing of 5 m/s. Sources: Chu and Hastenrath (1982) and Chu (1985).

tinuity axis and the convergence band over the western equatorial Atlantic. During this time, the band of highest SST reaches the southernmost position (Fig. 3.5). The relatively warm surface water off the Nordeste coast promotes heat and moisture fluxes, which create unstable conditions in the atmospheric boundary layer. For the Southern Nordeste, the rainy season spans from November to March, peaking in amounts in November/December. Frontal systems from the midlatitudes of the Southern Hemisphere may account for the maximum rainfall in these two months (Ratisbona, 1976; Kousky, 1979; Chu, 1983). A third rainfall regime is narrowly confined along the east coast, where the rainy season is in winter. Land–sea breeze circulation and westward propagating cloud clusters from the South Atlantic play a role in winter rainfall maximum in the eastern Nordeste coast (Ramos, 1975; Yamazaki

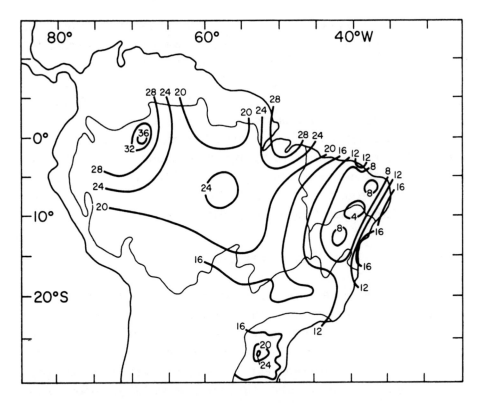

Fig. 3.8 Mean annual rainfall (in hundreds of mm) in Brazil. Isohyets in southern Brazil are adopted from Ratisbona (1976).

and Rao, 1977). Since the circulation features related to rainfall anomalies in Northern Nordeste are distinct, only this region in Nordeste will be included in the further discussion.

Rainfall variations in Amazonia are very complicated, and at least three major rainfall regimes can be recognized. In the area between the equator and 5°N (Rio Negro catchment), rainfall is abundant year-round. In a zonally oriented band extending approximately from the equator to 5°S along the Amazon River to its mouth, precipitation is greatest in March–May and tapers off toward August–October. The rhythm of the annual cycle is believed to act in concert with the displacement of the equatorial convergence (Ratisbona, 1976). In southern Amazonia, the rainy season extends from October through April. During this period, northerly flows with a slightly easterly component predominate over the basin at low levels while southerly flows prevail at high levels (Figs. 3.6 and 3.7). As the South Atlantic high intensifies and extends farther inland in winter, southeasterlies and northeasterlies appear to the north and south of 5°S, respectively (Fig. 3.6). Southern Amazonia, located in the area where the easterlies bifurcate, experiences its driest season during winter.

In southern Brazil, the rainfall distribution is fairly even from month to month, with the largest amounts in January and February. Concomitant with this steady annual cycle of rainfall is the prevailing northerly flow in the lower troposphere and the persistent subtropical westerlies in the upper troposphere (Figs. 3.6 and 3.7). The midlatitude synoptic systems and the associated precipitation bands from the South Atlantic are mainly responsible for rainfall in this region (Ratisbona, 1976).

Interannual variability of regional rainfall

The interannual variability of rainfall in Nordeste, Amazonia, and southern Brazil are presented in time series plots of yearly indices (Fig. 3.9). For the Northern Nordeste, the rainfall records from September to August for 32 stations are combined to yield a regional index. The indexed year refers to the later calendar year (e.g., 1958 denotes September 1957 to August 1958). After 1978, rainfall records for the aforementioned 32 stations are not complete and the regional index for Northern Nordeste is represented by records from Quixeramobim alone, which is located in the core region of the Northern Nordeste and has a high correlation with the regional index over the earlier time period (Fig. 3.1; Hastenrath, 1985).

As shown in Fig. 3.9, a run of deficient rainfall in the Northern Nordeste occurred in the 1950s, from 1951 to 1956 and again from 1958 to 1960. The severe 1958 drought in particular caused socioeconomic repercussions. In more recent years, rainfall was below average from 1981 to 1983, then replaced by abundant rainfall in 1984–86. There is a weak correspondence between low rainfall and ENSO events since the 1930s.

The complex rainfall regimes and scanty rainfall network in Amazonia preclude the construction of a regional index for the entire basin. A further problem is due to the lack of long-term and complete rainfall records in this vast, tropical humid lowlands, with the exception of Manaus. Therefore, only historical records at Manaus are used. In a strict sense, rainfall variations at Manaus may represent only the local climate regime or, at most, a limited

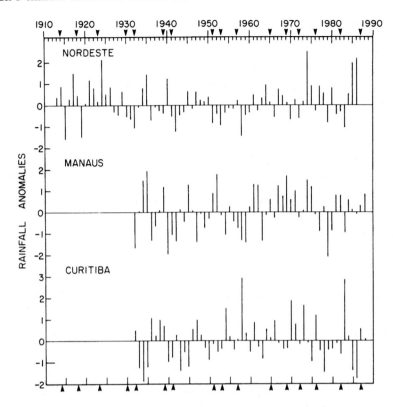

Fig. 3.9 Indices of annual rainfall anomalies for the Northern Nordeste (September–August), Manaus in Amazonia (September–August), and Curitiba in southern Brazil (July–June). Values are ascribed to the later calendar year. For the Northern Nordeste, regional rainfall index after 1978 is substituted with records for a single station, Quixeramobim. The inception of each ENSO event is shown by an arrow.

zonal band along the axis of the Amazon River as described in the previous section. Rainfall records at Manaus are analyzed by season from September through August, so that the highest rainfall occurs in the middle of the year (December through May). The indexed year again refers to the later calendar year. An examination of the raw data reveals that about 99 percent of the monthly totals are available from various sources. The few missing observations were filled with the long-term, monthly means only in order to construct the yearly rainfall indices.

Along the Amazon River, a run of low rainfall occurred in the 1950s (i.e., from 1953 to 1959, except 1955), similar to the rainfall pattern in the Northern Nordeste. Rainfall tended to be high during 1960–75 and 1981–88 (except in 1983 and near normal rainfall in 1986). Roughly speaking, low rainfall years in the heart of Amazonia did not coincide with the onset of ENSO events, but rather lagged by one year. These were exemplified by the ENSO years in 1939, 1941, 1953, 1957, 1965, 1976, and 1982, and the respective deficient rainfall years in 1940, 1942, 1954, 1958, 1966, 1977, and 1983. The year immediately following an ENSO event, however, was not always dry, as demonstrated by the abundant rainfall years in 1952, 1970, and 1988. Moreover, some low rainfall years (i.e., 1947, 1963, and 1979) were not preceded by ENSO events. Whether low rainfall years documented by records in Manaus are a local or a basin-wide phenomenon remains to be further investigated.

For southern Brazil, one station – Curitiba – with long-term records was chosen for study. The yearly rainfall index (July to June) again refers to the later calendar year. As with Manaus, only about 1 percent of the data is absent from the records, and the missing observations were filled with the long-term, monthly averages.

In southern Brazil, the nature of the interannual rainfall variability is different from that exhibited in tropical Brazil. For instance, southern Brazil in the 1950s apparently did not experience the persistent dry spells that occurred in the Nordeste and Amazonia. The moderately dry episode persisting for six consecutive years, 1977–82, in Curitiba was also not experienced at stations in the north. More importantly, rainfall tended to be unusually high immediately after the onset of ENSO, as demonstrated by the abundant rainfall years in 1954, 1958, 1970, 1973, and 1983. In fact, since 1940, eight out of the 10 years following the inception of an ENSO event were characterized by wet episodes in southern Brazil.

ENSO teleconnections, Atlantic circulation, and regional climate anomalies in Brazil

Nordeste

The remarkable coincidence of drought in Nordeste and the anomalous warming of the eastern equatorial Pacific Ocean (El Niño) has long been recognized (Walker, 1924, 1928). The physical causes of the spatial coherence between these two phenomena appear to be related to the Southern Oscillation (SO) (Berlage, 1966; Doberitz, 1969; Caviedes, 1973; Hastenrath, 1976). Using surface data over the global tropics, Kidson (1975) demonstrated that the SO is a preferred mode in the general circulation of the tropical atmosphere. Of particular interest in the manifestation of the SO is the approximately inverse pressure fluctuations between the South Atlantic and the southeast Pacific (Hastenrath and Heller, 1977). Thus, when pressures are unusually low over the South Pacific during the El Niño years, they tend to be high in the South Atlantic. The anomalously strong South Atlantic high is unfavorable for rainfall production in Nordeste, because of the enhancement of subsidence and, possibly, the reduction of moisture content in the air.

To illustrate the ENSO teleconnections on Nordeste's rainfall, Fig. 3.10a shows a scatter diagram based on the summer Southern Oscillation Index (SOI) and rainfall anomalies at the height of the rainy season in Quixeramobim. The SOI is obtained as the difference in normalized sea level pressure between Tahiti and Darwin, Australia. Details on the data source and data processing of the SOI are given in Chu and Katz (1985). The extreme low phase of the SO, as reflected by the large negative SOI, generally (but not always) corresponds to El Niño events.

In Fig. 3.10a, 12 observations occur in the lower left quadrant and 11 in the upper right quadrant. Taken together, they account for less than 50 percent of the total number of observations ($n = 51$), implying a weak phase linkage between the antecedent conditions in large-scale pressure seesaw and rainfall in Nordeste. However, as a summer SOI reached a value of -1.5 or below, the rainfall anomaly in the subsequent season tended to be negative. This occurred in six out of eight cases.

The scatter plot for autumn has 27 observations in the lower left and upper right quadrants, and thus accounts for more than 50 percent of the total observations (Fig. 3.10b). As a consequence, the correlation coefficient between the SOI and rainfall anomalies increases noticeably from 0.20 during the preseason to 0.31 during the rainy season. Although rainfall variations during the peak rainy season are better teleconnected to the concurrent fluctuations in the SO, this association is not statistically significant at the 95 percent level, when the reduction of the effective number

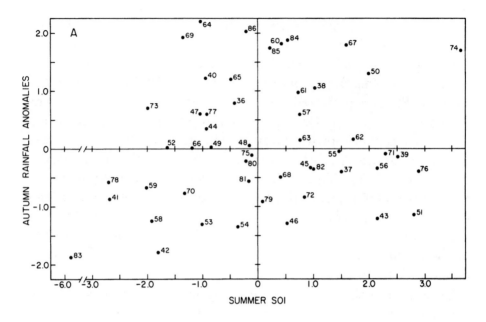

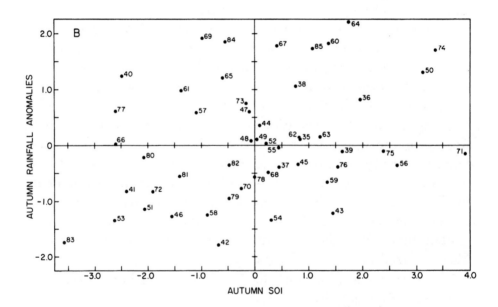

Fig. 3.10 Scatter diagrams based on the SOI and rainfall anomalies at Quixeramobim, Northern Nordeste during 1935–86. (a) Summer SOI and
autumn rainfall; (b) autumn SOI and autumn rainfall. Season refers
to the Southern Hemisphere season (e.g., autumn is March–May). The
numbers denote years (e.g., 58 denotes 1958).

of degrees of freedom due to persistence is considered (Quenouille, 1952). In particular, some drought years (1946, 1979) have occurred in the absence of ENSO phenomena. Rogers (1988) also noted a low correlation (0.15) between the seasonalized rainfall at Fortaleza, Ceará, and the SOI (Fig. 3.1). Thus, other circulation mechanisms, in addition to the SO, must be sought to explain climatic hazards in Nordeste.

In this regard, the mechanisms of extreme rainfall anomalies in the Northern Nordeste were explored in terms of the large-scale departure patterns over the adjacent tropical Atlantic. Departure maps of major meteorological variables have been compiled from composites of the 10 driest and wettest years covering the base period from 1912 to 1972. The variables considered for representation in the maps are sea level pressure (SLP), wind, and SST.

The SLP departure map for the composite of 10 extremely dry years shows positive values on the equatorward side of the South Atlantic high, a further northward position of the trough of low pressure over the western equatorial Atlantic, and a poleward retraction of the North Atlantic high (Figs. 3.11a and 3.2). The map of the zonal wind shows a band of increased easterlies stretching from the eastern equatorial Atlantic to the coast of the Northern Nordeste (Figs. 3.11b and 3.3). The patterns of meridional wind exhibit stronger southerlies in the South Atlantic and weaker northerlies in the North Atlantic, resulting in a more northerly position of the wind discontinuity axis in the equatorial area (Figs. 3.11c and 3.4). The South Atlantic is marked by anomalously cold water, and a major part of the North Atlantic by anomalously warm water (Figs. 3.11d and 3.5). Markham and McLain (1977) also found a high correlation between Nordeste rainfall and South Atlantic SST. The general circulation departure patterns during the wet years are nearly the reverse of those of the dry years, and they will not be shown here.

Departure patterns were further subjected to statistical testing and evaluated with respect to the reference years (1911–72), excluding the 10 driest and 10 wettest years. The conventional five percent confidence limit is reached within a large area of negative SLP anomalies over the subtropical North Atlantic and a pocket of positive anomalies off the Angola coast in Fig. 3.11a. In Fig. 3.11b, an area of stronger than normal easterlies that is statistically significant at the five percent level appears on the Northern Nordeste

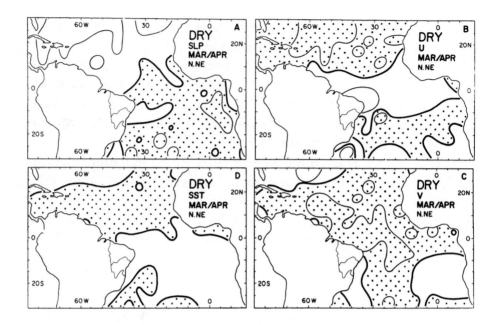

Fig. 3.11 March/April composite of the ten extremely dry years in the Northern Nordeste expressed as departure from the 60-year mean. (a) Sea-
level pressure; (b) zonal wind component; (c) meridional wind component; and (d) sea surface temperature. Positive areas are shaded
with heavy solid line denoting zero departure. Thin solid line refers
to the 5 percent significance level. Source: Chu (1983).

coast. The region of statistically significant weak easterlies over
the North Atlantic is consistent with the weakened meridional
pressure gradient depicted in Fig. 3.11a. The strong southerlies
to the south and weak northerlies to the north of the wind discontinuity axis are also statistically significant (Fig. 3.11c). Thus,
during the Nordeste drought, the enhanced southeast trade winds
along the Northern Nordeste coast are indeed statistically significant. The departure pattern of the SST has not been subjected
to the test because the SST series possess strong persistence from
year to year. It is recognized, however, that the effect of persistence can be adjusted by means of the method described earlier
for the correlation coefficient (Quenouille, 1952).

Other circulation anomaly mechanisms in the Atlantic have been
proposed for rainfall variations in Nordeste. Namias (1972) suggested a teleconnection between rainfall in the Nordeste and cyclonic activity over the subpolar North Atlantic involving a process

known as the North Atlantic Oscillation (van Loon and Rogers, 1978). Moura and Shukla (1981) conducted a numerical experiment with the general circulation model and posited that the juxtaposition of a warm anomaly in the North Atlantic and cold anomaly in the South Atlantic, as demonstrated in Fig. 3.11d, would feature a thermally direct, local meridional circulation with subsidence over Nordeste and the adjacent equatorial Atlantic.

In contrast to the idea of the local meridional circulation, Kousky et al., (1984), Rao et al. (1986), and Kayano et al., (1988) have proposed that the strengthening of the zonal Walker-type circulation along the equator partially accounts for the severe 1983 drought (Fig. 3.9). The Walker-type circulation is primarily driven by the east–west SST gradient in the tropical oceans and is thus related to ENSO. In light of the 1987 drought in Nordeste possibly being associated with the recent ENSO (Wagner, 1987), further investigation should be conducted on the displacement of the convergence band over the equatorial western Atlantic, the variations in intensity and position of the South and North Atlantic highs, and the roles of the local meridional circulation, the zonal circulation, and other circulation features in the Atlantic.

Amazon

To illustrate the effect of ENSO on monthly rainfall in the Amazon basin, rainfall anomalies for Manaus and Belém are presented in a composite time series during the ENSO cycles (Fig. 3.12a). Data for Belém for 1965 and 1966 are missing. Accordingly, only five ENSO events have been entered into the composite chart for Belém and six events for Manaus during the common period from 1961 to 1988. The months with missing data were not used in the construction of the composite time series.

The beginning of each ENSO event, as shown by an arrow in Fig. 3.9, is set in the middle of the three-year period and is denoted as Yr(0). The years preceding and following Yr(0) are denoted in Fig. 3.12 as Yr(−1) and Yr(+1), respectively. As examples, the large and positive SST anomalies (>2°C) off the Peruvian coast occurred in about September 1982 (Arkin et al., 1983). Thus, the 1982–83 ENSO event is embedded in the 1981–83 period, with 1981 understood as being Yr(−1), 1982 as Yr(0), and 1983 as Yr(+1).

During 1987, the evolution of SST anomalies off the South American coast and in the equatorial central Pacific, together with the SOI, mimics the patterns described in Rasmusson and Carpenter's (1982) composite for a Yr(0) of an ENSO event (Lander, 1989). Hence, monthly rainfall anomalies for both Manaus and Belém during the three-year period (1986–88) were also employed in constructing the composite chart shown in Fig. 3.12a.

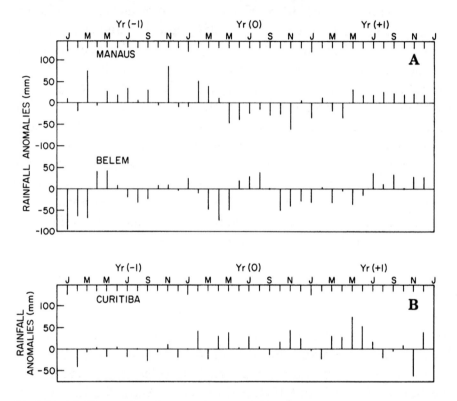

Fig 3.12 (a) Time series of composite monthly rainfall anomalies (mm) during ENSO events for Manaus and Belém in the Amazon Basin. Composites run from January of the year preceding [Yr(−1)] the ENSO event to December of the following year (Yr(+1)). Six ENSO events are included in the composite for Manaus and five for Belém since 1961. (b) Same as (a), but for Curitiba in southern Brazil.

Prior to the Yr(0), rainfall tends to be high in the interior of Amazonia (Fig. 3.12a). Deficient rainfall in Manaus begins to occur in May of the Yr(0) and continues into April of the Yr(+1). This low rainfall episode extends 12 months, only interrupted by a slightly above average rainfall in December Yr(0) and February Yr(+1). An examination of monthly rainfall anomalies for each

The scatter diagrams of observed versus predicted rainfall indices are displayed in Fig. 3.13. Forecasts are correct for the extreme drought year of 1958, as well as for the dry years of 1959 and 1966, and the wet year of 1964. The outlier occurs in 1960, mainly due to a sudden shift of the circulation patterns in the tropical Atlantic from January to March or April 1960. If our goal is limited to the prediction of the severe drought, then the results in Fig. 3.13 are quite encouraging. One cautionary note, however, is that the time scale associated with climate prediction should not be too far in advance, because the SO, which is the most dominant mode in short-term climate variations in the tropics, loses its predictability beyond two to three future seasons (Chu and Katz, 1987).

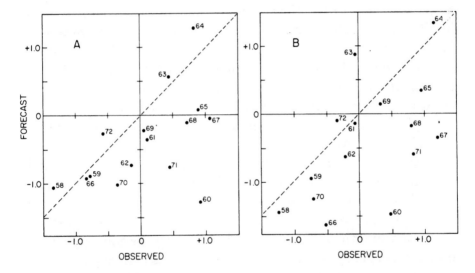

Fig. 3.13 Scatter diagrams of the Northern Nordeste rainfall index based on observations through January and predicted values for 1958–72. (a) March/April rainfall index; (b) March/September rainfall index. Numbers indicate the years. Source: Hastenrath, et al., (1984).

Although the drought in Northeast Brazil is generally predictable two to three months ahead (Fig. 3.13), there is a noticeable bias in rainfall forecasts for the period 1958–72 based on the model developed using an earlier period of record (1921–40 plus 1946–56). As revealed in Table 3.1(A), for instance, only the zonal wind component in January and the preseason rainfall (October to January) are used as predictors for March/April rainfall index. An examination of the time series of the preseason rainfall indicates a downward trend from the late 1940s to 1955, and this trend would contribute to the negative bias of the forecast (Fig. 3.13). Possible sources of bias could be the insufficient number of ship observations in the early period of record used in the model development, and a trend toward strengthening winds over the oceans since the 1950s (Cardone et al., 1990).

Summary and conclusions

This chapter has discussed research activities on rainfall anomalies in three regions of Brazil (Nordeste, Amazonia, and southern Brazil) known to be sensitive to ENSO phenomena.

Severe droughts in the Northern Nordeste occur frequently in the presence of ENSO. Nevertheless, the teleconnections between the SOI and rainfall are not particularly strong when examined statistically. In fact, the climatic hazards of the Northern Nordeste are better defined in terms of circulation patterns in the adjacent tropical Atlantic, although these patterns may be part of the manifestations of the planetary-scale ENSO phenomena.

Rainfall variations in the Amazon basin, at least in a narrow band along the Amazon River, are modulated by ENSO events. A composite of six recent events indicates that deficient rainfall in Manaus begins to appear around May during the year in which ENSO has occurred and persists for almost 12 months until April of the following year. A similar ENSO-related response also exists at the mouth of the river. The first dry period appears to run from February or March to May during an ENSO event and the second dry period commences in October during an event through June of the year following an event. A hypothesis involving ocean–atmosphere interaction in the realm of the tropical western Atlantic (e.g., wind-induced mixing, evaporative cooling, stability of the boundary layer), in response to ENSO phenomena, is suggested for the low rainfall episode in the Amazon basin.

For southern Brazil, as represented by Curitiba, two notable sequences of abundant rainfall accompany an ENSO event: the first occurs in April–December during an event and the second occurs in March–July following an event. The increased rainfall in this region may be related to the preferred location of the subtropical jet stream in association with ENSO.

Because the circulation anomaly patterns generally characteristic of dry and wet years in the Northern Nordeste are prominent, the severe drought in Nordeste is predictable two to three months ahead by using a physically based statistical scheme. However, an efficient data collection and dissemination system are required for near real-time, operational, long-range forecasting. Currently, such a system does not exist and presents a further challenge to be overcome in the future.

References

Aceituno, P. (1988). On the functioning of the Southern Oscillation in the South American sector. *Monthly Weather Review*, **116**, 505–24.

Arkin, P.A., Kopman, J.D. & Reynolds, R.W. (1983). 1982–1983 El Niño/Southern Oscillation event quick look atlas. Washington, DC: Climate Analysis Center, NOAA.

Berlage, H.P. (1966). The Southern Oscillation and World weather. *Koninklijk Nederlands Meteorologisch Instituut, Mededelingen en Verhandelingen*. No. 88.

Cardone, V.J., Greenwod, J.G. & Cane, M.A. (1990). On trends in historical marine wind data. *Journal of Climate*, **3**, 113–27.

Caviedes, C.N. (1973). Secas and El Niño: two simultaneous climatical hazards in South America. *Proceedings, Association of American Geographers*, **5**, 44–9.

Chu, P.-S. (1983). Diagnostic studies of rainfall anomalies in Northeast Brazil. *Monthly Weather Review*, **111**, 1655–64.

Chu, P.-S. (1984). Time and space variability of rainfall and surface circulation in the Northeast Brazil–tropical Atlantic sector. *Journal of the Meteorological Society of Japan*, **26**, 363–9.

Chu, P.-S. (1985). A contribution to the upper-air climatology of tropical South America. *Journal of Climatology*, **5**, 403–16.

Chu, P.-S. & Hastenrath, S. (1982). Atlas of upper-air circulation over tropical South America. Madison: Department of Meteorology, University of Wisconsin.

Chu, P.-S. & Katz, R.W. (1985). Modeling and forecasting the Southern Oscillation: A time-domain approach. *Monthly Weather Review*, **113**, 1876–88.

Chu, P.-S. & Katz, R.W. (1987). Measures of predictability with applications to the Southern Oscillation. *Monthly Weather Review*, **115**, 1542–9.

da Cunha, E. (1979). *Os Sertões*. 28th ed., Rio de Janeiro: Francisco Alves.

Doberitz, R. (1969). Cross spectrum and filter analysis of monthly rainfall and wind data in the tropical Atlantic region. *Bonner Meteorologische Abhandlungen*, **11**, 1–43.

Hastenrath, S. (1976). Variations in low-latitude circulation and extreme climatic events in the tropical America. *Journal of the Atmospheric Sciences*, **33**, 202–15.

Hastenrath, S. (1985). *Climate and Circulation of the Tropics*. Dordrecht: D. Reidel.

Hastenrath, S. & Heller, L. (1977). Dynamics of climatic hazards in Northeast Brazil. *Quarterly Journal of the Royal Meteorological Society*, **103**, 77–92.

Hastenrath, S. & Lamb, P. (1977). *Climatic Atlas of the tropical Atlantic and Eastern Pacific Oceans*. Madison: University of Wisconsin Press.

Hastenrath, S. & Kaczmarczyk, E.B. (1981). On spectra and coherence of tropical climate anomalies. *Tellus*, **33**, 453–62.

Hastenrath, S., Wu M.-C. & Chu, P.-S. (1984). Towards the monitoring and prediction of north-east Brazil droughts. *Quarterly Journal of the Royal Meteorological Society*, **110**, 411–25.

Kayano, M.T., Rao, V.B. & Moura, A.D. (1988). Tropical circulations and the associated rainfall anomalies during two contrasting years. *Journal of Climatology*, **8**, 477–88.

Kidson, J.W. (1975). Tropical eigenvector analysis and the Southern Oscillation. *Monthly Weather Review*, **103**, 187–96.

Kousky, V.E. (1979). Frontal influences on Northeast Brazil. *Monthly Weather Review*, **107**, 1140–53.

Kousky, V.E. (1987). The global climate for December 1986–February 1987. El Niño returns to the tropical Pacific. *Monthly Weather Review*, **115**, 2822–38.

Kousky, V.E. & Chu, P.-S. (1978). Fluctuations in annual rainfall for Northeast Brazil. *Journal of the Meteorolological Society of Japan*, **57**, 457–65.

Kousky, V.E., Kagano, M.T. & Cavalcanti, I.F.A. (1984). A review of the Southern Oscillation: Oceanic–atmospheric circulation changes and related rainfall anomalies. *Tellus*, **36A**, 490–504.

Lander, M. (1989). A comparative analysis of the 1987 ENSO event. *Tropical Ocean–Atmosphere Newsletter*, **49**, 3–6.

Markham, C.G. (1974). Apparent periodicities in rainfall at Fortaleza, Ceará, Brazil. *Journal of Applied Meteorology*, **13**, 176–9.

Markham, C.G. & McLain, D.R. (1977). Sea surface temperature related to rain in Ceará, Northeast Brazil. *Nature*, **265**, 320–3.

Moura, A.D. & Shukla, J. (1981). On the dynamics of droughts in Northeast Brazil: Observations, theory and numerical experiments with a general circulation model. *Journal of the Atmospheric Sciences*, **38**, 2653–75.

Namias, J. (1972). Influence of Northern Hemispheric general circulation on drought in Northeast Brazil. *Tellus*, **24**, 336–42.

Nobre, C.A. & de Oliveira, A.S. (1986). Precipitation and circulation anomalies in South America and the 1982–83 El Niño/Southern Oscillation episode. Second International Conference on Southern Hemisphere Meteorology, Wellington, New Zealand. Boston: American Meteorological Society, 442–445.

Quenouille, M.H. (1952). *Associated Measurements*. London: Butterworths.

Ramos, R.P.L. (1975). Precipitation characteristics in the Northeast Brazil dry region. *Journal of Geophysical Research*, **80**, 1665–78.

Rao, V.B., Satyarmurty, P. & de Brito, J.I.B. (1986). On the 1983 drought in Northeast Brazil. *Journal of Climatology*, **6**, 43–51.

4

Australasia

R.J. ALLAN

Climate Impacts Group
CSIRO Division of Atmospheric Research
Mordialloc, Victoria, Australia

Concerted research into the ENSO phenomenon and its physical manifestations in the Australasian region has occurred only recently (Australian locations mentioned in the text are shown in Appendix A). However, much of the historical pattern of ENSO research in Australasia has followed research developments overseas (Allan, 1985, 1988). Prior to the fusion of the atmospheric Southern Oscillation (SO) and oceanic El Niño (EN) phenomena into an interrelated entity (ENSO), they had been examined separately and each one was the subject of much debate before gaining acceptance by the scientific community.

Early historical evidence

As early as 1888, Sir Charles Todd (1888; p. 1456) stated:

> *Comparing our records* [Australia] *with those of India, I find a close correspondence or similarity of seasons with regard to the prevalence of drought, and there can be little or no doubt that severe droughts occur as a rule simultaneously over the two countries.*

The full impact of this statement has had to wait until recent findings showing strong correlations between the SO, drought, river discharges, and lake levels in both Australasia and India (McBride and Nicholls, 1983; Allan, 1985; Nicholls, 1985a; Williams et al., 1986; Ropelewski and Halpert, 1987, 1989; Adamson et al., 1987; Whetton and Baxter, 1987; Whetton et al., 1990). It has prompted Nicholls (1988a) to investigate colonial records for reports of protracted droughts and to compare dates with ENSO events deduced from northern Peruvian rainfall and documentary evidence by Quinn et al. (1978, 1987) and Hamilton and Garcia (1986).

His results, shown in Table 4.1, indicate a long period of linkages between ENSO and Australian droughts prior to instrumental records. There has also been research at the Australian Institute of Marine Sciences (AIMS) aimed at extending SO indices via coral cores (Isdale, 1984).

Table 4.1 Australian droughts and ENSO events 1788–1841. Dots indicate nonoccurrence of specific phenomena. Neither drought nor an ENSO event was experienced during the 28 unlisted years.

Australian drought	ENSO event according to:	
	Hamilton & Garcia	Quinn et al.
1790–91	1791	1791
1796–97	.	.
1798–99	.	.
1803–04	1804	1803–04
1810–11	.	.
1813–15	1814	1814
1817–18	.	1817
1819–20	.	1819
.	.	1821
1823–24	.	1824
1826–29	1828	1828–29
.	.	1832
1837–39	.	1837

Source: Nicholls (1988a).

In the Indonesian region, the records of the Dutch colonial period are a potential source of information on SO influences via prolonged regional droughts and frosts. Quinn et al. (1978) used salt production from the island of Madura (near Java) along with previously compiled historical data to produce chronologies of Indonesian droughts back to the mid-1800s (see Tables 4.2 and 4.3). Recent work by Murphy and Whetton (1989), re-examining Javan tree ring series back to the 1500s, indicates varying links with the SO and El Niño phenomena over time. It is difficult to determine whether this is due to problems with the data or reflects real instabilities in the climate system and ENSO events. Other studies by Allen (1989), Allen et al. (1989), and Brookfield (1989) have examined SO links with both Indonesian and New Guinean frost

Table 4.2 Association of east monsoon droughts in Java with El Niño type events.

Drought years	El Niño type event years	Notes	Drought years	El Niño type event years	Notes
1844	1844		1913	1914	
1845	1845–46		1914		
1850	1850		1918	1918–19	
1853	None	Event in 1852	1919	1923	
1855	1855		1923	1925–26	
1857	1857		1925	1929–30	
1864	1864		1926	1932	
1873	1873		1929	None	Slight lowering of index
1875	1875		1932	1939–40	
1877	1877–78		1935	1941	
1881	1880	Index low 1880–81	1940	1943–44	
1883			1941	1946	
1884	1884–85		1944	1953	
1885			1945	1954–75	
1888	1887–89		1946	1976	
1891	1891		1953		
1896	1896		Data unavail.		
1902	1902		1976		
1905	1905				

28 (separate events) ÷ 30 (east monsoon drought situations) = 0.93.
93 percent of east monsoon droughts can be associated with El Niño type events.
Source: Quinn et al. (1978).

and drought patterns through time. Although there is a close correspondence between strong SO and El Niño events and instances of widespread frost and drought in this region, there are many mismatches in the historical records. Once again, it is difficult to establish whether this is the result of data reliability or is a real consequence of the variability of physical manifestations of ENSO events over the region, particularly those of weak to moderate character.

Table 4.3 Association of El Niño type events with east monsoon droughts in Java.

El Niño type event years	Drought years	Notes	El Niño type event years	Drought years
1844	1844		1905	1905
1845–46	1845		1911–12	None
1850	1850		1914	1913–14
1852	None	Drought in 1853	1917	None
1855	1855		1918–19	1918–19
1857	1857		1923	1923
1864	1864		1925–26	1925–26
1868	None		1929–30	1929
1871	None		1932	1932
1873	1873		1939–40	1940
1875	1875		1941	1941
1877–78	1877		1943–44	1944
1880	1881	Index low 1880–81	1946	1945–46
1884–85	1883–85		1948	None
1887–89	1888		1951	None
1891	1891		1953	1953
1896	1896		Data unavail.	1954–75
1899–1900	None		1976	1976
1902	1902			

28 (east monsoon drought situations) ÷ 36 (separate events) = 0.78.
78 percent of El Niño type events can be associated with east monsoon droughts.
Source: Quinn et al. (1978).

The earliest attempts to use instrumental data to examine meteorological influences on Australian rainfall, and in particular their possible forecasting potential, followed the lead of Sir Gilbert Walker and colleagues in the Indian Meteorological Service (Hunt et al., 1913; Quayle, 1918, 1929; Taylor, 1918, 1920; Kidson, 1925;

Treloar, 1934; Rimmer and Hossack, 1939). In fact, Walker (1910) and Walker and Bliss (1930) found significant correlations between low (high) mean sea level pressure at Darwin (Honolulu and in South America) and high (low) rainfall in the following summer over north-northeastern Australia. Major Australian initiatives by Quayle (1918, 1929) indicated that mean sea level pressure (MSLP) at Darwin in winter was negatively correlated with the following spring rainfall over eastern inland Australia. Experiments with variants of Walker's algorithms for forecasting northern Australian monsoonal rainfall were undertaken by Treloar (1934) with mixed results. Similar work was done by Rimmer and Hossack (1939) using MSLP and temperature at northern Australian stations and various locations in the Indo-Pacific basin to forecast summer rainfall in Queensland.

Prior to the Second World War, other important studies aimed at forecasting atmospheric fluctuations in Australasia were undertaken by Dutch colonial workers in Indonesia (see a comprehensive review of these studies in Berlage (1957, 1966)). In particular, research by Braak (1919, 1921–29) and Berlage (1927, 1934) indicated that high (low) spring–early summer rainfall over Indonesia, and particularly Java, was significantly correlated with low (high) MSLP at Darwin in the previous autumn–winter. However, a general absence of physical mechanisms to explain SO related correlations stifled further concerted studies for some 20 years.

The resurgence

Renewed interest in the SO phenomenon in the Australasian region began with the works of Schell (1947), Loewe (1948), Reesinck (1952), Berlage (1957, 1961, 1966), Berlage and de Boer (1959), Troup (1965, 1967), Priestley (1962), O'Mahony (1961), and Priestley and Troup (1966). With longer data sets and more observations of parameters in space and with height in the atmosphere, these studies both showed the wider distribution and coherency of the correlations in the Indo-Pacific region and began to indicate something of the dynamics underlying the SO and its possible relationship to oceanic variables. One of the earliest compilations of the spatial pattern of correlations over the globe indicative of SO influences is shown in Fig. 4.1. This diagram shows that the northern Australian–Indian subcontinent region is in the core of a

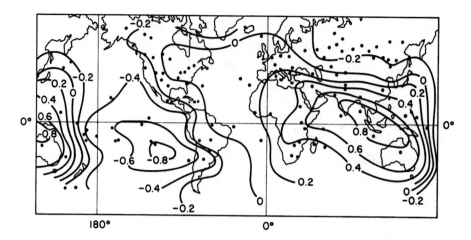

Fig. 4.1 Walker's Southern Oscillation as shown in Berlage (1957). Correlations are of annual pressure anomalies over the globe with simultaneous anomalies at Jakarta, Indonesia (from Rasmusson, 1984).

coherent positive correlation region with respect to pressure fluctuations.

However, it was the works of Bjerknes (1961, 1966, 1969, 1972) that have generally been seen as providing the basis for a decisive leap forward. Not only did they indicate a vital role for large–scale ocean atmosphere interactions and a link with El Niño, but they also suggested physical mechanisms for SO operation in terms of teleconnections and thus provided the focus for the evolution of the term ENSO. In Australasia, this led to a strong emphasis on rainfall patterns and climatic impacts under SO conditions (Nicholls, 1973, 1977, 1979a,b; Pittock, 1973, 1975; Streten, 1975; Pittock et al., 1978; Quinn et al., 1978). The basis for a number of these papers was a re-examination of the early findings of Quayle, Braak and Berlage and the comparison of their results with more recent findings using extended data sets.

ENSO, rainfall (and drought) correlations and associations

One of the major developments in understanding SO and rainfall links in Australasia was a paper by Pittock (1975). It showed that the continental distribution of correlations between annual values of the SO and Australian district rainfall was significant and coherent over much of northern and eastern Australia. Associations between Indonesian droughts and El Niño events were examined by Quinn et al. (1978), and extended into New Guinea by Allen (1989), Allen et al. (1989), Brookfield (1989), and McGregor (1989). In general, they note that although the best matchings are with regard to very strong SO and El Niño events, many discrepancies still occur. These may be the result of fluctuations in the spatial distribution of convective and subsident regions (teleconnections) during different events and time periods (Allan, 1985). This is evident in the spatial nature of SO correlations over Indonesia and New Guinea in Nicholls (1981). Building on the previous studies of Braak and Berlage, Nicholls indicated a general continuity in the spatial distribution of correlations between Darwin MSLP in winter and Indonesia spring rainfall during earlier and more recent data sets (Fig. 4.2a,b). However, further east into what is now Irian Jaya, the correlations at stations such as Manokwari change from –0.336 for the period 1883–1940 to –0.767 for the more recent 1941–65 period.

The study of McBride and Nicholls (1983) generated further understanding of the seasonal nature of SO and rainfall patterns in Australia. Using both simultaneous and lagged correlations between seasonal values of the SO Index (SOI) and district rainfall, they showed that the most (least) significant and coherent patterns occurred during the winter–spring (summer–autumn) period over the eastern portion of the continent. Correlation patterns over southwestern Australia were weak in all seasons. The strongest lag correlations supported the earlier studies of Quayle, with the winter SOI leading spring rainfall over northern and eastern inland Australia (Fig. 4.3a–d). Other relevant contributions were made by Nicholls and Woodcock (1981), Nicholls et al. (1982) and Nicholls (1983) in terms of long-range SO precursors and Victorian and northern Australian rainfall; Nicholls (1979b, 1984e, 1985a) and Hastings (1990) with regard to tropical cyclone frequency and ENSO and by Nicholls (1984d), Holland and Nicholls (1985) and Holland (1986) in terms of northern Australian monsoon onset and ENSO episodes.

These studies found that lower (higher) MSLP at Darwin in winter and warmer (colder) SSTs in northern Australian waters in spring were significantly correlated with more (less) tropical cyclone numbers in the following summer. The large-scale link

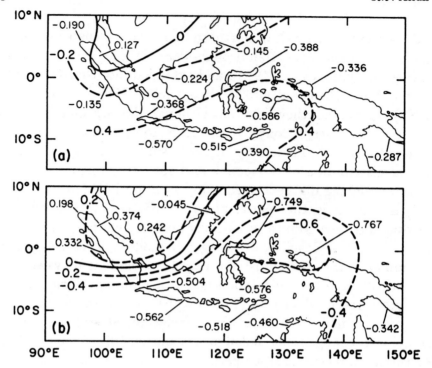

Fig. 4.2 (a) Lag correlation between Darwin pressure averaged for austral
winter (JJA) and Indonesian austral spring (SON) rainfall totals.
Data from 1883 to 1940. (b) As in (a), except from 1941 to 1965
(from Nicholls, 1981).

with the Indo-Australasian monsoon system and the ENSO phenomenon followed from the close interplay between the ocean and
atmosphere, with ENSO events often being preceded by an early
onset to the monsoon and cooler SSTs in northern Australian
waters, and coincident with a late monsoon onset and warmer
northern Australian SSTs. On longer time frames, Allan (1985)
has shown strong associations between persistent positive (negative) SOI periods and above (below) average spatial extent of
Australian annual rainfall from 1885 to 1976 (Fig. 4.4). These periods have been examined further in a recent study by Zhang and
Casey (1990).

When taken together with the findings of Williams et al. (1986)
and Nicholls (1988a), there is evidence for strong links between
Australian rainfall and ENSO over the last 200 years, although
some fluctuations in the relationship on decadal time scales are
also apparent. This is reinforced by the studies of Allan (1985),
Adamson et al. (1987), Whetton and Baxter (1987) and Whet-

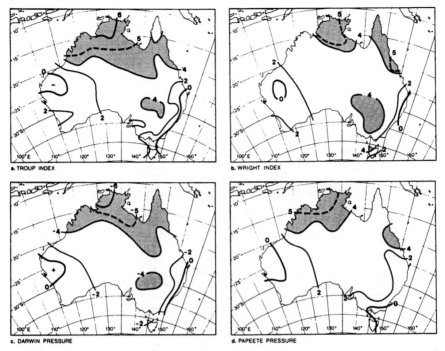

Fig. 4.3 (a–d) Patterns of lag correlation coefficients between austral winter (JJA) mean values of Southern Oscillation Indices, and austral spring (SON) district rainfall over Australia, 1932–74 (from McBride and Nicholls, 1983).

ton et al. (1990), which collectively have indicated links and correlations between ENSO (anti-ENSO) and low (high) river discharges in southeastern Australia and a dry (flooded) Lake Eyre in South Australia. Recent works by Isdale and Kotwicki (1987) and Kotwicki and Isdale (1990) have examined the links between Lake Eyre flooding history and the climatic reconstructions for northeastern Australia from Great Barrier Reef coral cores. They suggest that these records show a strong ENSO modulation through time.

Attempts have also been made to extend relationships between ENSO and Australian rainfall fluctuations through the use of oceanic variables alone. Nicholls (1981, 1984a, b) examined seasonal correlations between northern Australian SSTs and district rainfall. The most important long-range relationship found was a negative correlation between northern Australian summer–autumn SST and the following winter rainfall over eastern Australia. Recent work by Allan et al. (1990) has indicated that a robust and more significant relationship can be found between

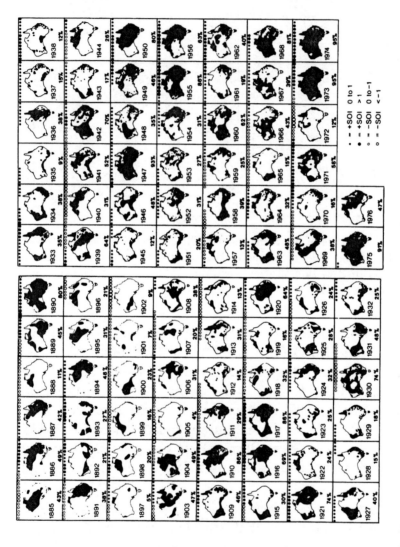

Fig. 4.4 Spatial distribution of above-average annual rainfall over Australia (shaded), compared with Wright (1977) seasonal Southern Oscillation Index (SOI), 1885–1976. The percentage of the continent's land area experiencing above-average annual rainfall is shown below each map. The range of + and – SOI values is shown in the key with open circles and dots (from Allan, 1985).

north-northwestern Australian (Darwin combined with Port Hedland) sea level in spring and the next winter rainfall over south-
southeastern and northern Australia (Fig. 4.5).

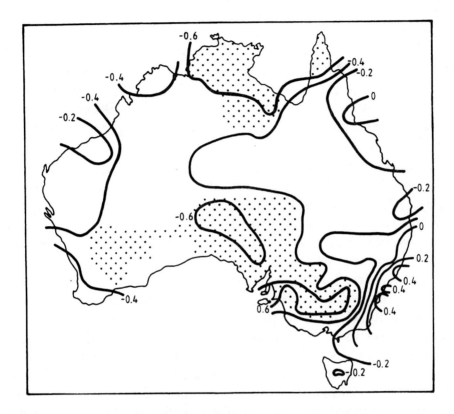

Fig. 4.5 Pattern of lagged cross correlations between austral spring (SON)
sea level anomalies at Darwin and Port Hedland (combined and averaged) and the following austral winter (JJA) district rainfall anomalies, 1966–82 data. Regions where the correlations are significant at
the 95 percent confidence level are stippled (from Allan et al., 1989).

A specific analysis of relationships between ENSO and drought
in Australia was carried out by Nicholls (1985b). Again building
on many of the results and findings reported above, this study developed an index of large-scale continental drought using the first
principal component of winter–spring district rainfall which was
predicted by the previous summer–autumn SST and air temperature in northern Australia. The ramifications in terms of economic
and agricultural impacts are examined in later sections.

The overall complexity of ENSO/anti-ENSO relationships in
Australasia can be seen in the extent of correlations between

oceanic and atmospheric variables in the region (Hackert and Hastenrath, 1986; Allan and Pariwono, 1990; Allan et al., 1990). Major features are the coherency of parameters during the winter–spring period and their generally poor coherence in autumn, the dominance of ENSO/anti-ENSO influences and, from a forecasting perspective, significant lag correlations in the range of three to nine months. Most recently, Drosdowsky (1990) has pointed to a possible ENSO precursor in midlatitude Australasia. This involves a negative pressure difference between the New Zealand (Chatham Island) and southern central Indian Ocean (Amsterdam Island) regions in the austral winter–spring months prior to an ENSO episode.

However, there have been concerns about the long-term stability of some of the correlation patterns detailed in the literature (Ramage, 1983; Pittock, 1984; Allan, 1985). In the Australasian region, Treloar and Grant (1953) identified fluctuations in the magnitude and sign of correlations suggested by Quayle (1918, 1929), Treloar (1934) and Rimmer and Hossack (1939) (Tables 4.4, 4.5, and 4.6). From these results, they have questioned the viability of such correlations from a forecasting perspective. In particular, they show major fluctuations in many relationships when comparisons are made between various time periods. With regard to Quayle's link between Darwin MSLP in winter and northern Victorian rainfall in spring, they show an increase in correlation magnitude between the 1889–1908 and 1909–28 periods and a marked decrease in the correlation values between the 1909–28 and 1929–48 periods. Nicholls and Woodcock (1981) report strong lag correlations for this link when using data from the more recent 1952–78 period. Overall, this suggests long-term fluctuations in this teleconnection. Other correlation changes were found with regard to relationships with northern Australian (Treloar, 1934) and Queensland (Rimmer and Hossack, 1939) rainfall when contrasting the 1886–1918 and 1919–41 to 1950 periods.

Major fluctuations in the global SO pattern shown earlier in Fig. 4.1 were detailed for two time periods in Berlage (1961) (Fig. 4.6a,b). Although these periods are relatively short, there is a noticeable increase in the spatial extent of the positive correlation "core region" covering low latitude Australasia, India and the equatorial Indian Ocean. Using spectral analysis, Wright (1977) has indicated that the SO power spectra for four contiguous

Table 4.4 Lag correlations between MSLP at Darwin, Alice Springs, Adelaide, Hobart and Alice Springs + Adelaide + 2×Darwin during the austral winter months of June and July and northern Victorian rainfall in the following austral springs months of August and September for the 1889–1908, 1909–1928, 1929–1948 and 1889–1948 time periods.

	Pressure at				
Period	Darwin	Alice Springs	Adelaide	Hobart	Alice Springs + Adelaide + 2 Darwin
1889–1908	−0.62	−0.25	−0.40	−0.61	−0.49
1909–1928	−0.87	−0.71	−0.65	−0.47	−0.77
1929–1948	−0.22	−0.54	−0.50	−0.14	−0.53
1889–1948	−0.65	−0.57	−0.62	−0.41	−0.67

Source: Treloar and Grant (1953).

Table 4.5 Lag correlations between MSLP at Darwin during the austral winter months of June and July and rainfall at various stations in northern Victoria during the following austral springs months of August and September for the 1889–1908, 1909–1928, 1929–1948, and 1889–1948 time periods.

	1889–1908	1909–1928	1929–1948	1889–1948
Bendigo	−0.57	−0.87	−0.14	−0.64
Charlton	−0.52	−0.86	−0.03	−0.56
Dookie	−0.68	−0.84	−0.30	−0.67
Echuca	−0.46	−0.86	−0.14	−0.60
Horsham	−0.46	−0.75	−0.23	−0.52
St. Arnaud	−0.54	−0.82	−0.19	−0.56
Swan Hill	−0.60	−0.58	−0.11	−0.60
Warracknabeal	−0.54	−0.79	−0.12	−0.56
Yarrawonga	−0.57	−0.82	−0.41	−0.65
Northern Victoria	−0.62	−0.87	−0.22	−0.65

Source: Treloar and Grant (1953).

Table 4.6 Lag correlations between MSLP at Darwin and Honolulu during austral autumn–spring months and rainfall at various locations and stations in Queensland and the Northern Territory during the following austral spring–early summer months for the 1886–1918 and 1919–1950 time periods.

Rainfall District	Period	Darwin Pressure			Honolulu Pressure		
		Period	1886–1918	1919–50	Period	1886–1918	1919–41
Relationships used by Treloar (1934)							
Darwin	Sept.–May	July	−0.65	−0.31			
Pine Creek	Sept.–May	July	−0.40	−0.21			
Victoria River Downs	Sept.–May	July	−0.41	−0.06			
Coastal	Oct.–Apr.	July	−0.68	−0.13	Mar.–Aug.	−0.67	+0.16
Inland	Oct.–Apr.	July	−0.66	−0.25	Mar.–Aug.	−0.57	−0.33
Relationships used by Rimmer & Hossack (1939)							
St George	Nov.–Jan.				June–Aug.	+0.73	+0.24
Downs	Nov.–Jan.				June–Aug.	+0.68	+0.40
Ipswich	Nov.–Jan.	June–Aug.	−0.64	−0.13	June–Aug.	+0.68	+0.15
Roma	Nov.–Jan.				June–Aug.	+0.63	+0.18
Springsure	Nov.–Jan.				June–Aug.	+0.63	+0.20
Charters Towers	Dec.–Feb.	July–Sept.	−0.32	−0.09	July–Sept.	+0.64	+0.35
Georgetown	Dec.–Feb.	July–Sept.	−0.74	−0.27			
Springsure	Jan.–Apr.	Aug.–Oct.	−0.72	+0.14	May–July	+0.54	−0.32

Source: Treloar and Grant (1953).

(a)

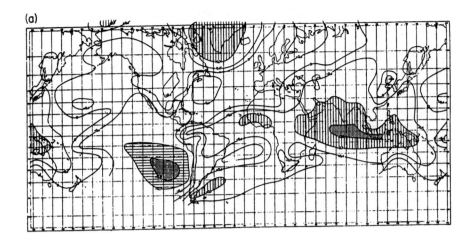

(b)

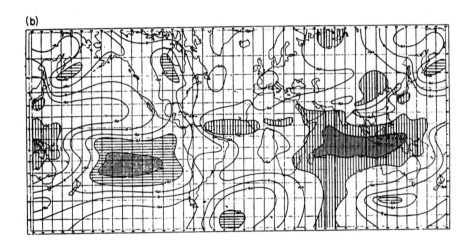

Fig. 4.6 (a) Simultaneous correlation between annual air pressure anomalies over the globe with anomalies at Jakarta, Indonesia for the period 1931–39 inclusive. (b) As in (a), except for the period 1949–57 inclusive. Regions of correlations above + (−) 0.7 are shaded by vertical (horizontal) lines, those above + (−) 0.9 are shaded by denser vertical (horizontal) lines (from Berlage, 1961).

32–year periods during 1851 to 1972 show changes in the dominant frequencies of the order of three to 12 years. Fluctuations in the spatial distribution of correlation patterns between SOIs and Australian district rainfall are also evident in McBride and Nicholls

(1983). This is shown in Fig. 4.7, where simultaneous correlation patterns between MSLP at Darwin and Australian district rainfall vary considerably over the central–western and east coastal portions of the continent when comparing the 1933–53 and 1954–74 periods. Only in the coastal margins of northern Australia are the correlation patterns of similar magnitude and sign. Perhaps this overall concern with correlation stability is best summed up by Pittock (1984; p. 80):

> ... *there is evidence that at least some of the patterns and sequences of correlations do not remain constant with time.*

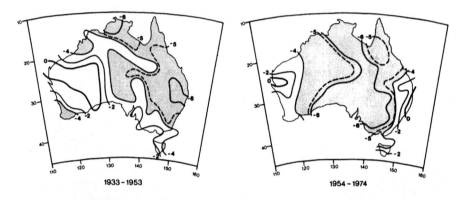

1933 – 1953   1954 – 1974

Fig. 4.7 Pattern of simultaneous correlations between austral spring (SON) MSLP at Darwin and district rainfall over Australia for the 1933–53 and 1954–74 time periods. Regions where the correlations are significant at the 95 percent confidence level are shaded (from McBride and Nicholls, 1983).

Dynamical, diagnostic and modeling studies

Physical support for statistical correlations relating ENSO and climatic events in Australasia has occurred only in the last 15–20 years. Apart from the early studies of Troup (1965, 1967), the vast majority of dynamical and diagnostic research on the atmosphere and ocean has been undertaken only since the mid-1970s.

A number of these studies have focused on large-scale cloudiness/convective changes in the Australasian region. Nicholls (1973, 1984a,b), Barry (1978) and Wright et al. (1985) have noted that there are reductions (increases) in cloudiness in the eastern Indonesian–New Guinean region during ENSO (anti-ENSO) episodes. This is associated with cloudiness changes further east, where Streten (1973, 1975) suggests that the South Pacific Convergence Zone (SPCZ) is displaced further to the east (west) during ENSO (anti-ENSO) years.

Evidence for major changes in circulation and tropical to higher latitude interactions in the region have been reported by Allan (1983, 1984, 1985), Williams (1987), Wright (1987), Drosdowsky (1988) and Whetton (1988, 1990a,b), and can also be seen in the broader, Southern Hemisphere studies of Pittock (1984), van Loon (1984), van Loon and Rogers (1981), Rogers and van Loon (1982), van Loon and Madden (1981), van Loon and Shea (1985) and Kiladis and van Loon (1988). These studies tend to support the indications of seasonal changes in persistence found in correlation analyses, with higher (lower) MSLP occurring over northern (central–southern) Australia during summer (winter) months of ENSO (anti-ENSO) events. More specific links between northern Australian SSTs and southeastern Australian rainfall have been found to be a consequence of extensive tropical to midlatitude cloud bands, resulting from convective activity generated by anomalously warm SSTs in the Timor Sea–North West Shelf and Coral Sea regions and often linked to midlatitude frontal systems during anti-ENSO phases (Wright, 1987; Wright, 1988a,b; Whetton, 1988, 1990b). General relationships involving tropical circulation fluctuations in Southern Hemisphere summers during ENSO and anti-ENSO episodes were explored by Allan (1983, 1984, 1985). These studies would tend to support the correlation patterns relating SOIs to eastern Australian rainfall, with evidence for marked subsidence (convective activity) in these longitudes during ENSO (anti-ENSO) events. They also indicate that broad scale rainfall patterns tend to be out of phase between the eastern and western portions of the Australian continent. At present, there are no strong correlations between ENSO or anti-ENSO episodes and rainfall patterns over southwestern Australia.

Substantial support has come more recently from a better understanding of ocean–atmosphere coupling processes. In particular, studies by Nicholls (1984a,b), Streten (1981, 1983), Hastenrath and Wu (1982, 1983), Wright (1984), Hackert and Hastenrath (1986) and Allan and Pariwono (1990) have shown that ocean–

atmosphere interactions in low-latitude Australasia vary seasonally and exhibit responses very closely linked to ENSO and anti-
ENSO events. Such fluctuations involve changes in SSTs, wind
fields, sea level, MSLP, cloudiness and oceanic circulation. SST–
wind relationships appear to be particularly important in "controlling" convective sources for tropical "moist infeed" that can
penetrate into the midlatitudes of the continent via cloud bands.
In fact, the location, strength and persistence of SST anomalies
appears to be linked closely with the seasonal nature of Australian
rainfall and ENSO and anti-ENSO episodes (Nicholls, 1989). Some
aspects of the sea level response to ENSO events have been examined in Pariwono et al. (1986), Pariwono (1987) and Allan and
Pariwono (1990). In general, the sea level response is an integration of a number of parameter forcings related to ENSO and
anti-ENSO extremes. As such, it is bound closely to the pattern
of responses. The nature of ocean–atmosphere fluctuations during
strong ENSO and anti-ENSO events in Australasia is shown in
Fig. 4.8.

Numerical modeling studies have provided additional support,
especially in terms of northern Australian SST and MSLP links
with wider Australian rainfall patterns. These studies have used
fixed SST anomalies in tropical Indo-Pacific locations as a forcing on atmospheric conditions (Voice and Hunt, 1984; Simmonds
and Smith, 1986). Recent work by Simmonds and Trigg (1988)
suggests that the location of SST anomalies is crucial to the response of the model atmosphere in terms of MSLP and rainfall
responses over Australasia. In particular, the presence of warm
or cold SSTs in the Timor Sea–North West Shelf, Indonesian and
Coral Sea regions was important for the distribution of above-
and below-average rainfall over the Australian continent. Ongo-
ing efforts to use coupled ocean–atmosphere models with fully interactive oceans and atmosphere are likely to improve our understanding of these interactions. This is especially true with regard
to developments in physical oceanography, and the dynamics of
oceanic models and their treatment of features such as Indonesian
"throughflow" (the mass transport of water between the Pacific
and Indian Oceans via the Indonesian archipelago), the Circumpo-
lar Current and the poleward flowing Leeuwin Current on the continental shelf margins of Western Australia (Allan, 1988). From
an atmospheric perspective, there are now efforts to understand

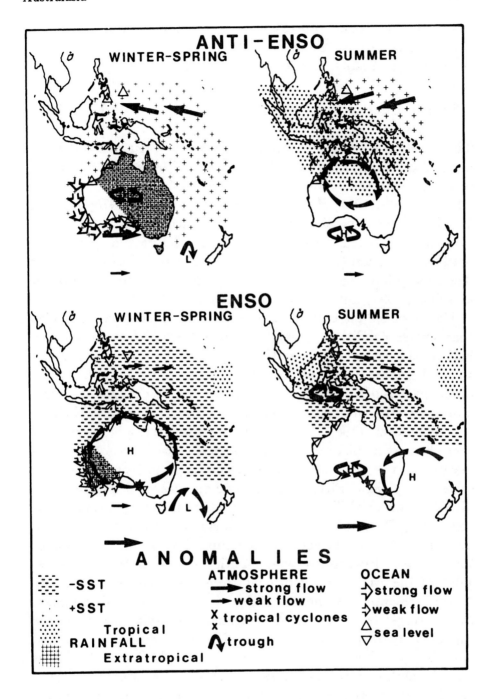

Fig. 4.8 Schematic representation of Australasian ocean-atmosphere responses during "typical" strong anti-ENSO and ENSO events. The seasons shown are for the Southern Hemisphere, winter–spring (JJASON) and summer (DJF) (from Allan, 1988).

the nature of physical processes due to the nonlinear nature of the
governing equations, those related to ocean–atmosphere interactions, such as ENSO, and to various low frequency signals.

Climatic fluctuations, low frequency signals and ENSO

The question of long-term trends in historical climatic patterns,
particularly Australian rainfall, has been addressed by a number
of studies over the last 35 years (Kraus, 1955; Das, 1956; Deacon,
1953; Gentilli, 1971; Pittock, 1975, 1983; Tucker, 1975; Russell,
1981). However, possible links between such trends and fluctuations in correlation patterns involving ENSO have only recently
received attention (Allan, 1987, and Zhang and Casey, 1990, for
Australia; Elliott and Angell, 1988; Quinn et al., 1987; Trenberth
and Shea, 1987; Enfield, 1988, for more global scales).

The evidence for climatic fluctuations in Australasia indicates
that wetter conditions prevailed across the northern to eastern
half of the continent from the 1860s to the 1910s and again
since the late 1940s, with a drier intervening period. Such
broad patterns are coincident with the major changes in correlation patterns outlined in Treloar and Grant (1953) and Nicholls
and Woodcock (1981). Attempts to find atmospheric circulation changes in historical meteorological data have focused primarily on the position of the anticyclonic pressure belt over
eastern Australia (Pittock, 1975) and the seasonal nature of individual station MSLP data over various time periods (Deacon, 1953; Das, 1956). A more conclusive result was obtained
by Allan (1987), through the examination of the location of
the Southern Hemisphere winter anticyclone over the central regions of the Australian continent from the 1880s and 1890s up
to the present. As shown in Fig. 4.9a, this analysis suggests
that across the center of the continent the winter anticyclone
was steadily displaced further to the north during the period
1920–50, and has tracked back to the south since. If this is
taken together with the 10-year average Adelaide minus Sydney MSLP differences since 1861 (Fig. 4.9b), there are suggestions of seasonal changes in the meridional wind field over southeastern Australia as pressure differences wax and wane in magnitude. The most notable features are the stronger (weaker)
southerlies in early mid-Southern Hemisphere winter during the

1870s–80s and again in the 1940s–50s (1900s–10s and 1970s), and the stronger northerlies (southerlies) during Southern Hemisphere summer months of the 1870s, 1890s–1910s and since the 1950s–60s (1920–40s). Such changes suggest stronger winds from the major Southern Ocean moisture source during the winters of the 1870s–80s and the 1940s–50s and from tropical moisture sources allied with the summer monsoon in the 1870s, 1890s–1910s and the 1950s–70s. Overall, this indicates that the higher rainfall regimes over eastern Australia in the 1870s–1910s resulted from a combination of stronger tropical summer and midlatitude winter influences, while the wetter phase since the 1940s began with a period of increased midlatitude southerlies in winter and was then followed by a period of enhanced tropical northerlies in summer. Weaker winter influences occurred in the 1900s–20s, with weaker summer influences during the 1920s–40s.

Links between ENSO and these broad climatic fluctuations were also examined by Allan (1987), Elliott and Angell (1988), and Zhang and Casey (1990). Using 10-year MSLP transects contoured for data stations from Adelaide to Alice Springs during Southern Hemisphere winters since the 1880s, Allan (1987) has shown a broad pattern of latitudinal shifts in the winter anticyclone (Fig. 4.10a). These patterns were then juxtaposed with a similar transect compositing the MSLP for winters with very strong ENSO events (Quinn et al., 1978) in each 10-year period (Fig. 4.10b). A comparison of the two diagrams suggests that increased anticyclonicity over the Australian region during the winter months of ENSO events is linked closely with the longer-term pattern of circulation. Certainly, such relationships may explain many of the observed changes in the spatial distributions of correlations over time.

Elliott and Angell (1988) have examined wider SO relationships using data sets covering the last 100 years. Their findings indicate that correlations were generally weakest in the period from the 1920s–50s and strongest in the 1880s–1910s and particularly since the 1950s. They suggest that such results could be interpreted as showing changes in the locations and magnitudes of the "centers of action" across the Indo-Pacific region, and when contrasting the 1883–1940

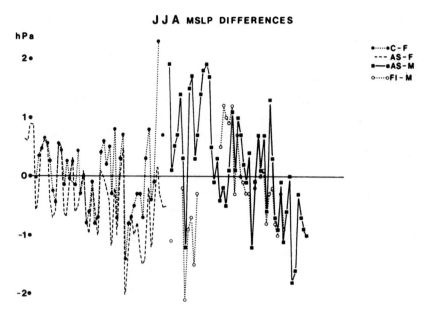

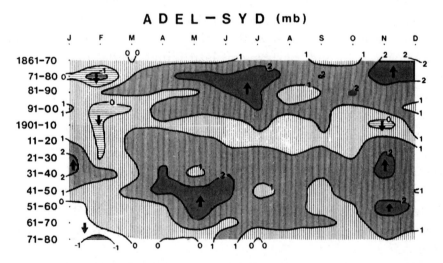

Fig. 4.9 (a) Austral winter (JJA) MSLP differences in hPa between
         stations from 24°S to 31°S across the center of Australia
         from 1888 to 1987.   Stations used are C=Charlotte Wa-
         ters (25°56′S, 134°55′E); F=Farina (30°4′S, 138°16′E); AS=Alice
         Springs (23°49′S, 133°54′E); M=Marree (29°39′S, 138°3′E) and
         FI=Finke (25°35′S, 134°34′E). Positive (negative) differences indi-
         cate that the high pressure belt is displaced northward (southward).
         (b) Adelaide minus Sydney 10-year MSLP differences in hPa from
         1861 to 1980 (from Allan, 1987).

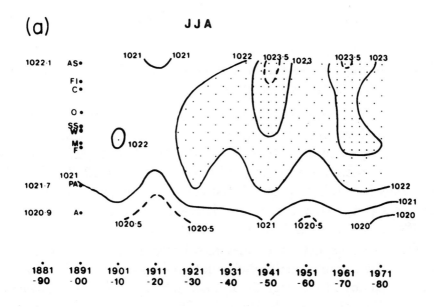

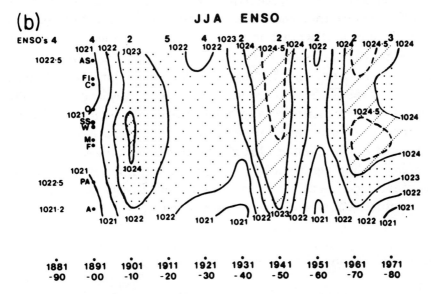

Fig. 4.10 (a) Transects from Adelaide to Alice Springs of 10 year MSLP for the austral winter (JJA) period from 1881 to 1980. MSLP is in hPa, with pressures above 1022 hPa shaded. Stations used are AS=Alice Springs; FI=Finke; C=Charlotte Waters; O=Oodnadatta; SS=Strangways Springs; W=William Creek; M=Marree; F=Farina; PA=Port Augusta and A=Adelaide. (b) As in (a), except for composites of ENSOs during each 10-year period. The number of ENSO events composited in each decade is shown along the top of the diagram. Data for all stations is not available before 1890, so only the MSLP for AS, PA and A are shown (from Allan, 1987).

and 1941–84 periods note that (Elliott and Angell, 1988, p. 737):

> ... *the SO was certainly weaker or operating in a different mode during the earlier period.*

Recently, Zhang and Casey (1990) have provided a finer resolution of these periods using Darwin MSLP. They find four predominantly positive (1884–1906, 1911–15, 1939–47, 1976–88) and three predominantly negative (1906–11, 1915–39, 1947–76) MSLP anomaly intervals that are statistically significant since the 1880s. It is suggested that these periods are marked by a dominance of ENSO and anti-ENSO events, respectively.

One other aspect that must also be considered in the context of climatic fluctuations and circulation changes is the influence of low frequency MSLP signals propagating through the Indo-Pacific region (Barnett, 1984a,b, 1985a,b, 1988). In fact, Krishnamurti et al. (1986) report that these features display decadal fluctuations between Asian and Antarctic to southern African source regions. Possible ramifications for ENSO links with such features in Australasia follow from the findings in Allan and Pariwono (1990). They note that ENSO-like conditions in Australasia during 1967 occurred in the absence of any wider ENSO signal in the Pacific basin, and attribute them to the passage of an Antarctic to southern African low frequency MSLP feature through low-midlatitude Australasia. Further research is required to improve the current understanding of possible links between the ENSO phenomenon, climatic fluctuations and low frequency features.

Ecological impacts and manifestations

Research into ENSO impacts on the ecology of terrestrial and marine environments in Australasia has expanded rapidly in recent years. Some of the manifestations can be linked directly to changes in rainfall and atmospheric and oceanic circulations. However, in general, many are indirect influences, with ENSO impacts often modulated by, or in concert with, other forcing factors.

Terrestrial ENSO and anti-ENSO influences have generally been sought in the area of economically important agricultural products and crops. Several papers by Nicholls (1984c, 1985b,c, 1988b) have attempted to quantify possible ENSO relationships with cereal crops, particularly wheat and sorghum (Figs. 4.11a,b, 4.12, 4.13). In Fig. 4.11a, the scores of the first principal component of Australian district average June to November rainfall (used as a drought index) are aligned closely with Australian wheat yields since 1951. As such rainfall patterns have been related to ENSO, it is suggested that forecasting major ENSO and anti-ENSO manifestations in the Southern Hemisphere winter–spring period would be most beneficial to the rural economy (McKeon et al., 1990). This is evident in Fig. 4.11b, where the total value of Australian crops is linked strongly to the index of drought.

The potential for forecasting crop yields from ENSO and anti-ENSO precursors has been shown to be particularly relevant to sorghum yields (Nicholls, 1985c, 1988b) (Figs. 4.12 and 4.13). In Fig. 4.12, sorghum yield anomalies are shown to be aligned closely to the Darwin MSLP trend from the previous January–March to June–August months since 1954. The statistical significance of lag correlations between monthly MSLP at Darwin and sorghum yield is shown in Fig. 4.13. There would appear to be a case for examining the negative correlations from May to September and even the earlier positive correlations in January and February with the ENSO phenomenon in terms of likely yields in the next year, or the viability of planting in a particular year.

Links between weather and insect populations and migrations have received increasing attention over the last 40 years. This has included insect responses to most scales of atmospheric structure and motions. The possibility of ENSO influence on insect life cycles has been suggested by Drake and Farrow (1988). These influences appear to be derived from major changes in wind systems and rainfall patterns. The implications for Australian crops and insect pests require further investigation.

In Indonesia, ENSO events have been seen to influence fluctuations in wet and dry season rice crops through the prevalence of drought conditions. This has been reported as having occurred during the 1972 and 1976 ENSO events (Malingreau, 1987). However, more concerted studies of the impact of the 1982–83 ENSO on the Indonesian rice crops suggest that the situation is more complex. Given the varied topography and geography of the Indonesian archipelago and the nature of wet and dry season rice growing regions, this is not surprising. Basically, the Malingreau

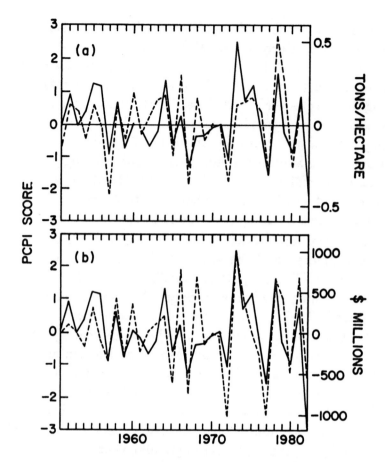

Fig. 4.11 (a) Time-series of the first principal component scores (PCP1) (full
lines) of Australian winter-spring (JJA-SON) district rainfall and
anomalies of Australian wheat yield in metric tons/hectare (broken
line). (b) As in (a), except that PCP1 scores are plotted against
the total value of Australian crops in $A millions (broken line) (from
Nicholls, 1985b).

(1987) study suggests that the growth rate rather than the production of rice was most affected, particularly in eastern Indonesia.
For the dry season crop there was a 10 percent reduction in the
area harvested in 1982, while the wet season crop suffered from de-
lays in planting due to the late onset of the Australasian summer
monsoon in 1982–83. Otherwise, the Indonesian rice crop fared
reasonably well under the impact of the 1982–83 ENSO. Thus,
the most direct climatic impacts are from low and/or late monsoon rainfall, which generally occurs in the broad longitudes of
eastern Indonesia to New Guinea during ENSO events.

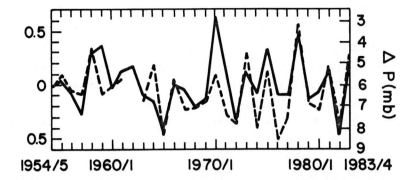

Fig. 4.12 Time-series of sorghum yield anomalies per hectare (full line) and
the trend in Darwin MSLP from January–March up to June–August
(austral summer to winter) prior to planting (broken line). Scale for
MSLP trend is reversed. MSLP data for 1962–63 are missing (from
Nicholls, 1985c).

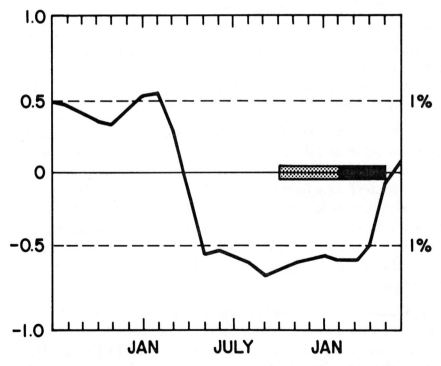

Fig. 4.13 Correlations of Australian sorghum yield (metric tons per hectare)
with monthly means of Darwin station-level pressure during and before the sorghum growing season. Data from 1954–83. The planting
period is shown as a hatched box and the harvest period as a black
box. The 1 percent significance levels for the correlation coefficients
are shown as thin dashed lines (from Nicholls, 1988b).

On a larger scale, and of more indirect nature, are instances of widespread bushfires and possible links to ENSO in Australasia. Many factors have been seen to act towards creating dangerous bushfire conditions. Such factors include many non-meteorological or climatological elements that can act on different scales leading up to, and during, bushfire events. It would seem that the most important factors linking bushfires to ENSO in southeastern Australia are the prevalence of a wet winter–spring to summer period in the year before ENSO and a drier period during the corresponding phase of the following ENSO year. In the absence of other influences, such conditions would favor a buildup in potential "fuel" throughout the growing season in an anti–ENSO year. The most extensive research into such relationships has only occurred since the massive bushfires associated with the 1982–83 ENSO (Allan and Heathcote, 1987). More studies of these potential links are required. Closer links between major ENSO events and major fires in Kalimantan are noted by Malingreau (1987).

Surprisingly, most research on ENSO linkages with ecological variables in Australasia has focused on the marine environment. This has included associations with tuna, rock lobster, prawns, green turtles and corals. Most links have been found with aspects of the food chains or life cycles of these organisms.

Examining southern blue fin tuna catches in the western Tasman Sea between 1966 and 1974, Rochford (1981) concluded that warmer eastern Australian SSTs (especially in the Southern Hemisphere winter) were found to coincide with both ENSO events and low tuna catches in 1969, 1973, and 1976 (Fig. 14). It was suggested that the tuna did not tolerate temperatures above 20°C and were sensitive to oceanic temperature fronts. However, the study also found that it was difficult to separate local effects from large-scale ENSO influences.

In southwestern Australian waters, there have been concerted studies of the local rock lobster fishery. Possible relationships with climatic and oceanic factors have been examined by Phillips (1981) and Phillips and Pearce (1988). This research has suggested links between various stages in the rock lobster life cycle and both the SO and sea level fluctuations at Fremantle. In Fig. 4.15 there are indications of a rapid, in-phase response in terms of "settlement" of the larval stage and both the SOI and the Fremantle sea level.

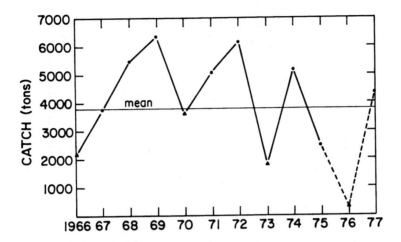

Fig. 4.14 Changes in total catch of southern blue fin tuna by the New South Wales (NSW) pole fishery 1966–77. The 1976 catch was so poor that only an approximate value was recorded. Years with anomalously warm austral winter SST in eastern Australian waters are shown by a solid triangle (from Rochford, 1981).

From a physical perspective, it is thought that large offshore eddies provide the impetus for the early lobster stage to be shifted thousands of kilometers offshore and to return and resettle later in the inshore reefs. Why so much time is required to complete this process is still unknown. One possible explanation may involve the time and size of the warm eddies, as they are a feature of the warm, coastal Leeuwin Current which flows to the south bringing low salinity tropical waters to southwestern Australia. Relationships between the Leeuwin Current, oceanic throughflow via Indonesian waters and ENSO have been postulated by some researchers. Generally, they suggest that reduced (increased) throughflow and Leeuwin Current flow are a response to ENSO (anti–ENSO) influences on the Pacific and Indian Ocean trade wind fields. However, such links have yet to be reliably established (Allan, 1988; Allan and Pariwono, 1990; Prata et al., 1989, 1990).

Another living marine resource, whose productivity has been associated with ENSO conditions in Australasia, is the banana prawn in the Gulf of Carpentaria (Staples, 1983; Vance et al., 1985; Love, 1987). The study of Vance et al. (1985) attempted to correlate a number of atmospheric variables with banana prawn catch data from 1970–79. Basically, the best correlations with prawn

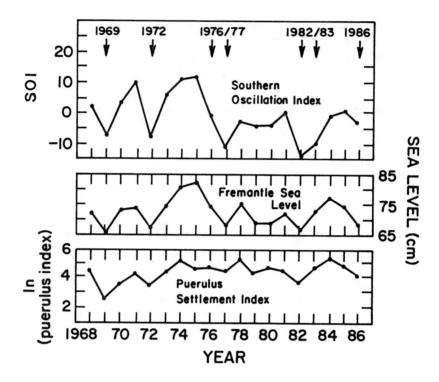

Fig. 4.15 Annual mean data for the Southern Oscillation Index (SOI), Fremantle (near Perth, Western Australia) sea level, and an index of *puerulus*
settlement at Seven-Mile Beach, Dongara (just south of Geraldton,
28°42′S, 114°42′E). ENSO events are indicated by arrows (there was
an El Niño off Peru in 1976, but the SOI was in fact lowest in 1977).
The SOI used is that of Troup (1965) (from Phillips and Pearce,
1988).

catch were as follows: positive correlations with spring–autumn
rainfall in southern regions of the gulf; a negative correlation with
summer rainfall in one northern region and positive correlations
with wind components and air temperature in most northern regions (Table 4.7). The highest correlations were with rainfall and
banana prawn catches in the southern Gulf of Carpentaria, e.g.,
the Karumba region (Fig. 4.16). These relationships have been updated yearly and used to predict catches up to six weeks prior to
the fishing season. CSIRO research has shown that the main causal
mechanism underlying the relationships is the effect of rainfall on
the emigration process of prawns from their estuarine nursery areas into the coastal fishing grounds. Staples (1983) linked these

Table 4.7 Simultaneous correlations between mean seasonal and annual rainfall and regional prawn catch in the Gulf of Carpentaria for the years 1970–79 inclusive. Weipa, Mitchell River and Groote Eylandt are denoted as northern locations, while Karumba, Mornington Island and Limmen Bight are on, or near, the southern coast of the Gulf.

Region	Winter	Spring	Summer	Autumn	Annual
Weipa	−0.46	0.46	−0.41	0.11	−0.02
Mitchell River	0.89†	0.14	−0.10	0.56	0.14
Karumba	−0.12	0.76†	0.80†	0.37	0.86†
Mornington Island	0.78†	0.34	0.59	0.03	0.74†
Limmen Bight	0.82†	0.57	0.46	0.49	0.77†
Groote Eylandt	0.75†	0.69*	−0.22	0.15	0.22

*0.01 < P < 0.05
†P < 0.01
Source: Vance et al. (1985).

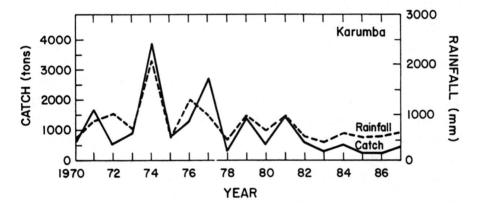

Fig. 4.16 Relationship between rainfall and banana prawn catch in the Karumba region of the Gulf of Carpentaria (from Staples, 1983).

correlations with the SOI and described the effect of the ENSO on the Gulf of Carpentaria banana prawn catches.

Love (1987) has extended this analysis and suggested the use of the SOI as a possible predictor of total banana prawn catch. Using a more extensive data set of prawn catch from the entire Gulf of Carpentaria from 1968 to 1984 and Troup's SOI (Tahiti minus Darwin normalized MSLP), this study indicated that the spring (especially November) SOI could be used to forecast total season banana prawn catch (Fig. 4.17). The correlation coefficients

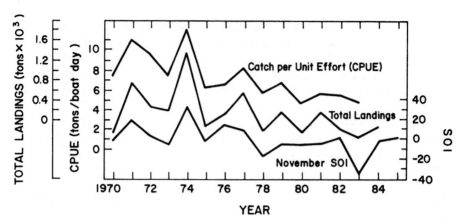

Fig. 4.17 Time-series of November Southern Oscillation Index (SOI), total landings of banana prawns (metric tons) and the catch per unit effort (metric tons/boat day) in the Gulf of Carpentaria from 1970 to 1985. The SOI used is that of Troup (1965) (from Love, 1987).

between the variables shown in Fig. 4.17 were of the order of 0.8 and are significant at the 0.001 confidence level.

Northern Australian green turtle numbers have been related to ENSO events by Limpus and Nicholls (1988) and Nicholls (1988). This study suggested that the numbers of breeding turtles on northern Australian and Indonesian islands lag the ENSO phenomena by two years. In particular, data collected from two northern Queensland rookeries on Heron and Raine Islands have been related to the Darwin MSLP from November to January two years earlier. Despite the small data samples involved, significant correlations were reported between Darwin MSLP and turtle numbers on both islands some 20 to 30 months later. The biological factors causing such responses could not be resolved by this study, although some influence on turtle breeding habits was suspected.

Attempts to relate a number of fishery responses to interannual climatic fluctuations in the Tasmanian region have been reported by Harris et al. (1988). The extent of correlations between a number of biological, atmospheric and oceanic variables in this study is shown in Table 4.8. In general, the results suggest a wide range of marine and aquatic organisms are related to climatic fluctuations and especially ENSO influences. Some of the most significant simultaneous correlations suggesting larger scale influences are between: Darwin MSLP and the spring bloom at Maria Island and also Tasmanian lobster catch; Hobart MSLP and both Tasmanian

Table 4.8 An assemblage of statistically significant simultaneous correlations between various climatological, limnological, oceanographical and fisheries data in the Australian region.

	DAR	HOB	MAC	ZWW	MAX	RAI	GLL	DEL	TRO	SPR	MAR	TAS	NSW	NZ	TUN	SEA
DAR	—									0.40		0.32‡				
HOB	0.41‡	—		-0.65†	0.73†			-0.33				0.72‡	-0.57†			
MAC	-0.69		—	-0.67			0.47			-0.40	0.43	-0.36‡			-0.47	-0.33
ZWW				—	-0.28	0.30		0.43		0.50		-0.35‡	0.62†	0.30		
MAX					—				0.53*			0.66†				
RAI						—		0.46								
GLL							—	0.31	0.40	-0.50	0.31					
DEL								—								
TRO									—		0.40	0.59*				
SPR										—	-0.70†					
MAR											—			0.41		
TAS												—	-0.32†			
NSW													—		-0.43†	
NZ														—		
TUN															—	0.37
SEA																—

* Data showing long-term trend.
† Data analyzed from 1950 to 1980.
‡ Two lowest MAC years deleted as outliers from trends DAR, HOB, MAC.

Statistical analyses were performed by GENSTAT and SYSTAT. Series were rendered stationary by differencing.
DAR, HOB, MAC, atmospheric pressure at Darwin, Hobart and Macquarie Island. ZWW, annual total of zonal westerly winds over Tasmania. MAX, maximum temperature at Shannon. RAI, annual rainfall at Shannon. GLL, Great Lake level. DEL, change in Great Lake level. TRO, percentage of 3-year-old female brown trout. SPR, timing of spring bloom on Maria Island. MAR, maximum summer sea-surface temperature at Maria Island. TAS, NSW, NZ, total lobster catch (weight), Tasmania, New South Wales, New Zealand (Otago). TUN, New South Wales tuna catch (weight). SEA, abundance of leopard seals Macquarie Island.
Source: Harris et al. (1988).

and New South Wales lobster catch; Macquarie Island MSLP and both the spring bloom and SST at Maria Island, plus the Tasmanian lobster catch, the New South Wales southern blue fin tuna catch and leopard seal numbers at Macquarie Island; and the number of days of westerlies over Tasmania and the spring bloom at Maria Island and Tasmanian, New South Wales and New Zealand rock lobster catches. Overall, such relationships suggest that major interannual fluctuations in large-scale oceanic and atmospheric circulation, modulated by ENSO and anti-ENSO events, have profound effects on many living marine resources.

Specific relationships with ocean–atmosphere variables are particularly revealing. The rock lobster catch in Tasmania, New South Wales and New Zealand and the spring bloom at Maria Island seem to be influenced by MSLP and zonal wind effects on oceanic water transport. The spring bloom appears to be influenced by westerly winds driving colder sub-Antarctic waters, which are rich in nutrients, northwards along the eastern coast of Tasmania. Fluctuations in rock lobster catch are thought to be affected by the location of the surface subtropical convergence of oceanic waters, which is displaced further to the north in the Tasman Sea during ENSO events when the westerly wind field is enhanced in more southern latitudes. The regions of greatest lobster catch are on the southern flank of the subtropical convergence. Fluctuations in the long-range transport of various larval stages of some marine species under ENSO and/or anti-ENSO influences on the west wind drift have also been suggested. Some modulations of scallop catch in the Tasmanian region by such processes were also discussed. The ramifications for southern blue fin tuna (as reported earlier by Rochford, 1981) are thought to be more in terms of major shifts in SST fields (especially the critical 20°C isotherm) associated with circulation fluctuations. Further south at Macquarie Island, it was suggested, occurred another possible consequence of the above changes, namely variations in leopard seal numbers.

Relationships between ENSO and coral growth have been reported from a number of locations in tropical–subtropical regions of the globe (see Allan, 1988). In the Australasian region, research in this area has focused on coral coring on the Great Barrier Reef and in northern Australian waters. Aspects of this work are reported in Isdale (1984) and Isdale and Kotwicki (1987). The major thrust of these studies is the resolution of historical climatic signals from coral responses to environmental fluctuations. Coral cores are being used to reconstruct environmental conditions prior to instrumental records and European settlement. In this context, the evaluation of ENSO and anti-ENSO episodes is being seen as providing insights into wider Australian rainfall patterns and climatic fluctuations over time (Isdale and Kotwicki, 1987; Kotwicki and Isdale, 1990).

However, it must be remembered that all of these relationships are pertinent to certain regions and/or time frames. Furthermore, many ecological responses result from the integrated effect of a number of influences, and may not be readily linked to simply one forcing in space or time. Finally, the extent of ecological data on many organisms is short and may be difficult to validate, especially with regard to crop production or marine resource landings where accurate information on yields can be a closely held secret or deliberately falsified.

ENSO forecasting

At present, there are no major institutional bodies making ENSO forecasts on an operational basis in Australia. Three large groups have indicated that they are making significant progress in this general field: the National Climate Center in the Bureau of Meteorology (Melbourne), the Applied Climate Research Unit at the University of Queensland in conjunction with the Queensland State Department of Primary Industry (Brisbane), and the Flinders Institute for Atmospheric and Marine Sciences National Tidal Facility (Adelaide). Of these groups, the first two are issuing, or are in the process of issuing, regular ENSO outlook statements. However, these are not ENSO forecasts.

The National Climate Centre's approach is based on lagged correlations between the Troup SOI and area average rainfall, analogue periods from historical and contemporary records, and general SOI trends. This technique focuses on factors such as the winter–spring persistence of features in Australasia and links between Australasian monsoon onset and anti-ENSO to ENSO sequences. The result is a series of regions on the continent which wax and wane seasonally in terms of their relationships to the SOI as a rainfall predictor (Fig. 4.18). This information is being made available to the general community at no cost. More detailed studies for specific regions and specific needs will be available but here a charge will be levied.

SOI BASE PERIOD	OUTLOOK PERIOD	AREA
1. April – May	June – July	Inland Northern N.S.W.
2. April – June	July – September	Southeastern Australia
3. May – July	August – October	Inland Southeastern
4. June – August	September – November	and Tropical
5. July – September	October – December	North
6. September – October	November – December	Tropical North

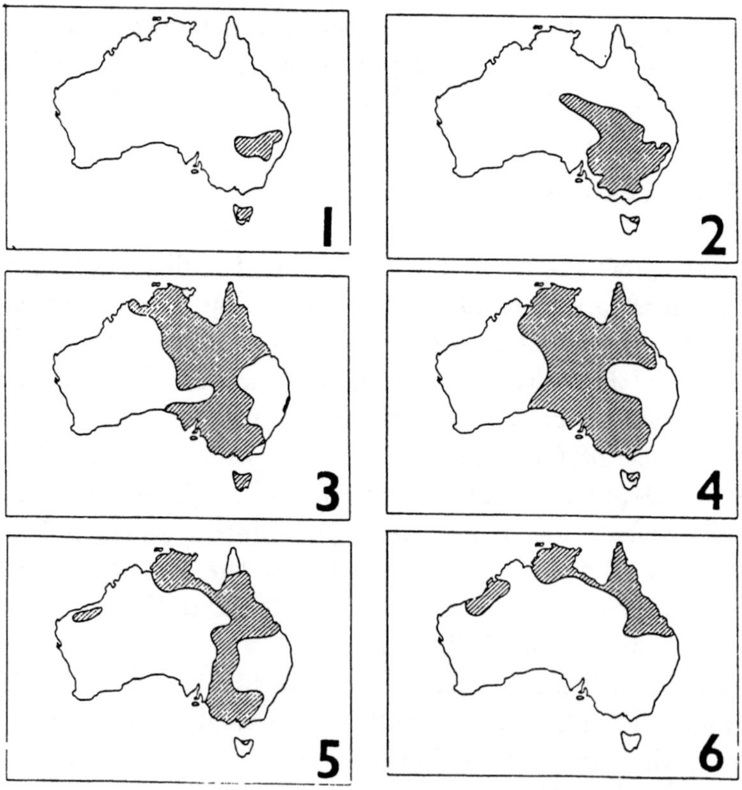

Fig. 4.18 Isopleths of correlation between the SOI (Troup, 1965), and predicted district rainfall for the periods indicated. Areas where the correlation coefficient is greater than or equal to 0.4 are shaded. Outlooks are prepared at the beginning of each indicated period (from National Climate Centre, 1988).

With the Queensland group, the approach has been to use rainfall probabilities based on 10 years of climatological data, SOI values and circulation features, such as the position of the subtropical ridge. Prognoses of Queensland rainfall are being issued on a monthly basis by the Applied Climate Research Unit, usually aimed specifically at rural and industrial needs. Such outlooks have been issued regularly since 1988.

The Flinders Institute for Atmospheric and Marine Sciences group has been interested in ENSO forecasting using atmospheric and oceanic variables, with a strong focus on sea level data from ports in Australasia. Several regions have come under close scrutiny by this group: the North West Shelf region in Western Australia and the Timor Sea, the southern coast of the continent and the Southern Ocean. The latter has been "targeted" as a key region which requires further urgent study, as it is thought to play a vital role in the onset of ENSO events. Such an approach requires the establishment of tide gauges on Southern Ocean islands, with new telemetering gauges around the Australian continent. To date, gauges have been maintained for short periods on Macquarie and Heard Islands to the south and southwest of Australia in the Southern Ocean.

In summary, ENSO forecasting for the Australasian region on a regular basis has not yet been achieved. Many factors have led to this situation, most of them common to all countries engaged in such research. The current lack of an ENSO forecasting scheme has as much to do with a lack of funding and human resources being dedicated to such a task, as it has to do with the question of the dissemination and nature of probabilistic forecasts for the wider community.

Conclusion

ENSO and anti-ENSO influences in Australasia are evident over the period of historical and instrumental records. Manifestations have now been found to encompass the full range of physical, ecological, biological and agricultural concerns. The bulk of these relationships have been identified by statistical correlation studies.

Significant correlation patterns have been reported between ENSO and physical variables such as rainfall, wind/circulation,

MSLP, sea level, sea surface and air temperature, and cloudiness. In general, correlation responses suggest the importance of seasonally modulated fluctuations in ocean–atmosphere interactions, with the greatest persistence centered on the winter–spring period. The predominance of anti-ENSO to ENSO sequences in this region indicates that some measure of forecast skill out to nine to 12 months for certain variables and locations is likely. Of particular significance are possible precursors involving changes in other environmental elements such as agricultural crop production and marine and terrestrial organisms. Thus, there is a potential for outlooks or forecasts of impacts on yields, catches, distributions and the viability of primary products in a number of industries.

More recently, many correlation links have received strong physical support from diagnostic, dynamical, and modeling studies of oceanic and atmospheric circulation patterns during ENSO and anti-ENSO phases. However, there are indications that the longer-term stability of ENSO correlations and teleconnections may be linked to climatic fluctuations on decadal time frames. Further complexity is suggested by the presence of low frequency MSLP features propagating through the Australasian region and their links to the ENSO phenomenon. As noted earlier, each of these aspects requires further study.

Appendix A

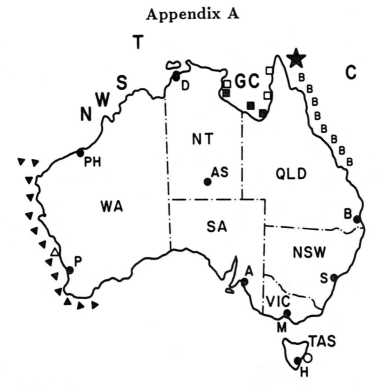

Map of Australia showing the locations mentioned in the text:

States of Australia are:
NT=Northern Territory,
WA=Western Australia,
SA=South Australia,
QLD=Queensland,
NSW=New South Wales,
VIC=Victoria, and
TAS=Tasmania.

Locations mentioned are: **D**=Darwin, **PH**=Port Hedland, **P**=Perth, **AS**=Alice Springs, **A**=Adelaide, **B**=Brisbane, **S**=Sydney, **M**=Melbourne, **H**=Hobart, **T**=Timor Sea, **NWS**=North West Shelf, **C**=Coral Sea, **GC**=Gulf of Carpentaria.

Symbols used are:
solid triangles down the Western Australian coast=the Leeuwin Current;
the open triangle north of Perth=Seven-Mile Beach, Dongara;
open circle just north of Hobart=Maria Island;
open (solid) squares in the Gulf of Carpentaria show northern (southern) locations of regions where rainfall and prawn catch are compared in Vance et al. (1985);
the solid star off the top of Queensland=Heron and Raine Islands;
and the line of Bs down the Queensland coast=the Great Barrier Reef.

References

Adamson, D., Williams, M.A.J. & Baxter, J.T. (1987). Complex late Quaternary alluvial history of the Nile, Murray–Darling and Ganges Basins: three rivers presently linked to the Southern Oscillation. *International Geomorphology 1986 Part 11*. New York: John Wiley and Sons, 875–87.

Allan, R.J. (1983). Monsoon and teleconnection variability over Australasia during the Southern Hemisphere summers of 1973–77. *Monthly Weather Review*, **111**, 113–42.

Allan, R.J. (1984). Variations in low latitude circulation and rainfall over Australasia during the Southern Hemisphere summer monsoon regime. Adelaide: University of Adelaide (unpublished thesis).

Allan, R.J. (1985). The Australasian summer monsoon, teleconnections, and flooding in the Lake Eyre Basin. *South Australian Geographical Papers No. 2*, Royal Geographical Society of Australasia (South Australian Branch, Inc.).

Allan, R.J. (1987). ENSO and climatic fluctuations in Australasia. In *CLIMANZ III Extended Abstracts*, University of Melbourne, 28–29 November, 29–38.

Allan, R.J. (1988). El Niño Southern Oscillation influences in the Australasian region. *Prog. Physical Geography*, **12**, 4–40.

Allan, R.J. & Heathcote, R.L. (1987). The 1982–83 drought in Australia. In *The Societal Impacts Associated with the 1982-83 Worldwide Climate Anomalies*, ed M.H. Glantz, R. Katz & M. Krenz. Report based on the Workshop on Economic and Societal Impacts Associated with the 1982–83 Worldwide Climate Anomalies, 11–13 November 1985, Lugano, Switzerland. Boulder, Colorado, U.S.A.: NCAR, 18–23.

Allan, R.J., Beck, K. & Mitchell, W. (1990). Sea level and rainfall correlations in Australia. Accepted for publication by the *Journal of Climate*, **3**.

Allan, R.J. and Pariwono, J.I. (1990). Ocean-Atmosphere interactions in low latitude Australasia. *International Journal of Climatology*, **10**, 145–79.

Allen, B.J. (1989). Frost and drought through time and space, part I: The climatological record. *Mountain Research and Development*, **9**, 3, 252–78.

Allen, B.J., Brookfield, H. & Byron, Y. (1989). Frost and drought through time and space, Part II: The written, oral and proxy records and their meaning. *Mountain Research and Development*, **9**, 3, 279–305.

Barnett, T.P. (1984a). Interaction of the monsoon and Pacific trade wind system at interannual time scales. Part II: The tropical band. *Monthly Weather Review*, **112**, 2380–7.

Barnett, T.P. (1984b). Interaction of the monsoon and Pacific trade wind system at interannual time scales. Part III: A partial anatomy of the southern oscillation. *Monthly Weather Review*, **112**, 2388–400.

Barnett, T.P. (1985a). Variations in near-global sea level pressure. *Journal of the Atmospheric Sciences*, **42**, 478–501.

Barnett, T.P. (1985b). Three-dimensional structure of low-frequency pressure variations in the tropical atmosphere. *Journal of the Atmospheric Sciences*, **42**, 2798–803.

Barnett, T.P. (1988). Variations in near-global sea level pressure: Another view. *Journal of Climate*, **1**, 225–30.

Barry, R.G. (1978). Aspects of the precipitation characteristics of the New Guinea mountains. *Journal of Tropical Geography*, **47**, 13–30.

Berlage, H.P. (1927). *East-Monsoon Forecasting in Java.* Verhandelingen Koninklijk Magnetisch en Meteorologisch Observatorium te Batavia, No. 20. Jakarta, Indonesia.

Berlage, H.P. (1934). *Further Research into the Possibility of Long Range Forecasting in Netherlands-India.* Verhandelingen Koninklijk Magnetisch en Meteorologisch Observatorium te Batavia, No. 26. Jakarta, Indonesia.

Berlage, H.P. (1957). *Fluctuations in the General Atmospheric Circulation of more than One Year, their Nature and Prognostic Value.* Mededlingen en Verhandelingen No. 69, Koninklijk Meteorologische Instituut, Staatsdrukkerijs-Gravenhage, The Netherlands.

Berlage, H.P. (1961). Variations in the general atmospheric and hydrospheric circulation of periods of a few years duration affected by variations of solar activity. *Annals New York Academy of Sciences*, **95**, 1, 354–67.

Berlage, H.P. (1966). *The Southern Oscillation and World Weather.* Medelingen en Verhandelingen No. 88, Koninklijk Meteorologische Instituut, Staatsdrukkerijs-Gravenhage, The Netherlands.

Berlage, H.P. & de Boer, H.J. (1959). On the extension of the southern oscillation throughout the world during the period July 1, 1949 to July 1, 1957. *Geofisica Pura e Applicata*, **44**, 287–95.

Bjerknes, J. (1961). El Niño study based on analyses of ocean surface temperatures 1935–1957. *Bulletin of the Inter-American Tropical Tuna Commission*, **5**, 217–303.

Bjerknes, J. (1966). A possible response of the atmospheric Hadley circulation to equatorial anomalies of ocean temperature. *Tellus*, **18**, 820–9.

Bjerknes, J. (1969). Atmospheric teleconnections from the equatorial Pacific. *Monthly Weather Review*, **97**, 163–72.

Bjerknes, J. (1972). Large-scale atmospheric response to the 1964–65 Pacific equatorial warming. *Journal of Physical Oceanography*, **2**, 212–7.

Braak, C. (1919). *Atmospheric Variations of Short and Long Duration in the Malay Archipelago.* Verhandelingen Koninklijk Magnetisch en Meteorologisch Observatorium te Batavia, No. 5. Jakarta, Indonesia.

Braak, C. (1921–29). *The Climate of the Netherlands Indies.* Verhandelingen Koninklijk Magnetisch en Meteorologisch Observatorium te Batavia, Vol. 1, No. 8, Jakarta, Indonesia.

Brookfield, H. (1989). Frost and drought through time and space, Part III: What were conditions like when the high valleys were first settled? *Mountain Research and Development*, **9**, 3, 306–21.

Das, S.C. (1956). Statistical analysis of Australian pressure data. *Australian Journal of Physics*, **9**, 394–9.

Deacon, E.L. (1953). Climatic change in Australia since 1880. *Australian Journal of Physics*, **6**, 209–18.

Drake, V.A. & Farrow, R.A. (1988). The influence of atmospheric structure and motions on insect migration. *Annual Review of Entomology*, **33**, 183–210.

Drosdowsky, W. (1988). Lag relations between the Southern Oscillation and the troposphere over Australia. BMRC Research Report No. 13, Bureau of Meteorology Centre, Melbourne, Australia.

Drosdowsky, W. (1990). A simple index of the second POP component of Southern Oscillation. *Tropical Ocean–Atmosphere Newsletter*, **54**, 13–5.

Elliott, W.P. & Angell, J.K. (1988). Evidence for changes in Southern Oscillation relationships during the last 100 Years. *Journal of Climate*, **1**, 729–37.

Enfield, D.B. (1988). Is El Niño becoming more common? *Oceanography Magazine*, **1**, 2, 23–37.

Gentilli, J. (Ed.) (1971). *Climates of Australia and New Zealand, World Survey of Climatology Volume 13*. Amsterdam: Elsevier.

Hackert, E.C. & Hastenrath, S. (1986). Mechanisms of Java rainfall anomalies. *Monthly Weather Review*, **114**, 745–57.

Hamilton, K. & Garcia, R.R. (1986). El Niño/Southern Oscillation events and their associated midlatitude teleconnections 1531–1841. *Bulletin of the American Meteorological Society*, **67**, 1354–61.

Harris, G.P., Davies, P., Nunez, M. & Meyers, G. (1988). Interannual variability in climate and fisheries in Tasmania. *Nature*, **333**, 754–7.

Hastenrath, S. & Wu, M.C. (1982). Oscillations of upper-air circulation and anomalies in the surface climate of the tropics. *Archiv für Meteorologie, Geophysik und Bioklimatologie Serie B*, **31**, 1–37.

Hastenrath, S. & Wu, M.C. (1983). On upper-air mechanisms of the Southern Oscillation. *Tropical Ocean–Atmosphere Newsletter*, **17**, 2–4.

Hastings, P.A. (1990). Southern Oscillation influences on tropical cyclone activity in the Australian/Southwest Pacific region. *International Journal of Climatology*, **10**, 291–8.

Holland, G.J. (1986). Interannual variability of the Australian summer monsoon at Darwin 1952–82. *Monthly Weather Review*, **114**, 594–604.

Holland, G.J. & Nicholls, N. (1985). A simple predictor of El Niño. *Tropical Ocean–Atmosphere Newsletter*, **30**, 8–9.

Hunt, H.A., Taylor, G. & Quayle, E.T. (1913). *The Climate and Weather of Australia*. Melbourne: Commonwealth Bureau of Meteorology.

Isdale, P. (1984). Fluorescent bands in massive corals record centuries of coastal rainfall. *Nature*, **310**, 578–9.

Isdale, P. & Kotwicki, V. (1987). Lake Eyre and the Great Barrier Reef: A paleohydrological ENSO connection. *South Australian Geographical Journal*, **87**, 48–55.

Kidson, E. (1925). Some periods in Australian weather. *Bureau of Meteorology, Bulletin* **17**, 5–33.

Kiladis, G.N. & van Loon, H. (1988). The Southern Oscillation. Part VII: Meteorological anomalies over the Indian and Pacific sectors associated with the extremes of the Oscillation. *Monthly Weather Review*, **116**, 120–36.

Kotwicki, V. & Isdale, P. (1990). Hydrology of Lake Eyre: El Niño link. Submitted to *Palaeogeography, Palaeoclimatology, Palaeoecology*.

Kraus, E.B. (1955). Secular changes of east coast rainfall regimes. *Quarterly Journal of the Royal Meteorological Society*, **81**, 430–9.

Krishnamurti, T.N., Chu, S.H. & Iglesias, W. (1986). On the sea level pressure of the Southern Oscillation. *Archiv für Meteorologie, Geophysik und Bioklimatologie Serie A*, **34**, 385–425.

Limpus, C.J. & Nicholls, N. (1988). The Southern Oscillation regulates the annual numbers of green turtles (*Chelonia mydas*) breeding around northern Australia. *Australian Journal of Wildlife Research*, **15**, 157–61.

Loewe, F. (1948). Variability and periodicity of meteorological elements in the Southern Hemisphere with particular reference to Australia: Some considerations regarding the variability of annual rainfall in Australia. *Bulletin* **39**, Melbourne, Australia: Commonwealth Meteorological Bureau, 1–13.

Love, G. (1987). Banana prawns and the Southern Oscillation Index. *Australian Meteorological Magazine*, **35**, 47–9.

Malingreau, J-P. (1987). The 1982–83 drought in Indonesia: Assessment and monitoring. In *The Societal Impacts Associated with the 1982–83 Worldwide Climate Anomalies*, eds. M.H. Glantz, R. Katz, & M. Krenz. Report based on the Workshop on Economic and Societal Impacts Associated with the 1982-83 Worldwide Climate Anomalies, 11–13th November 1985, Lugano, Switzerland. Boulder, CO: NCAR, 11–8.

McBride, J.L. & Nicholls, N. (1983). Seasonal relationships between Australian rainfall and the Southern Oscillation. *Monthly Weather Review*, **111**, 1998–2004.

McGregor, G.R. (1989). An assessment of the annual variability of rainfall: Port Moresby, Papua New Guinea. *Singapore Journal of Tropical Geography*, **10**, 1, 43–54.

McKeon, G.M., Day, K.A., Howden, S.M., Mott, J.J., Orr, D.M., Scattini, W.J. & Weston, E.J. (1990). Management for pastoral production in northern Australian savannas. *Journal of Biogeography*, **17**, in press.

Murphy, J.O. & Whetton, P.H. (1989). A re-analysis of a tree ring chronology from Java. *Proceedings of the Koninklijke Nederlandse Akademic van Weterschappen*, **B92**, 3, 241–57.

National Climate Center (1988). *Seasonal Outlooks (Based on El Niño/Southern Oscillation (ENSO) Relationships)*. Melbourne: Bureau of Meteorology.

Nicholls, N. (1973). The Walker Circulation and Papua New Guinea rainfall. *Technical Report No. 6, 40/169*, Bureau of Meteorology, Australia.

Nicholls, N. (1977). Tropical–extratropical interactions in the Australian region. *Monthly Weather Review*, **105**, 826–32.

Nicholls, N. (1979a). A simple air–sea interaction model. *Quarterly Journal of the Royal Meteorological Society*, **105**, 93–105.

Nicholls, N. (1979b). A possible method for predicting seasonal tropical cyclone activity in the Australian region. *Monthly Weather Review*, **107**, 1221–4.

Nicholls, N. (1981). Air–sea interaction and the possibility of long-range weather prediction in the Indonesian Archipelago. *Monthly Weather Review*, **109**, 2435–43.

Nicholls, N. (1983). Predictability of the 1982 Australian drought. *Search*, **14**, 154–5.

Nicholls, N. (1984a). El Niño Southern Oscillation and North Australian sea surface temperature. *Tropical Ocean–Atmosphere Newsletter*, **24**, 11–2.

Nicholls, N. (1984b). The Southern Oscillation and Indonesian sea surface temperature. *Monthly Weather Review*, **112**, 424–32.

Nicholls, N. (1984c). ENSO and Australian crops. Scientific Paper presented at the *Report of the First Australian TOGA Meeting*, July 9–10, Hobart, CSIRO Division of Oceanography.

Nicholls, N. (1984d). A system for predicting the onset of the North Australian wet-season. *Journal of Climatology*, **4**, 425–35.

Nicholls, N. (1984e). The Southern Oscillation, sea-surface-temperature, and interannual fluctuations in Australian tropical cyclone activity. *Journal of Climatology*, **4**, 661–70.

Nicholls, N. (1985a). The Southern Oscillation and Australian tropical cyclones. *Tropical Ocean–Atmosphere Newsletter*, **29**, 2.

Nicholls, N. (1985b). Towards the prediction of major Australian droughts. *Australian Meteorological Magazine*, **33**, 161–6.

Nicholls, N. (1985c). Impact of the Southern Oscillation on Australian crops. *Journal of Climatology*, **5**, 553–60.

Nicholls, N. (1986). Use of the Southern Oscillation to predict Australian sorghum yield. *Agricultural and Forest Meteorology*, **38**, 9–15.

Nicholls, N. (1988a). More on early ENSOs: Evidence from Australian documentary sources. *Bulletin of the American Meteorological Society*, **69**, 4–6.

Nicholls, N. (1988b). El Niño–Southern Oscillation impact prediction. *Bulletin of the American Meteorological Society*, **69**, 173–6.

Nicholls, N. (1989). Sea surface temperatures and Australian winter rainfall. *Journal of Climate*, **2**, 965–73.

Nicholls, N. & Woodcock, F. (1981). Verification of an empirical long-range weather forecasting technique. *Quarterly Journal of the Royal Meteorological Society*, **107**, 973–6.

Nicholls, N., McBride, J.L. & Ormerod, R.J. (1982). On predicting the onset of the Australian wet-season at Darwin. *Monthly Weather Review*, **110**, 14–7.

O'Mahony, G. (1961). Time series analysis of some Australian rainfall data. *Meteorological Studies No. 14*, Melbourne: Bureau of Meteorology.

Pariwono, J.I. (1987). Wind stress, mean sea level and inter-ocean transport in the Australasian region. Flinders Institute for Atmospheric and Marine Sciences *Research Report*, **43**.

Pariwono, J., Bye, A.T. & Lennon, G.W. (1986). Long period variations of sea level in Australasia. *Geophysical Journal of the Royal Astronomical Society*, **87**, 43–54.

Phillips, B.F. (1981). The circulation of the southeastern Indian Ocean and the planktonic life of the western rock lobster. *Oceanography of Marine Biology Annual Review*, **19**, 11–39.

Phillips, B.F. & Pearce, A. (1988). ENSO events, the Leeuwin current, and larval recruitment of the western rock lobster. *Journal du Conseil International pour l'Exploration de la Mer*, **45**, 13–21.

Pittock, A.B. (1973). Global meridional interactions in stratosphere and troposphere. *Quarterly Journal of the Royal Meteorological Society*, **99**, 424–37.

Pittock, A.B. (1975). Climatic change and the patterns of variation in Australian rainfall. *Search*, **6**, 11–12, 498–504.

Pittock, A.B. (1983). Recent climatic change in Australia: Implications for a CO_2 warmed earth. *Climatic Change*, 5, 4, 321–40.

Pittock, A.B. (1984). On the reality, stability, and usefulness of Southern Hemisphere teleconnections. *Australian Meteorological Magazine*, 32, 75–82.

Pittock, A.B., Frakes, L.A., Jenssen, D., Peterson, J.A.& Zillman, J.W. (eds.) (1978). *Climatic Change and Variability – A Southern Perspective*. Melbourne: Cambridge University Press.

Prata, A.J., Pearce, A.F., Wells, J.B., Hick, P.T., Carrier, J.C. & Cechet, R.P. (1989). A satellite sea surface temperature climatology of the Leeuwin Current, Western Australia. Report to the Marine Sciences and Technology Council, Canberra, Australia.

Prata, A.J., Pearce, A.F. & Wells, J.B. (1990). An index for the Leeuwin Current from satellite derived sea surface temperatures. Submitted to *Journal of Geophysical Research* (Oceans).

Priestley, C.H.B. (1962). Some lag associations in Darwin pressure and rainfall. *Australian Meteorological Magazine*, 38, 32–42.

Priestley, C.H.B. & A.J. Troup (1966). Droughts and wet periods and their association with sea surface temperature. *Australian Journal of Science*, 29, 2, 56–7.

Quayle, E.T. (1918). Tropical control of Australian rainfall. *Bulletin No. 15*, Melbourne: Bureau of Meteorology.

Quayle, E.T. (1929). Long range rainfall forecasting from tropical (Darwin) air pressures. *Proceedings of the Royal Society of Victoria*, 41, 160–4.

Quinn, W.H., Zopf, D.O., Short, K.S. & Kuo Yang, R.T.W. (1978). Historical trends and statistics of the Southern Oscillation, El Niño and Indonesian droughts. *Fishery Bulletin*, 76, 663–78.

Quinn, W.H., Neal, V.T. & Antunez de Mayolo, S.E. (1987). El Niño occurrences over the past four and a half centuries. *Journal of Geophysical Research*, 92, 14449–61.

Ramage, C.S. (1971). *Monsoon Meteorology*. New York: Academic Press.

Ramage, C.S. (1983).Teleconnections and the siege of time. *Journal of Climatology*, 3, 223–31.

Rasmusson, E.M. (1984). El Niño: The ocean/atmosphere connection. *Oceanus*, 27, 5–12.

Reesinck, J.J.M. (1952). *Some Remarks on Monsoon Forecasting for Java*. Verhandelingen No. 44. Jakarta, Indonesia: Kementerian Perhubungan Djawatan Meteorologi Dan Geofisik.

Rimmer, T. & Hossack, A.W.W. (1939). Forecasting summer rain in Queensland. *University of Queensland Papers, Department of Physics*, 1.

Rochford, D.J. (1981). Anomalously warm sea surface temperatures in the western Tasman Sea, their causes and effects upon the southern bluefin tuna catch, 1966-1977. *CSIRO Division of Fisheries and Oceanography, Report 114*.

Rogers, J.C. & van Loon, H. (1982). Spatial variability of sea level pressure and 500 mb height anomalies over the southern hemisphere. *Monthly Weather Review*, 110, 1375–92.

Ropelewski, C.F. & Halpert, M.S. (1987). Global and regional scale precipitation patterns associated with the El Niño/Southern Oscillation. *Monthly Weather Review*, **115**, 1606–26.

Ropelewski, C.F. & Halpert, M.S. (1989). Precipitation patterns associated with the high index phase of the Southern Oscillation. *Journal of Climate*, **2**, 268–84.

Russell, J.S. (1981). Geographic variation in seasonal rainfall in Australia – An analysis of the 80-year period 1895–1974. *Journal of the Australian Institute of Agricultural Sciences*, **47**, 59–66.

Schell, I.I. (1947). Dynamic persistence and its applications to long-range forecasting. *Harvard Meteorological Studies No. 8*, Milton, MA: Blue Hill Observatory.

Simmonds, I. & Smith, I.N. (1986). The effect of the prescription of zonally-uniform sea surface temperatures in a general circulation model. *Journal of Climatology*, **6**, 641–59.

Simmonds, I. & Trigg, G. (1988). Global circulation and precipitation changes induced by sea surface temperature anomalies to the north of Australia in a General Circulation Model. *Mathematics and Computers in Simulation*, **30**, 99–104.

Staples, D.J. (1983). Gulf of Carpentaria prawns and rainfall. *Proceedings of the Colloquium on the Significance of the Southern Oscillation–El Niño Phenomena and the Need for a Comprehensive Ocean Monitoring System in Australia*. Melbourne: AMSTAC, Department of Science and Technology.

Streten, N.A. (1973). Some characteristics of satellite-observed bands of persistent cloudiness over the Southern Hemisphere. *Monthly Weather Review*, **101**, 486–95.

Streten, N.A. (1975). Satellite derived inferences to some characteristics of the South Pacific atmospheric circulation associated with the Niño Event of 1972–73. *Monthly Weather Review*, **103**, 989–95.

Streten, N.A. (1981). Southern Hemisphere sea surface temperature variability and apparent associations with Australian rainfall. *Journal of Geophysical Research*, **86**, 485–97.

Streten, N.A. (1983). Extreme distributions of Australian annual rainfall in relation to sea surface temperature. *Journal of Climatology*, **3**, 143–153.

Taylor, G. (1918). The Australian environment. *Advisory Council of Science and Industry Memoir No. 1*, Melbourne, Australia.

Taylor, G. (1920). *Australian Meteorology*. Oxford: Clarendon Press.

Todd, C. (1888). *The Australasian*, 1456.

Treloar, H.M. (1934). Foreshadowing monsoonal rains in Northern Australia. *Bulletin 18*, Melbourne, Australia: Bureau of Meteorology.

Treloar, H.M. & Grant, A.M. (1953). Some correlation studies of Australian rainfall. *Australian Journal of Agricultural Research*, **4**, 424–9.

Trenberth, K.E. & Shea, D.J. (1987). On the evolution of the Southern Oscillation. *Monthly Weather Review*, **115**, 3078–96.

Troup, A.J. (1965). The Southern Oscillation. *Quarterly Journal of the Royal Meteorological Society*, **91**, 490–506.

Troup, A.J. (1967). Opposition of anomalies of upper tropospheric winds at Singapore and Canton Island. *Australian Meteorological Magazine*, **15**, 32–7.

Tucker, G.B. (1975). Climate: Is Australia's changing? *Search*, **6**, 323–8.

Vance, D.J., Staples, D.J. & Kerr, J.D. (1985). Factors affecting year-to-year variation in the catch of banana prawns (Penaeus merguiensis) in the Gulf of Carpentaria. *Journal du Conseil International pour l'Exploration de la Mer*, **42**, 83–97.

van Loon, H. (1984). The Southern Oscillation Part III: Associations with the trades and with the trough in the westerlies of the South Pacific Ocean. *Monthly Weather Review*, **112**, 947–54.

van Loon, H. & Madden, R.A. (1981). The Southern Oscillation Part I: Global associations with pressure and temperature in northern winter. *Monthly Weather Review*, **109**, 1150–62.

van Loon, H. & Rogers, J.C. (1981). The Southern Oscillation Part II: Associations with changes in the middle troposphere in the northern winter. *Monthly Weather Review*, **109**, 1163–8.

van Loon, H. & Shea, D.J. (1985). The Southern Oscillation Part IV: The precursors south of 15°S to the extremes of the Oscillation. *Monthly Weather Review*, **113**, 2063–74.

Voice, M.E. & Hunt, B.G. (1984). A study of the dynamics of drought initiation using a global General Circulation Model. *Journal of Geophysical Research*, **89**, D6, 9504–20.

Walker, G.T. (1910). Correlations in seasonal variations in weather II. *Memoirs of the Indian Meteorological Department*, **21**, Part II, 45 pp.

Walker, G.T. & Bliss, E.W. (1930). World weather IV. Some applications to seasonal foreshadowing. *Memoirs of the Royal Meteorological Society*, **3**, 81–95.

Whetton, P.H. (1988). A synoptic climatological analysis of rainfall variability. *Journal of Climatology*, **8**, 155–77.

Whetton, P.H. (1990a). Relationships between monthly anomalies of sea surface temperature and mean sea-level pressure in the Australian region. *Australian Meteorological Magazine*, **38**, 17–30.

Whetton, P.H. (1990b). Relationships between monthly anomalies of Australian region sea-surface temperature and Victorian rainfall. *Proceedings, Geological Society of Australia*. *Australian Meteorological Magazine*, **38**, 31–41.

Whetton, P.H. & Baxter, J.T. (1987). The Southern Oscillation and river behaviour in South-East Australia. In *CLIMANZ III Extend Abstracts*, University of Melbourne, 28–29th November, 39–43.

Whetton, P.H., Adamson, D.A. & Williams, M.A.J. (1990). Rainfall and river flow variability in Africa, Australia and East Asia, linked to El Niño–Southern Oscillation events. *Proceedings, Geological Society of Australia*, **1**, in press.

Williams, M. (1987). Relations between the Southern Oscillation and the troposphere over Australia. BMRC Research Report No. 6, Bureau of Meteorology Research Centre, Melbourne, Australia.

Williams, M.A.J., Adamson, D.A. & Baxter, J.T. (1986). Late quaternary environments in the Nile and Darling Basins. *Australian Geographical Studies*, **24**, 128–44.

Wright, P.B. (1977). The Southern Oscillation – Patterns and mechanisms of the teleconnections and the persistence. *Hawaii Institute of Geophysics Report*, HIG-77-13.

Wright, P.B. (1984). Relationship of SST near northern Australia to the Southern Oscillation. *Tropical Ocean–Atmosphere Newsletter*, **27**, 17–8.

Wright, P.B., Mitchell, T.P. & Wallace, J.M. (1985). *Relationships between surface observations over the global oceans and the Southern Oscillation.* Data Report ERL PMEL-12, Seattle: National Oceanic and Atmospheric Administration.

Wright, W.J. (1987). Synoptic and climatological factors influencing winter rainfall variability in Victoria. Melbourne: Meteorology Department, University of Melbourne (unpublished Ph.D. thesis).

Wright, W.J. (1988a). The low latitude influence on winter rainfall in Victoria, south-eastern Australia – I. Climatological aspects. *Journal of Climatology*, **8**, 437–62.

Wright, W.J. (1988b). The low latitude influence on winter rainfall in Victoria, south-eastern Australia – II. Relationships with the Southern Oscillation and Australian region circulation. *Journal of Climatology*, **8**, 547–76.

Zhang, X.G. & Casey, T.M. (1990). Long term variation in the Southern Oscillation and the relationship with rainfall in Australia. Submitted to the *Australian Meteorological Magazine*.

5

West Africa

PETER J. LAMB* and RANDY A. PEPPLER
Climate and Meteorology Section
Illinois State Water Survey
Champaign, IL 61820

Introduction

Much of the rest of this book is concerned with the mechanisms and characteristics of the Southern Oscillation (SO) and its associated El Niño (EN) phenomenon (collectively termed ENSO) and their teleconnections with diverse and remote atmospheric, oceanic, and biological phenomena. The material presented attests to both the pervasiveness of ENSO and its wide-ranging and considerable impacts upon society. In this chapter, however, we are concerned with the highly seasonal rainfalls of two contrasting parts of West Africa that are extremely important for agriculture, water supply, and hence for society, but which are not strongly related to ENSO on more than an occasional basis.

The regions concerned are the nation of Morocco, which occupies the northwestern extremity of the African continent (Figs. 5.1 and 5.2), and the Subsaharan zone that spans West Africa between the approximate latitudes of 11° and 18°N, extending from Mauritania, Senegal, and Gambia in the west to Niger and Nigeria in the east (Fig. 5.3). The Subsaharan zone is a transition belt of natural grassland (savanna in the south, sparser towards the north), nomadic herding, and rain-fed agriculture (e.g., millet, grain sorghum, groundnuts) between the essentially rainless Sahara desert to the north and the area of much more abundant rainfall and vegetation to the south. Morocco possesses rain-dependent agricultural areas in the vicinity of its Mediterranean

* Also affiliated with Departments of Atmospheric Sciences and Geography, the University of Illinois.

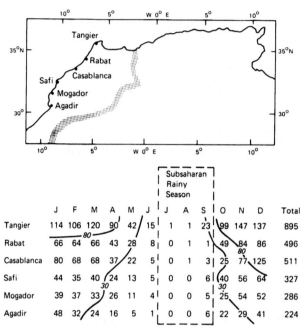

	J	F	M	A	M	J	J	A	S	O	N	D	Total
							Subsaharan Rainy Season						
Tangier	114	106	120	90	42	15	1	1	23	99	147	137	895
Rabat	66	64	66	43	28	8	0	1	1	49	84	86	496
Casablanca	80	68	68	37	22	5	0	1	3	25	77	125	511
Safi	44	35	40	24	13	5	0	0	6	40	56	64	327
Mogador	39	37	33	26	11	4	0	0	5	25	54	52	286
Agadir	48	32	24	16	5	1	0	0	6	22	29	41	224

Fig. 5.1 Annual and latitudinal variation of rainfall (mm) for the Atlantic coast of Morocco. The rainfall data were obtained from Griffiths (1972, Tables VIII, XIV, XXII, XXIV, XXIX). Note that Mogador is now known as Essaouira. Taken from Lamb and Peppler (1987).

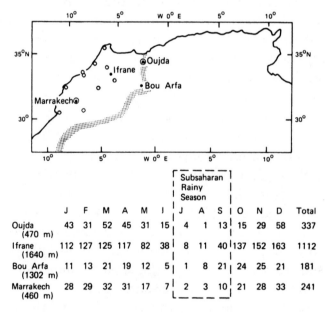

	J	F	M	A	M	J	J	A	S	O	N	D	Total
							Subsaharan Rainy Season						
Oujda (470 m)	43	31	52	45	31	15	4	1	13	15	29	58	337
Ifrane (1640 m)	112	127	125	117	82	38	8	11	40	137	152	163	1112
Bou Arfa (1302 m)	11	13	21	19	12	5	1	8	21	24	25	21	181
Marrakech (460 m)	28	29	32	31	17	7	2	3	10	21	28	33	241

Fig. 5.2 Same as Fig. 5.1 except for inland Morocco. The rainfall data were obtained from Griffiths (1972, Tables VIII, XVI, XXI, XXVII). The Moroccan rainfall indices used in Figs. 5.4, 5.5, and 5.9 are based on the 12 circled stations. Taken from Lamb and Peppler (1987).

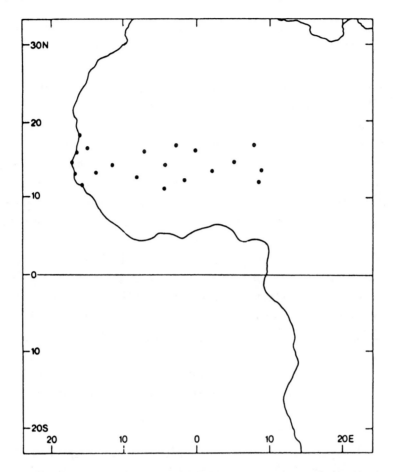

Fig. 5.3 Dots locate 20 rainfall stations in Subsaharan West Africa whose 1941–88 data were used to construct the Subsaharan index used in Figs. 5.4 and 5.5. The stations lie between 11° and 18°N.

and northern Atlantic coasts (wheat, barley, citrus fruits, vegetables) beyond which, to the southeast and south, lie the imposing Atlas mountain ranges and the Sahara desert.

This chapter, therefore, illustrates that not all regional climate fluctuations, including those of extreme societal importance, are strongly ENSO-related. This circumstance may have become obscured by the considerable publicity that ENSO has attracted during the past decade. More importantly, however, this chapter also reveals that even regional climate fluctuations that are not strongly ENSO-related can be the products of other atmosphere–ocean interactions and their associated teleconnections. Because of this situation, some potential exists for the long-range prediction of Moroccan and Subsaharan rainfall on an individual year basis. However, substantial progress is still needed in that regard.

We commence by providing some important background climatological information concerning the above two regions. This includes documenting and explaining their contrasting annual rainfall marches and indicating how each region's rainfall has varied substantially over time, both during the most recent 50 years and throughout much longer periods. We then show that the rainfall variations experienced in the two regions during the past 50 years have been largely unrelated to each other and the SO. With this as background, we proceed to demonstrate that the interannual variability of Moroccan rainfall is relatively strongly related to the concurrent state of a major atmospheric circulation fluctuation known as the North Atlantic Oscillation (NAO). However, this relationship is found to be less strongly foreshadowed by the state of the NAO during the months immediately preceding the rainy season. A key remaining challenge is therefore the identification of the components of the climate system that possess advance information on the likely NAO state for individual Moroccan rainy seasons.

The subsequent section dealing with Subsaharan West Africa shows that most of the very deficient recent rainy seasons there have coincided with a distinctive pattern of tropical Atlantic sea surface temperature (SST) anomalies. That pattern has been accompanied by physically consistent changes in the overlying pressure and wind fields that likely reduce the penetration of the moisture-bearing southwest monsoon flow into Subsaharan West Africa. It is also revealed that, for some of those deficient Subsaharan rainy seasons, the tropical Atlantic SST anomalies were part of a global pattern whose tropical Pacific component probably further contributed to the suppression of rainfall in Subsaharan West Africa. We then indicate how the evolutions of the tropical Atlantic and global SST fields are providing the basis for attempts to predict the quality of individual Subsaharan rainy seasons several months in advance. However, since the success of these predictive efforts has been mixed, it is clear that we need to further enhance our understanding of the links between SSTs and Subsaharan rainfall.

We conclude by identifying, for both Morocco and Subsaharan West Africa, the challenges that now exist as a result of the considerable progress made during the last 10–15 years.

Climatological background and relation to Southern Oscillation

Annual march of Moroccan and Subsaharan rainfall

As Figs. 5.1 and 5.2 and Table 5.1 show, both Subsaharan West Africa and Morocco experience very pronounced and well defined rainy seasons that are strongly out of phase with each other. Eighty percent of the year's rainfall in the Subsaharan zone typically occurs during July–September (Table 5.1). Small amounts of Subsaharan rain fall in the adjacent months of May, June, and October, with virtually no rainfall occurring in the half-year period from November to April. In contrast, approximately 85 percent of the rainfall along the Atlantic coast of Morocco (Fig. 5.1), and around 70 percent in inland Morocco (Fig. 5.2), typically occur during the November–March period that is almost totally dry in Subsaharan West Africa. Morocco experiences small amounts of rain in May, June, and October, but little or none (particularly along the Atlantic coast) in the July–September period that is the Subsaharan rainy season.

Table 5.1 Percentage of annual average Subsaharan rainfall occurring in individual calendar months. Rainfall data used are from the 20 stations located in Fig. 5.3 for the entire period of available records up to 1974. Taken from Lamb (1980, 1985a).

May	June	July	August	September	October	Nov–April
3	9	22	37	21	5	3

These contrasting annual rainfall marches are the *immediate* products of very different components of the global atmospheric circulation pattern. Subsaharan rainfall is produced by distinctive *tropical* weather systems – westward propagating lines of thunderstorms ("West African Disturbance Lines" or "Lignes de Grains") with north–south orientations – that affect this zone only during

Northern Hemisphere summer. These systems are displaced to the south of the Subsaharan zone during the remainder of the year, when Saharan high pressure and its associated subsidence are responsible for the general absence of Subsaharan rainfall. Most Moroccan rainfall, on the other hand, is associated with the *extratropical* cyclonic storm systems that move eastward off the North Atlantic and onto Western Europe and the Mediterranean during the Northern Hemisphere winter half-year. The tracks of those systems are too far north during the rest of the year for them to affect Morocco, whose generally rainless weather in those months is the result of the extension of the North Atlantic subtropical high pressure cell onto northwestern Africa. Fuller discussions of the annual marches of Subsaharan and Moroccan rainfall, along with their global atmospheric circulation causes, appear in Lamb (1980, 1985a) and Lamb and Peppler (1988). Those treatments are intended for multidisciplinary audiences.

Despite the foregoing discussion, there is one very strong commonality between Moroccan and Subsaharan rainfall. Because both are so strongly seasonally concentrated, a deficient rainy season in either region severely handicaps its agriculture and water supplies, and hence socioeconomic life, for an entire year. Successive rainy season failures compound those problems in a nonlinear manner, as recent history (see below) has graphically demonstrated for both areas.

Secular variation of Moroccan and Subsaharan rainfall

Figure 5.4 shows that rainfall has declined markedly in both Morocco and Subsaharan West Africa since about 1970, with the Subsaharan decline being the continuation of a trend that started almost 20 years earlier. Subsaharan West Africa experienced extremely deficient rainy seasons in 1972, 1977, and 1983–84, all of which were the culmination of progressive rainfall decreases during several preceding years. Subsaharan rainfall was also substantially below the 1941–82 average in many other years during 1968–88; indeed, only in 1969, 1975, and 1988 did it approach or slightly exceed that average. In contrast to this post-1967 situation, rainfall in Subsaharan West Africa was generally close to the 1941–82 mean during 1959–67, rather abundant between 1950–58, and much more varied from 1941–49.

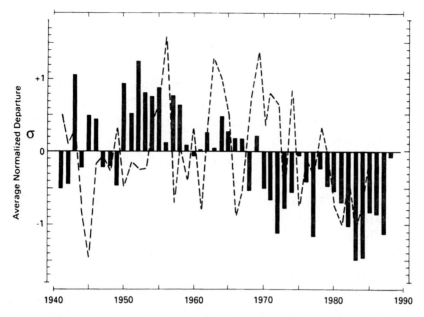

Fig. 5.4 Time series plots of yearly rainfall index values for Morocco (November–April, dashed line, 1941–85) and Subsaharan West Africa (April–October, vertical bars, 1941–88). The years indicated pertain to the ends of the seasonal periods. The Moroccan index was computed using data for all 12 stations circled in Fig. 5.2; its yearly values are averages of the normalized rainfall departures (with respect to 1941–84 base period) for those stations. The Subsaharan index has the same formulation (except for 1941–82 base period) and was computed using data for all 20 stations shown in Fig. 5.3. See Lamb (1985b) and Lamb and Peppler (1988) for further information on the index formulation and evaluations.

Before turning to the secular variation of Moroccan rainfall, it is necessary to place the Subsaharan rainfall variations of the last few decades in their historical context. Historical records (e.g., Grove, 1973; Nicholson, 1978, 1980a), hydrological data for the rivers (e.g., Senegal, Niger) and lakes (e.g., Chad) of the zone, and the (fewer) available pre-1941 rainfall data (e.g., Jenkinson, 1973; Davy, 1974; Nicholson, 1979, 1980b) suggest that the years of the early 1910s, 1919, 1921, and 1926 were also particularly dry in Subsaharan West Africa. It seems probable that 1913 experienced a rainfall deficiency similar to those of 1972, 1977, and 1983–84 (Grove, 1973). However, 1900–40 apparently did not include distinct epochs when rainfall anomalies of a particular type were most persistent, like those Fig. 5.4 shows have occurred since, nor any long-term trend like the decline since the early 1950s. Going further back, there is evidence that severe Subsaharan drought prevailed between 1820–40 and 1736–58, during centuries that were otherwise wetter than the present one (Nicholson, 1978, 1980a). On an even longer time-scale, we now know from lake, alluvial, and pollen records that Subsaharan rainfall was greater between 10000 and 5000 years ago than it is today (Grove, 1972; Street-Perrott and Grove, 1979; Street-Perrott and Roberts, 1983). Recent climate modeling research has suggested that this resulted from a strengthening of the southwest monsoon flow into West Africa from the tropical Atlantic that was, in turn, the product of enhanced solar heating of the North African land surface during summer (Kutzbach and Otto-Bliesner, 1982; Kutzbach and Street-Perrott, 1985; Kutzbach and Guetter, 1986). The latter stemmed from the very slow variations that characterize the earth's orbit about the sun.

The most outstanding feature of Fig. 5.4 for Morocco is the prolonged and severe drought that was experienced during 1979–84. This followed and extended the aforementioned pronounced general rainfall decline of the 1970s. In contrast, the Moroccan rainfall of 1941–68 was dominated by strong short-period (2–5 year) variation, with no long-term trend being apparent. The extreme duration and intensity of the 1979–84 Moroccan drought so adversely affected that nation's agricultural production and water supplies that a special study to determine the maximum length and severity of previous droughts was commissioned in late 1984 by His Majesty King Hassan II (Stockton, et al., 1985; Stockton, 1988a,b). This involved in particular the development and analysis of a dendrochronological record for central Morocco for the last 1000 years. The work revealed that although droughts of comparable duration (6 years) and intensity have occurred in the past (between 1069–74 and 1626–31), they have been widely separated in time. Furthermore, droughts lasting four to five years were also relatively infrequent during the last millennium, having been confined to 1404–07 (particularly severe), 1714–17, and 1794–98. However, these results did establish that the 1979–84 drought was not totally unprecedented and that it accordingly should not be viewed as a harbinger of more permanent desiccation, as the Moroccan government had feared.

Relation between Moroccan and Subsaharan rainfall

The foregoing concern was prompted by (1) both Morocco and Subsaharan West Africa experiencing drought in the first half of the 1980s (Fig. 5.4) and (2) some scientists suggesting that the Subsaharan drought could be self-perpetuating because of vegetation or soil moisture depletion induced by the lack of rainfall. The latter hypotheses were based on the work of Charney (1975), Charney et al. (1977), Walker and Rowntree (1977), Sud and Fennessy (1982, 1984), and others. They involve possible biogeophysical feedback mechanisms that would operate within the Subsaharan zone, and are therefore essentially beyond the scope of this book dealing with teleconnections (see the background section on Subsaharan rainfall for further discussion). However, because of (1) and (2) above, the Moroccan government was interested in the question of the relation between the rainfall of the two regions. This issue is addressed in Figs. 5.4 and 5.5. They suggest that the interannual variation of Moroccan November–April rainfall since 1941 has been essentially unrelated to the April–October rainfall of Subsaharan West Africa (Fig. 5.5), and that there was also little agreement at longer time-scales prior to the late 1970s (Fig. 5.4). As already noted, however, both areas have experienced general rainfall declines since 1970 and severe drought during the early 1980s. Overall, Figs. 5.4 and 5.5 and information presented in the section on Moroccan rainfall and the NAO suggest that Moroccan rainfall is much more strongly related to the large-scale atmospheric circulation of the subtropical and extratropical North Atlantic than to Subsaharan rainfall and, by implication, the tropical circulation systems that contribute to Subsaharan rainfall. However, the aforementioned pronounced change in the relation between Subsaharan and Moroccan rainfall that occurred in the late 1970s implies that the North Atlantic subtropical atmospheric circulation may have played different roles in producing the severe Subsaharan drought conditions of the early 1970s as opposed to the early 1980s. This subject is currently being investigated.

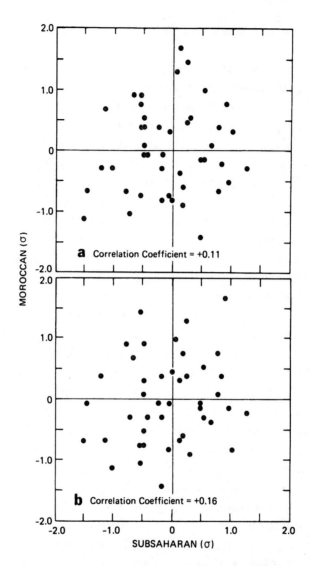

Fig. 5.5 Relation between the yearly rainfall index values for Morocco and
        Subsaharan West Africa shown in Fig. 5.4 for 1941–84. The compar-
        ison in a/b is between Moroccan rainfall for individual November–
        April periods and Subsaharan rainfall for the following/prior April–
        October periods.

Relation of Moroccan and Subsaharan rainfall
to Southern Oscillation

Table 5.2 summarizes the results of a comprehensive investigation
into the question of whether teleconnections existed between the

Table 5.2 Correlation coefficients relating Southern Oscillation (SO) state (for 3-month running seasons) to rainfall in Subsaharan West Africa (April–October; 1941–88) and Morocco (November–April; 1933–85). SO state was represented by SO index of Ropelewski and Jones (1987); Moroccan rainfall by the three indices portrayed in Fig. 5.9; and Subsaharan rainfall by index shown in Fig. 5.4. Correlation coefficients within ±0.02 were obtained using the SO index of Nicholls (1989). SO Lag 0 correlations are for SO seasons ending in same year as rainfall period ends; SO Lag −1 correlations are for SO seasons ending in year prior to year in which rainfall period ends. One asterisk indicates that the individual correlations were significant at the 95 percent level (two-sided test). None were significant at the 99 percent level, and none were significant using the Bonferroni test with $\alpha_o = 0.05$ and $K = 268$ unique variable pairings ($\alpha = \alpha_o/K$), as recommended by Brown and Katz (1989).

| | Subsaharan rainfall | | Moroccan rainfall | | | | | |
| | | | National | | Atlantic coast | | Tangiers | |
SO Season	SO Lag −1	SO Lag 0	SO Lag −1	SO Lag 0	SO Lag −1	SO Lag 0	SO Lag −1	SO Lag 0
Nov–Jan	+0.06	+0.14	+0.03	+0.28*	+0.02	+0.17	+0.09	+0.08
Dec–Feb	+0.10	+0.23	+0.03	+0.21	+0.04	+0.11	+0.06	+0.04
Jan–Mar	+0.14	+0.25	+0.02	+0.26	−0.02	+0.19	+0.07	+0.02
Feb–Apr	+0.14	+0.28	+0.03	+0.29*	−0.03	+0.23	+0.03	+0.04
Mar–May	+0.04	+0.23	+0.08	+0.27	+0.04	+0.20	+0.12	+0.01
Apr–June	+0.07	+0.27	+0.21	+0.09	+0.16	+0.01	+0.08	−0.08
May–July	+0.09	+0.27	+0.23	−0.04	+0.18	−0.12	+0.06	−0.10
June–Aug	+0.16	+0.34*	+0.24	−0.08	+0.16	−0.12	+0.04	−0.13
July–Sept	+0.12	+0.23	+0.23	−0.09	+0.16	−0.14	+0.05	−0.11
Aug–Oct	+0.09	+0.19	+0.22	−0.07	+0.12	−0.12	+0.04	−0.12
Sept–Nov	+0.02	+0.13	+0.28*	−0.05	+0.15	−0.10	+0.08	−0.10
Oct–Dec	+0.14	+0.19	+0.28*	−0.06	+0.15	−0.15	+0.10	−0.13

SO and Subsaharan and Moroccan rainfall during 1933–88. This inquiry considered not only the relationship between the rainfall and the concurrent state of the SO, but also whether the evolution of the SO prior to the respective rainy seasons could provide a basis for the skillful prediction of rainy season quality several months or more in advance. Unfortunately, the results are rather negative on both counts. The maximum correlations obtained (Table 5.2) between the SO and Subsaharan (+0.34) and Moroccan (+0.29) rainfall account for only 8–12 percent of the rainfall variance.

For Subsaharan (April–October) rainfall, the correlation with the SO is strongest for June–August of the same year – i.e., very close to (but slightly leading) the July–September core of the concurrent rainy season. The SO states during the preceding December–February, January–March, ..., May–July periods correlate even more weakly with Subsaharan April–October rainfall (+0.23 to +0.28). There is also a progressive decrease in the correlation magnitude as the SO evolves from June–August (+0.34) to July–September (+0.23) to August–October (+0.19) to September–November (+0.13) of the same year as the Subsaharan rainy season. Table 5.2 further shows that the SO state during a given calendar year contains no predictive information about the Subsaharan rainfall of the subsequent calendar year.

Similar results were obtained for Moroccan (November–April) rainfall (Table 5.2). Its correlation with the SO is strongest (+0.27 to +0.29) for much of the rainy season itself, after which the correlation magnitudes weaken even further (rainfall variance explained approaches zero). Even though the pre-rainy season (June–November) SO state correlates almost as well with the ensuing Moroccan rainfall (+0.22 to +0.28) as the concurrent SO state, the evolution of the SO clearly contains little predictive information concerning that rainfall. Note, too, that the Moroccan correlation results in Table 5.2 are stronger for the country as a whole than for its Atlantic coastal region. Interestingly, this feature is counter to other (much stronger) teleconnection results presented in the section on Moroccan rainfall and the NAO.

Summary

The discussion thus far has established the following key points – the strongly seasonally concentrated (but out of phase) rainfalls of Subsaharan West Africa and Morocco have varied considerably during both the most recent 50 years and the more distant past; only in the late 1970s and early 1980s did the two regions experience similar rainfall anomalies during the past 50 years; for the post-1941 period as a whole, the rainfall in both regions has been only very weakly related to the concurrent state of the SO; and the SO, despite being a dominant and conservative mode of global atmosphere–ocean variation, does not in general provide a basis for the skillful prediction of the quality of Moroccan and Subsaharan rainy seasons several months in advance. Given the extreme dependence on rainy season quality of the agricultural production, water supplies, and general socioeconomic life of the two regions, the existence of such a predictive capability would clearly be of considerable value. The remainder of this chapter summarizes and evaluates the research effort that has been directed to that end. In both cases, the work was a response to a climate-triggered societal crisis.

Moroccan rainfall

Background

Prior to the 1979–84 drought and its adverse impacts on agricultural production and water supplies, there seems to have been little governmental or scientific interest in the causes or predictability of the interannual and longer time-scale variations of Moroccan rainfall. The awakening of the Moroccan government's interest in the problem in 1984 stemmed not just from the first-order impacts of the 1979–84 drought that have already been mentioned, but also, and perhaps more importantly, from the realization that resulting food and water shortages would ultimately threaten the government's stability if the drought persisted further. The dimensions of this situation are fully described in Seddon (1986) and Swearingen (1987).

The Moroccan government responded by launching three complementary initiatives. One resulted in the conducting of a five-year experimental cloud seeding program to investigate the potential for snowpack augmentation in the Atlas mountains (Bensari and Benarafa, 1988; Bensari et al., 1989). This program was partially funded by the U.S. Agency for International Development and received technical assistance from the U.S. Bureau of Reclamation. The second initiative was the commissioning of the dendrochronological investigation for the last millennium that has already been mentioned (Stockton et al., 1985; Stockton, 1988a,b). In addition to establishing that Morocco had previously experienced the other severe but widely separated four- to six-year droughts noted above, this investigation showed that "Shorter droughts, however, lasting about 1.5 years are not uncommon and their expected return period is about 11 years." The latter finding led to the third of the aforementioned initiatives. In the words of Stockton (1988a), "For this reason a conference was held in Agadir (Morocco, 21–24 November 1985) to discuss various options available to the Kingdom of Morocco to mitigate the impact of future droughts. Most of the options discussed centered around the development of water resources. However, a wide range of subjects were covered, ranging from large-scale atmospheric features associated with droughts in Morocco to weather modification as a water management tool." The complete proceedings of that conference were published by the Kingdom of Morocco (1988).

The senior author of the present chapter was invited to make the aforementioned conference presentation dealing with "Large-scale atmospheric features associated with drought in Morocco." We began our literature review for that presentation by searching for works that were directly motivated by a desire to understand the interannual variability of Moroccan rainfall. That initial effort encompassed all major atmospheric and oceanographic journals for the period 1960–85. It unearthed nothing of substance, which suggested that the subject had previously received little if any serious scientific attention. This finding also required that we adopt a different approach in our inquiry. We therefore reoriented the literature review to ensure identification of research that had the potential to increase our understanding of the interannual variability of Moroccan rainfall, irrespective of its particular motivation. The most promising result to emerge concerned the North Atlantic Oscillation. This atmospheric phenomenon has since provided the framework for our inquiry into the mechanisms and predictability of Moroccan rainfall fluctuations, which to date have been pursued using a classical correlation-based teleconnections approach.

North Atlantic Oscillation

The term North Atlantic Oscillation (NAO) refers to a large-scale alternation of atmospheric mass between the North Atlantic regions of subtropical high surface pressure (centered near the Azores) and subpolar low surface pressure (extending south and east of Greenland). Both of these pressure field features are clearly apparent in Fig. 5.6, particularly for January. The alternation of mass between them tends to be characterized by either (1) anomalously high subtropical surface pressure and anomalously low subpolar surface pressure or (2) the opposite surface pressure anomalies in those areas. These are the two extremes of the NAO, and are referred to as such below. The state of the NAO determines the strength and orientation of the poleward pressure gradient over the North Atlantic, and hence the speed and direction of the mid-latitude westerlies across that ocean. These, in turn, affect the tracks of European-sector low-pressure storm systems that were earlier noted to be responsible for most Moroccan rainfall.

The NAO was discovered in the 1920s by a well-known British meteorologist, Sir Gilbert Walker, during his encyclopaedic inquiry into world weather (Walker, 1924; Walker and Bliss, 1932) that included the identification of the SO and its initial application to regional climate variation. However, scientific interest in the NAO (like the SO) languished until quite recently, when it became a central concept in a series of papers dealing with winter temperature differences between Greenland and Northern Europe (van Loon and Rogers, 1978; Rogers and van Loon, 1979; Meehl and van Loon, 1979; Rogers, 1984, 1985). This research effort, which was clearly of much smaller scope than the last decade's SO-related endeavors reported in the remainder of this book, built on the earlier ideas of several well-known European scientists (e.g., Dove, 1839; Hann, 1890; Defant, 1924; Ångström, 1935; Loewe, 1937, 1966; Dannmeyer, 1948). As is illustrated in Fig. 5.7, it established that NAO extreme (1) coincides with cold winters in western Greenland and warm ones in Northern Europe (termed "Greenland below"), while NAO extreme (2) is associated with the opposite temperature anomaly pattern (termed "Greenland above"). In the former (latter) case, the westerly flow across the North Atlantic is strong (weak). Even more recently, Moses et al. (1987) and Barnston and Livezey (1987) have documented fur-

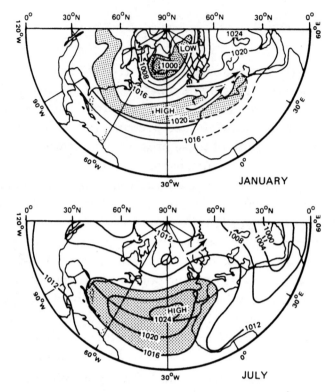

Fig. 5.6 Large-scale atmospheric surface conditions over the Northern Hemi-
         sphere during January and July. The isopleths give the mean sea-
         level pressure in mb; arrows indicate principal tracks of low-pressure
         systems in the European sector and are based on Klein (1957) and
         Whittaker and Horn (1984); the high-pressure cell is light-stippled
         and the low-pressure area is heavy-stippled. Taken from Byers (1959,
         pp. 265–6).

ther the NAO's importance for the climates of Europe and eastern
North America.

   During the course of the above work, Rogers (1984) developed
an index of the NAO for winter. Following Walker (1924) and
Walker and Bliss (1932), the NAO Index was defined as the difference between the normalized mean winter (December–February)
surface-pressure anomaly for Ponta Delgada (Azores) and that for
Akureyri (Iceland). A 1933–83 time series of this NAO index is
one of several features of Fig. 5.9, all of which are fully discussed
in the following section. Rogers' (1984) research also documented
the large-scale nature of the NAO, including its manifestation as
high in the atmosphere as the 500-mb level (see Fig. 5.8).

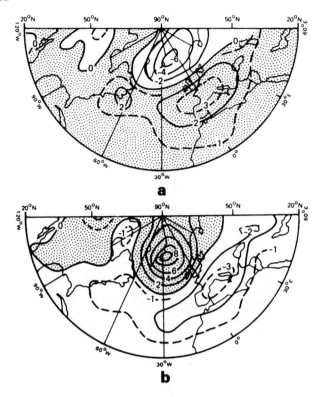

Fig. 5.7 The extremes of the NAO. Panel (a) shows the average January
    sea-level-pressure anomaly pattern (mb) for 16 years during 1899–
    1975 when the temperature at Jakobshavn (Greenland) was below
    the mean and that at Oslo (Norway) was above the mean, and the
    temperature anomalies were 4°C apart ("Greenland below"). Panel
    (b) is the same as (a) except that the temperature at Jakobshavn
    is above the mean and that at Oslo is below the mean ("Greenland
    above"), and the pattern is a 15-year average. Pressure anomalies
    were computed with reference to 1899–1975 means; positive values
    are shaded. Taken from van Loon and Rogers (1978).

Relation of North Atlantic Oscillation to interannual variation of Moroccan rainfall

Figure 5.9 relates Moroccan November–April rainfall to Rogers'
(1984) aforementioned winter NAO index for 1933–83. Three Moroccan rainfall indices are used. One is the 12-station national
index already portrayed in Figs. 5.4 and 5.5 for 1941–84/85; the
second is a counterpart index for the four southernmost coastal
stations in Fig. 5.2; and the third is for the extreme northern
coastal station in Fig. 5.2. It is clear from Fig. 5.9 that Moroccan winter semester rainfall is inversely related to the concurrent

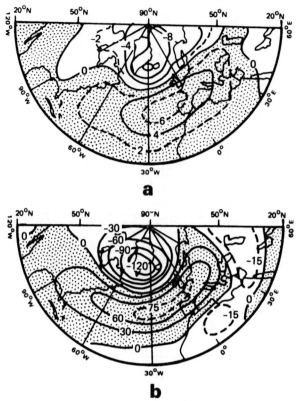

a

b

Fig. 5.8 Further characteristics of NAO. Panel (a) shows the difference in average sea-level pressure (mb) between two subsets of winters (December–February) for 1943–79 — (1) when the NAO index in Fig. 5.9 was above zero and (2) when it was below zero. Panel (b) gives the difference in average 500 mb height (geopotential meters) between the identically defined subsets of winters during 1947–79. Areas of positive difference (i.e., positive NAO pressure/height > negative NAO pressure/height) are shaded. Taken from Rogers (1984).

state of the NAO. High Moroccan rainfall tends to coincide with large negative values of the NAO index, where the latter result from anomalously low subtropical (Azores) and anomalously high subpolar (Icelandic) North Atlantic surface pressures. This situation corresponds to the NAO extreme (2) "Greenland-above" case defined earlier, one that is also characterized by relatively weak westerly flow across the North Atlantic. Low Moroccan rainfall tends to accompany a large-scale North Atlantic atmospheric circulation pattern that is the opposite in all respects to the one just described.

The inverse relationship between Moroccan rainfall and the NAO is strongest for that nation's Atlantic coast, and particularly

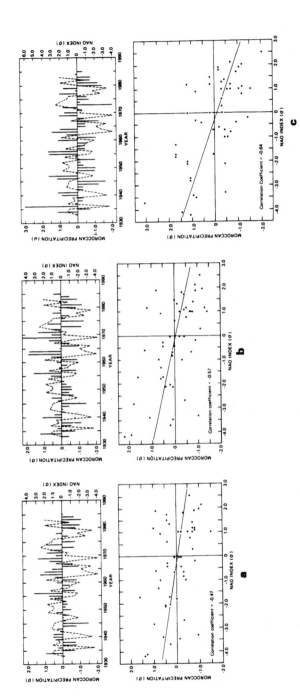

Fig. 5.9 Relation between Moroccan November–April rainfall and Rogers' (1984, Fig. 1) December–February NAO index for 1933–83. Both parameters are expressed as normalized departures; see text and Fig. 5.4 for computational details. The rainfall index in (a) is derived from all 12 circled stations in Fig. 5.2, that in (b) is from the four southernmost coastal stations in Fig. 5.2, and that in (c) is for the northernmost coastal station in Fig. 5.2. In the upper panels, bars denote Moroccan precipitation, broken line is NAO index, and years indicated pertain to ends of November–April and December–February periods. In lower panels, thin lines are least-squares fits and correlation coefficients were assessed to be significant at the 99.9 percent (b and c) and 99 percent (a) levels. This included use of Quenouille's (1952, p. 168) method to account for the reduction of effective degrees of freedom due to persistence. Moroccan rainfall information was obtained from Stockton (1988b) and Meko (1988). Taken from Lamb and Peppler (1987, 1988).

its northernmost station (Fig. 5.9). The correlation coefficients concerned are –0.64 (41 percent of rainfall variance explained) for the northernmost coastal station and –0.57 (32 percent explained variance) for the four coastal stations located further south, as opposed to –0.47 (22 percent explained variance) when all 12 Moroccan rainfall stations were considered. This finding is probably not surprising, given that the Atlantic coast is the part of Morocco that is most exposed to the North Atlantic atmospheric circulation and among the least affected by Morocco's considerable topographic influences on rainfall. However, it should be stressed that the magnitudes of *all* of the above correlation coefficients are moderate-to-large by the standards of recent research into the mechanisms of tropical and subtropical rainfall fluctuations, as is shown by Table 5.3 and comparison with the results presented elsewhere in this book. The comparison of Fig. 5.9 with Table 5.2 shows that Moroccan rainfall is much more strongly related to the concurrent state of the NAO than the SO. Furthermore, the fact that the concurrent Moroccan rainfall–NAO relationship is stronger for the Atlantic coast of Morocco (particularly the northernmost station) than for the country as a whole, runs counter to the much weaker Moroccan rainfall–SO results in Table 5.2.

The magnitudes of the above correlation coefficients also exceed counterparts obtained by separately relating each Moroccan rainfall index in Fig. 5.9 to the individual components of the NAO, namely the subtropical high pressure centered near the Azores and the subpolar low pressure extending south and east of Greenland. Time series of normalized surface-pressure anomalies for both of these components were extracted from Rogers (1984, Fig. 1). The magnitudes of the correlation coefficients obtained were 0.05–0.14 lower than those cited above, values which correspond to 5–16 percent less explained rainfall variance. No clear pattern emerged as to which NAO component has the greater influence on Moroccan rainfall.

These results suggest that Moroccan rainfall is related not only to the strength of the adjacent Azores high pressure area (negatively), but also to the much more removed low pressure region near Greenland (positively). As Fig. 5.10 indicates, the mass alternation between these two components of the NAO influences the immediate mechanism of Moroccan rainfall variation previously identified – the number of low pressure systems that cross the

North Atlantic to the west of Morocco and southwestern Europe. These are more plentiful in the NAO extreme (2) "Greenland-above" winters that were found to produce higher Moroccan rainfall, than in winters that are the opposite in all those respects. Interestingly, the difference shown in Fig. 5.10 is greater for the zone that affects Morocco than for more northern latitudes.

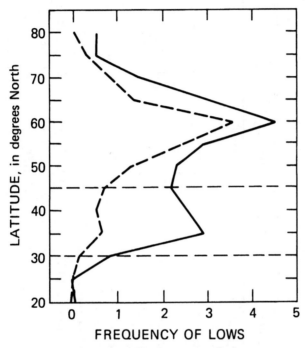

Fig. 5.10 Influence of NAO on number of January low-pressure systems in 10–30°W sector of the North Atlantic for 1899–1975. Solid line shows frequency of lows per January for the 15 "Greenland-above" years contributing to Fig. 5.7b; thick broken line does likewise for 16 "Greenland-below" years contributing to Fig. 5.7a. Latitude zone of particular relevance to Morocco is enclosed by thin broken lines. Taken from Rogers and van Loon (1979).

Prediction of Moroccan rainfall from NAO

The preceding section established that Moroccan winter semester rainfall is inversely related to the concurrent state of the NAO, and that the relationship is relatively strong by the standards of recent research into the mechanisms of tropical and subtropical rainfall fluctuations.

Table 5.3 Compilation of correlations between tropical/subtropical regional precipitation (or proxy) and various climate system indices. Except where indicated, the precipitation considered was in the form of average normalized departures for sizable sets of stations for lengthy single time periods. PC denotes Principal Component, SST is sea surface temperature, and SLP is sea level pressure. One, two, and three asterisks indicate the correlations were considered to be significant at the 95 percent, 99 percent, and 99.9 percent levels, respectively. See publications concerned for further details.

Regional precipitation or proxy	Climate system index	Linear correlation	Source
Northeast Brazil rainfall (September–August)	PC of tropical Atlantic calendar year mean SST field	+0.53***	Hastenrath (1978, Table 3)
Northeast Brazil rainfall (September–August)	PC of calendar year mean tropical Atlantic SLP field	+0.51***	Hastenrath (1978, Table 2)
Northeast Brazil rainfall (September–August)	North Atlantic minus South Atlantic subtropical SLP difference (July–June)	+0.43***	Hastenrath and Heller (1977, Table 3)
Northeast Brazil rainfall (March–April)	Southern Oscillation index for March–April (Tahiti minus Darwin SLP difference)	+0.34*	Hastenrath et al., 1987, Table 1)
Northeast Brazil rainfall (September–August)	North Atlantic subtropical SLP (July–June)	+0.33**	Hastenrath and Heller (1977, Table 3)
Northeast Brazil rainfall (September–August)	South Atlantic subtropical SLP (July–June)	+0.31**	Hastenrath and Heller (1977, Table 3)
Central American–Caribbean Sea rainfall (calendar year)	PC of tropical Atlantic calendar year mean SST field	-0.58***	Hastenrath (1978, Table 3)
Central American–Caribbean Sea rainfall (calendar year)	Southern Oscillation Index (defined above) for July–August	+0.58**	Hastenrath et al. (1987, Table 1)

Central American–Caribbean Sea rainfall (calendar year)	PC of tropical Atlantic calendar year mean SLP field	−0.41**	Hastenrath (1978, Table 2)
Indian rainfall (summer)	Latitude of 500-mb ridge along 75°E in preceding April	+0.72**	Hastenrath (1987, Table 2)
Indian rainfall (calendar year)	Southern Oscillation Index (defined above) for October	+0.61***	Wu and Hastenrath (1986, Figure 9)
Indian rainfall (calendar year)	Arabian Sea SLP (June–August)	−0.35**	Hastenrath (1985, Table 8.5:1)
Java rainfall (July–June)	Southern Oscillation Index (defined above) for preceding June	+0.49***	Wu and Hastenrath (1986, Figure 9)
Indonesian rainfall (September–November) treated separately for 13 stations for 2 time periods	Darwin SLP (preceding June–August)	Range from −0.77** to +0.37 with average of −0.28; 9 out of 26 lie between −0.50* and −0.77**	Nicholls (1981, Table 1)
Eastern Australian district rainfall (calendar year)	Darwin annual mean SLP	Largely range from −0.40** to −0.65**	McBride and Nicholls (1983, Figure 1)
Southeastern Australian district rainfall (calendar year)	Annual mean value of Wright's Southern Oscillation Index	Largely range from −0.40** to −0.60**	McBride and Nicholls (Figure 1)
Australian annual wheat production	Darwin June–August SLP	−0.56**	Nicholls (1985, Table 1)
Australian wheat production	June–August SST around Northern Australia	+0.67***	Nicholls (1985, Table 1)
Subsaharan African rainfall (April–October)	Southern Oscillation Index (defined above) for July–August	+0.31*	Hastenrath et al., 1987, Table 1)

This situation suggests that the NAO is of particular significance for the important issue of the long-range prediction of Moroccan winter rainfall. Reasonably accurate predictions issued as early as the preceding summer may be of considerable potential socioeconomic benefit to Morocco; see Lamb (1981) for a general discussion of climate prediction design issues. It is therefore critical that the following two questions be addressed: (1) can Moroccan winter rainfall be predicted from the state of the NAO during earlier seasons?; and (2) if not, can the winter NAO state (which we have shown to contain information on concurrent Moroccan rainfall) be forecast from earlier observations of other components of the global climate system? To the extent currently possible, these issues are now treated in turn.

Concerning question (1), we begin by summarizing the key results of earlier work dealing with the seasonal evolution of the NAO. Meehl and van Loon (1979) found that occurrences of the two winter (December–February) NAO extremes defined above were not foreshadowed in the North Atlantic sea-level pressure fields of the preceding summers (June–August periods). However, this is probably not surprising, given that one component of the NAO (the large area of low pressure to the south and east of Greenland) is typically only weakly apparent in summer (Fig. 5.6). The NAO may be a much less meaningful concept in summer than in winter. On the other hand, Meehl and van Loon (1979) did find that a sea-level "pressure anomaly pattern of the type seen during (NAO extreme) winters, but of smaller amplitude . . ., exists at all latitudes in autumns (September–November periods) preceding those . . . winters."

Because of this mildly encouraging finding, we used a time series approach to investigate the relation between Moroccan (November–April) rainfall and the NAO state during all possible overlapping and preceding three-month periods. This exploited newly available NAO index data for individual months (J.C. Rogers, 1989, personal communication). The results are summarized in Table 5.4. They reveal that the relatively strong relationships between Moroccan November–April rainfall and the December–February NAO state depicted in Fig. 5.9 are only weakly foreshadowed by the evolution of the NAO during preceding months. The correlations obtained for all three Moroccan rainfall indices decline moderately as the NAO period considered retreats from December–February (–0.47 to –0.64) to November–January (–0.40 to –0.50) and then decline precipitously to October–December (–0.17 to –0.34) and September–November (–0.03 to –0.18). Interestingly, this correlation decline is greatest for the rainfall index (single northernmost Atlantic coastal station) that is most strongly correlated with the December–February NAO state, and least for the 12-station national rainfall index that had the weakest such correlation. This intra-Moroccan variation bears some similarity to that of the SO–Moroccan rainfall correlation results presented in Table 5.2 and discussed earlier. It should also be noted that, for two of the three Moroccan rainfall indices, most of Table 5.4's September–November and October–December NAO correlations with Moroccan December–February rainfall are *weaker* than their SO counterparts in Table 5.2. The exception is the four-station Atlantic coastal index. However, all the correlations concerned are small (0–8 percent of rainfall variance explained) and statistically insignificant.

The analyses summarized in Table 5.4 were also separately performed using indices representing the two components of the NAO defined in the previous section. These indices were found to contain less or equal September–November and October–December predictive information concerning Moroccan November–April rainfall than the full NAO index.

These negative findings concerning question (1) above require that the follow-up question (2) be addressed. Unfortunately, the relevant information available is not very conclusive at present. The only nonatmospheric component of the global climate system considered so far is sea surface temperature (SST) (Rogers and van Loon, 1979; Meehl and van Loon, 1979). The results obtained (not shown) suggest that the SST anomaly patterns that occur in the tropical and subtropical North Atlantic during the two winter NAO extremes defined earlier already exist by the preceding summer (June–August) and persist through the intervening autumn. The SST in those zones tends to be higher during and prior to the NAO extreme (2) "Greenland-above" winters that were previously shown to produce high Moroccan rainfall, than in winters that are the opposite in all those respects. While this situation may prove to be important for the long-range prediction of Moroccan winter rainfall, it should be noted that neither Rogers and van Loon (1979) nor Meehl and van Loon (1979) discuss the possi-

Table 5.4 Correlation coefficients relating North Atlantic Oscillation (NAO) state (for 3-month running seasons) to rainfall in Morocco (November–April) for 1933–83. NAO state was represented by NAO index formulation of Walker (1924) and Walker and Bliss (1932) and evaluated using data supplied by J.C. Rogers (1989, personal communication); Moroccan rainfall was represented by the 3 indices portrayed in Fig. 5.9. NAO lag 0 correlations are for NAO seasons ending in the same year as the rainfall period ends; NAO lag −1 correlations are for NAO seasons ending in year prior to year in which rainfall period ends. One (two) asterisks indicate that the individual correlations were significant at the 95 percent (99 percent) level using a two-sided test. One (two) daggers indicate that the correlations were significant using the Bonferroni test with $\alpha_o = 0.05$ (0.01) and $K = 268$ unique variable pairings ($\alpha = \alpha_o/K$), as recommended by Brown and Katz (1989).

Moroccan rainfall

NAO season	National		Atlantic coast		Tangiers	
	NAO Lag −1	NAO Lag 0	NAO Lag −1	NAO Lag 0	NAO Lag −1	NAO Lag 0
Nov–Jan	−0.05	−0.42**	−0.10	−0.50***†	−0.10	−0.40**
Dec–Feb	−0.08	−0.47**	−0.11	−0.57***††	−0.08	−0.64**††
Jan–Mar	−0.04	−0.36*	−0.07	−0.45**	−0.01	−0.58**††
Feb–Apr	−0.07	−0.23	−0.06	−0.31*	−0.05	−0.47**
Mar–May	−0.16	+0.18	−0.16	+0.14	−0.23	+0.05
Apr–June	−0.16	+0.31*	−0.14	+0.31*	−0.23	+0.33*
May–July	+0.06	+0.27	+0.02	+0.28	+0.04	+0.33*
June–Aug	+0.19	−0.10	+0.10	−0.08	+0.30*	+0.06
July–Sept	+0.19	−0.06	+0.10	−0.02	+0.23	+0.06
Aug–Oct	+0.10	−0.13	−0.00	−0.06	+0.15	−0.09
Sept–Nov	−0.09	+0.10	−0.18	+0.10	−0.03	−0.01
Oct–Dec	−0.27	+0.13	−0.34*	+0.09	−0.17	+0.03

ble physical role of the above SST anomaly patterns for the NAO. However, the hemispheric modeling study of Rowntree (1976) obtained an extratropical atmospheric response to a warm tropical–subtropical North Atlantic SST anomaly (patterned after observations for December 1962–February 1963) that was consistent with NAO extreme (2) and high Moroccan rainfall (both of which occurred during the above period; see Fig. 5.9).

In contrast to the tropical and subtropical zones, the SST anomaly patterns in the middle latitudes of the North Atlantic during the two winter NAO extremes are not strongly evident in preceding seasons (Rogers and van Loon, 1979; Meehl and van Loon, 1979). Those winter NAO extremes apparently induce a substantial change in the midlatitude SST anomaly pattern that persists in the following summer. However, it should be noted that the modeling and observational research reported in Palmer and Zhaobo (1985) suggest "that it appears that some form of air–sea interaction could help both atmospheric and oceanic anomalies develop and persist over several (late autumn and early winter) months" in the extratropical North Atlantic.

Subsaharan rainfall

Background

Like the Moroccan situation described earlier, prior to the recent period of extended Subsaharan drought there seems to have been little governmental or scientific interest in the causes or predictability of the interannual and longer-time-scale variations of that region's rainfall. Such neglect was perhaps more understandable in the Subsaharan case, given that its rainfall had been adequate-to-abundant during the previous two decades, whereas Morocco had experienced some poor individual rainy seasons (Fig. 5.4). Consistent with this, and probably also because of lesser resources, the Subsaharan governments have been much less active than their Moroccan counterpart in promoting initiatives to address the drought problem. On the other hand, the scientific response of the international climate community to the widespread publicity of the disastrous societal consequences of the recent Subsaharan drought has been broad and strong. As outlined below, much progress has resulted.

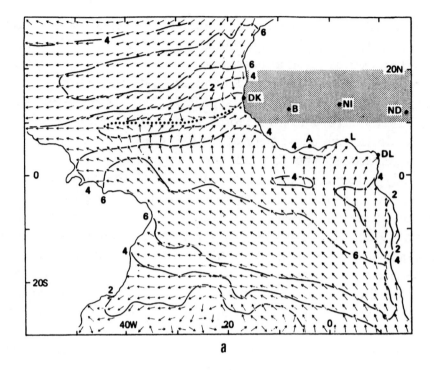

a

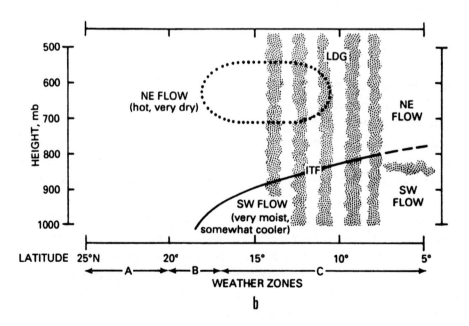

b

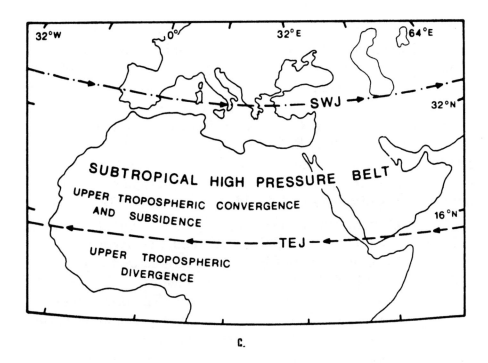

c.

Fig. 5.11 Large-scale atmospheric circulation influences on Subsaharan rainfall during July–September rainy season core. Shading in panels a
and c delineates 10°–20°N zone. a: Arrows and isotachs indicate
the 1911–70 resultant wind direction and speed (m/s) over the tropical Atlantic. Dotted line across ocean is discontinuity (kinematic
axis) between northerly and southerly wind directions. Dots over
continent locate the following aerological stations: Abidjan (A), Ba-
mako (B), Dakar (DK), Douala (DL), Lagos (L), Ndjamena (ND),
Niamey (NI). Taken from Lamb (1983). b: Schematic meridional-
vertical cross-section through West Africa along approximately 3°W.
NE and SW denote wind directions; ITF is the Intertropical Front
that separates the two airstreams; prevailing weather zones are designated by A (southern Sahara; cloudless), B (ITF vicinity; limited
cloud development, little rain), and C (south of ITF; greatest cloud
development, weather activity, and rainfall, especially in southern
two-thirds of zone); LDG denotes rain-producing Lignes de Grains
(i.e., Disturbance Lines); dotted envelope delineates approximate
location of 700–600-mb easterly wind maximum. Based on Lamb
(1980, 1985a), which contain more detailed descriptions. c: Upper
tropospheric flow over North Africa, Middle East, Mediterranean,
and Europe. TEJ is Tropical Easterly Jet (centered near 150-mb or
14-km level) and SWJ is Subtropical Westerly Jet (centered around
200–250 mb or 10–12 km level). Taken from Lamb (1980, 1985a).

The recent investigations into the interannual and longer-time-scale variability of Subsaharan rainfall have benefited from prior knowledge of the synoptic meteorology and climatology of the zone. Particularly valuable has been information on the nature of the rain-producing, westward-propagating West African Disturbance Lines ("Lignes de Grains") mentioned earlier (e.g., Hamilton and Archbold, 1945; Eldridge, 1957; Burpee, 1972), and their relation to the general climatic zonation of the region as summarized in Fig. 5.11b. Some of that information originated from the military needs of the Second World War (e.g., Hare, 1977).

The recent work has had two major "pulses" – one in the mid-1970s in response to the publicity afforded the particularly severe 1972–73 drought years (Fig. 5.4), and a second that began with the renewal of the severe drought conditions in the mid-1980s (Fig. 5.4) and continues to the present. The lack of interest in the problem in the intervening years was probably a consequence of the widespread but incorrect scientific perception that the drought conditions ended by 1974 or 1975 (reviewed in Lamb, 1982a; cf. Fig. 5.4). Most of the research in these two "pulses" has been conducted at major institutions in the United States (University of Wisconsin, Massachusetts Institute of Technology, Illinois State Water Survey, Goddard Institute for Space Studies) and the United Kingdom (University of East Anglia, Meteorological Office). Recently, however, there has been an encouraging appearance of studies by individual West African scientists (e.g., Bah, 1987; Adedoyin, 1989; and others). The accumulated body of work has by now used most of the principal modern techniques (empirical and numerical modeling) available for the investigation of interannual and longer-time-scale climatic variability.

The review of this work that follows is from the standpoint of the teleconnections theme of this book. It therefore does not consider the possibility that the local forcing of biogeophysical feedback mechanisms (vegetation or soil moisture depletion → less rainfall → further vegetation/soil moisture depletion, etc.) operating within the Subsaharan zone may have contributed to the initiation or persistence of the recent period of extended drought. A discussion of this work appears in the review paper by Druyan (1989), where it is concluded that such local forcing is probably of substantially subordinate importance to larger-scale influences of the type considered below. The modeling study of Owen and Folland

(1988) reached a similar conclusion. The pursuit of the teleconnections theme in the Subsaharan case has been substantially broader than the classical correlation-based approach used in the foregoing Moroccan work. However, it has not involved the NAO, which we found to be essentially uncorrelated with Subsaharan rainfall.

Relation of tropical Atlantic sea–air interaction to interannual variation of Subsaharan rainfall

Initial empirical investigations   Inquiry into the possible relationship between sea–air interaction and Subsaharan rainfall variability was initiated by Lamb (1976, 1978a,b, 1982b, 1983). This effort focused on the tropical Atlantic and the adjacent portion of West Africa to the south of the Sahara. It was prompted by the fact that the weather systems that produce Subsaharan rainfall have their base within, and receive most of their moisture from, the wedge-shaped southwest monsoon flow off the tropical Atlantic. This situation is illustrated in Fig. 5.11. It suggested that rainfall in Subsaharan West Africa may be sensitive to variations in not just the monsoon flow (including its interactions with the immediately overlying easterlies), but also the larger-scale tropical Atlantic sea–air interaction that gives rise to the monsoon flow and furnishes its moisture.

Investigation of this possibility began in the mid-1970s (Lamb 1976, 1978a,b) with the documentation of the large-scale atmospheric and oceanic patterns (pressure, wind, sea surface temperature, cloudiness, rainfall) at the surface of the tropical Atlantic during six moderate-to-severe Subsaharan drought years (1942, 1949, 1968, 1970–72; Fig. 5.4) and six years when the rainfall was slightly or substantially above the 1941–82 average (1943, 1950, 1952, 1954, 1957, 1967; Fig. 5.4). The marine meteorological data base used was adequate to support case study analyses for 1967 (slightly above average rainfall) and 1968 (moderate drought), which were presented in Lamb (1978a), but not the remaining years whose data were therefore combined into DRY and WET composite sets that were investigated in Lamb (1978b). The analyses for 1967, 1968, DRY, and WET involved comparison with 60-year (1911–70) average fields presented in Lamb (1977). The principal results obtained are summarized in Figs. 5.12 and 5.13.

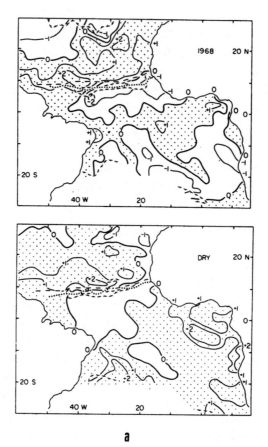

a

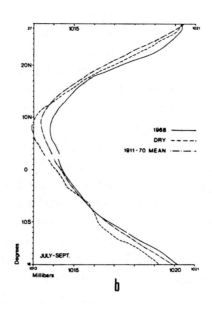

b

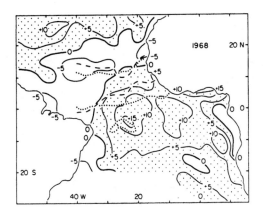

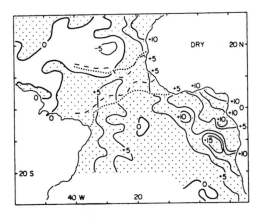

c

Fig. 5.12 Tropical Atlantic surface conditions during deficient Subsaharan rainy
season cores (July–September) of 1968 and a DRY composite data set
(produced by combining data for 1942, 1949, 1970, 1971, 1972 droughts).
Taken from Lamb (1978a,b). *a*: Resultant wind fields for 1968 and DRY
expressed as departures from 1911–70 average patterns. Solid lines are departure isotachs (m/s), positive values shaded; dotted lines locate discontinuity (kinematic axis) between northerly and southerly resultant wind
directions for 1968 and DRY, with broken lines doing likewise for 60-year
mean; barbed lines enclose resultant wind direction departures of more
than 30° in areas where the directional steadiness of wind exceeds 40 percent. *b*: Meridional transects of surface pressure for the north tropical
Atlantic (30°–1°N; 10°–60°W) and south tropical Atlantic (4°N–20°S;
13°E–52°W). Solid line is 1968, broken line is DRY, and dash-dot line the
1911–70 mean. *c*: SST fields for 1968 and DRY expressed as departures
from the 1911–70 average patterns. Solid lines are departure isotherms
(tenths of 1°C), positive values shaded; dotted lines enclose area of maximum SST east of 40°W for 1968 and DRY (>26.7°C), with broken lines
doing likewise for 1911–70 mean.

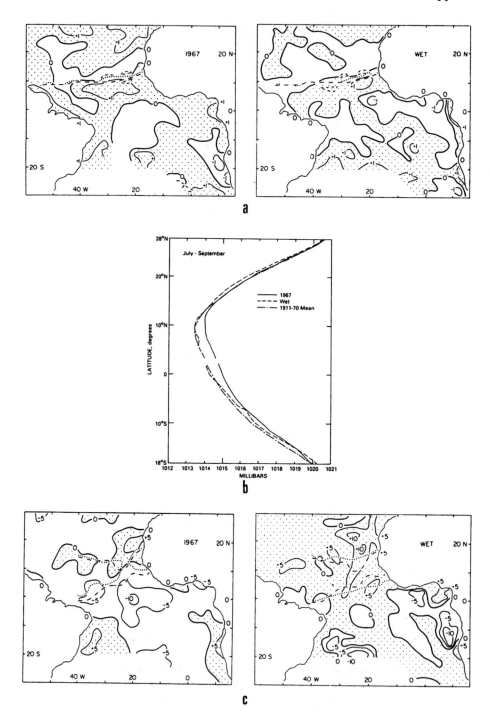

Fig. 5.13 Same as Fig. 5.12 except for Subsaharan rainy seasons of 1967 (near-
average) and a WET composite data set (produced by combining
data for excessively WET years of 1943, 1950, 1952, 1954, 1957).
Taken from Lamb (1978a,b).

This research established that the southwesterly surface monsoon flow did not extend as far north along the West African coast during the July–September cores of deficient Subsaharan rainy seasons as in the 60-year mean circulation pattern (by 200–300 km) or in more abundant rainy seasons (300–400 km) (cf. Figs. 5.11–5.13, a). Similar or slightly larger southward displacements were also characteristic of the following eastern tropical Atlantic surface features during poor rainy seasons – the kinematic axis between Northern and Southern Hemisphere trades (Figs. 5.12 and 5.13, a); the near-equatorial trough of minimum atmospheric pressure (Figs. 5.12 and 5.13, b); and the zone of maximum SST (Figs. 5.12 and 5.13, c). In addition, the North (South) Atlantic subtropical high pressure cell showed evidence of extending further (less) equatorward than normal during deficient rainy seasons (Fig. 5.12b); the opposite anomalies were characteristic of the near-average 1967 and WET rainy seasons (Fig. 5.13b). The location of the near-equatorial pressure trough during July–September probably controls the latitude of the kinematic axis to the west of West Africa and thus the northward extent of the moist surface southwesterly airflow into the western Subsaharan zone. Particularly relevant in this regard are the extensive zones of negative (positive) wind speed anomalies to the south (north) of the kinematic axis for 1968 and DRY (Fig. 5.12a). Note that the opposite wind speed anomaly pattern was characteristic of the wetter Subsaharan rainy seasons investigated (1967, WET; Fig. 5.13a).

Since surface pressure troughs over the tropical oceans are generally considered to be heat troughs that coincide with areas of maximum SST (e.g., see Ramage, 1974), the tropical Atlantic SST anomaly pattern that produced the aforementioned southward displacement of the zone of maximum SST is of particular interest. Figure 5.12c shows that this pattern contains warmer than average water to the south of 10°N and east of 35°W, a feature that extends to at least 20°S, and colder than average water to its immediate north and west. The latter anomaly occupies most of the 10–20°N zone, and extends westward into the Caribbean. It is the contiguity of these two anomalies near 10°N that produces the southward displacement of the zone of maximum SST noted above. The occurrence and significance of this drought SST pattern are discussed further below. Figure 5.13c shows that it was not characteristic of the rainy seasons of the WET composite,

for which the zone of maximum SST was expanded both northward and southward relative to the 60-year mean, or 1967 which
was very wet in the extreme western Subsaharan zone (Landsberg,
1973, 1975) if only near-average for Subsaharan West Africa as a
whole. In the case of 1967, the SST anomaly pattern was in fact
the opposite to the drought one described above, and accordingly
produced a northward displacement of the zone of maximum SST.

Although the foregoing displacements of the near-equatorial
kinematic axis, pressure trough, and zone of maximum SST are not
large, confirmation of their validity and significance is offered by
both their internal physical consistency and the very strong northward decrease of rainfall in Subsaharan West Africa. The latter,
which averages 1–3 mm/km depending on longitude (Lamb, 1978a,
1980, 1985a), has the important implication that drought-inducing
circulation displacements in the meridional direction need not be
large. This research also noted that the lower tropospheric mixing
ratios recorded over Subsaharan West Africa at the height of several very deficient rainy seasons (1968, 1970–72) were smaller than
during the more productive one of 1967 (Lamb, 1978a, Table 2).

The above results suggested that the wedge-shaped, moisture-
bearing southwest monsoon flow (Fig. 5.11b) may not penetrate as
far north in West Africa during Subsaharan droughts as in more
abundant rainy seasons. This interpretation was supported by
Kidson's (1977) finding that the July–August 850 mb flow over
West Africa south of approximately 8°N was from the east during
two severe Subsaharan drought years (1972–73; Fig. 5.4), whereas
it had a westerly component in two wetter years (1959, 1961;
Fig. 5.4). Because of this situation, Lamb (1983) investigated the
interannual and intraseasonal variability of the southwest monsoon
layer's thickness over West Africa, and its water vapor supply to
this region from the tropical Atlantic, for sets of deficient (1968,
1971, 1972; Fig. 5.4) and near-average (1967, 1969, 1975; Fig. 5.4)
Subsaharan rainy seasons. This used rawinsonde data for the three
Gulf of Guinea and four Subsaharan aerological stations located
in Fig. 5.11a, and also involved the estimation of the evaporation and surface specific humidity fields for the tropical Atlantic
domain treated in Figs. 5.12 and 5.13 (a and c).

The Lamb (1983) study had four principal findings. First, it
established that Subsaharan drought is not associated with suppressed tropical Atlantic evaporation into the southwest monsoon

flow, or with the northward supply of unusually dry surface air across the West African coast from the tropical Atlantic. Second, the extremely poor 1972 Subsaharan rainy season coincided with particularly shallow southwesterly flow across the Gulf of Guinea coast (approximately 5°N; Fig. 5.11a); thicker monsoon flows there were found characteristic of the less severe 1968 and 1971 droughts. However, the monsoon layer depth above the Gulf of Guinea coast during the non-drought study years did not always exceed that for some individual drought months, particularly the less severe ones. Third, the direction of the low-level water vapor flux above Dakar, a sensitive Subsaharan location (approximately 15°N; Fig. 5.11a) on the West African coast showed a stronger tendency to be from north of west during unproductive Subsaharan rainy seasons than in near-average ones. Fourth, in contrast to the foregoing Gulf of Guinea results, the monsoon layer over the interior of the Subsaharan zone at approximately 13°N (Fig. 5.11a) exhibited minimal variability for the drought study years; the better developed non-drought southwesterly flows here tended to be thicker than those characteristic of most drought months.

The above study thus provided some confirmation of our earlier ideas and findings concerning the role of the southwesterly monsoon flow for Subsaharan rainfall variability, but complete endorsement could not be claimed. It was concluded that, for some aspects such as the shallow monsoon layer over the Subsaharan zone, more definitive and physically instructive results would emerge from the analysis of individual rawinsonde observations, as opposed to the monthly mean data that were necessarily used. Such work is now being undertaken.

In the meantime, additional confirmation of the foregoing type was provided by Newell and Kidson's (1984) analysis of composite sets of African tropospheric wind data for two periods of contrasting Subsaharan rainy seasons (1958–62 = Wet; 1970–73 = Dry). This revealed that the southwest monsoon flow was stronger, produced greater water vapor convergence, and reached further north at 850 mb for the wet period, and that the immediately overlying Subsaharan 700-mb easterly wind maximum (Fig. 5.11b) was stronger and extended further equatorward for the dry period. These wind differences resulted in the 850–700-mb vertical wind shear over the Subsaharan zone being stronger for the dry period than the wet period; however, whether this was a cause or a consequence of the low 1970–73 Subsaharan rainfall was not assessed. Druyan (1989) further noted that the role of the 700-mb easterly wind strength for Subsaharan rainfall is ambiguous, since a negative thermal feedback may alter its initial characteristics once a rainfall (and cloud) regime is established. The Newell and Kidson (1984) study also confirmed Kidson's (1977) earlier finding that the upper tropospheric Tropical Easterly Jet (Fig. 5.11c) tends to be weaker during deficient as opposed to more abundant Subsaharan rainy seasons. The two easterly wind maxima just discussed determine, along with the southwesterly monsoon flow, the extent to which the Subsaharan atmospheric environment is conducive to the development of the rain-producing West African Disturbance Lines ("Lignes de Grains") described earlier (Fig. 5.11).

Subsequent empirical and numerical modeling investigations The possible relationship between tropical Atlantic sea–air interaction and Subsaharan rainfall variability has been the subject of several further empirical investigations during the 1980s.

The first, which is reported in Lough (1980, 1981, 1982, 1986), was restricted to SST and, to a lesser extent, sea level pressure. It included separate unrotated Principal Component (PC) analyses of tropical Atlantic SST for 1911–39 (except for 1915–20) and 1948–72. One of the major spatial modes of SST variation identified for both periods (Fig. 5.14a) corresponds closely to the drought SST anomaly pattern identified in Fig. 5.12c and its 1967 inverse (Fig. 5.13c). This mode features SST anomalies of opposite sign to the northwest and southeast of a line joining southwestern West Africa and northeastern Brazil, and essentially duplicates an earlier result obtained by Weare (1977). Inspection of the associated time series of PC scores (Fig. 5.14b) reveals that the positive variant of this mode (i.e., negative/positive anomalies to the northwest/southeast of the aforementioned West Africa–Brazil line) has characterized many of the deficient Subsaharan rainy seasons identified earlier (1913, 1921, 1949, 1968, 1972, and to a lesser extent 1971) and did not occur during two periods (1927–34, 1950–56) when Subsaharan West Africa was drought-free. Note that some of the other drought years identified earlier (e.g., 1919, 1941–42) were not considered in Lough's analyses because of war-time data deficiencies. Furthermore, the magnitude of this pattern's correlation with Subsaharan rainfall during 1948–72, as obtained using the aforementioned PC score time series, is in the range 0.51–0.66

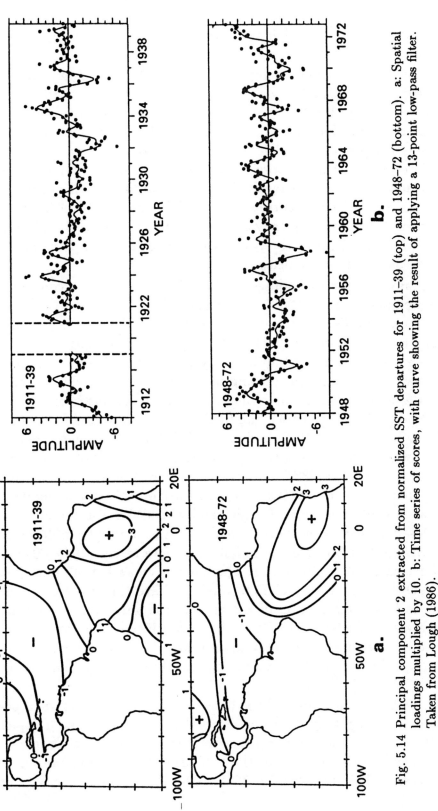

Fig. 5.14 Principal component 2 extracted from normalized SST departures for 1911–39 (top) and 1948–72 (bottom). a: Spatial loadings multiplied by 10. b: Time series of scores, with curve showing the result of applying a 13-point low-pass filter. Taken from Lough (1986).

(26–44 percent of rainfall variance explained), depending on the time-scale of the analysis.

However, this relationship was not found to be characteristic of the earlier 1911–39 period as a whole, despite the SST pattern occurring during some of its driest years, as already noted. This could be due to the weaker SST and rainfall data bases of the earlier period or a different functioning of the climate system. Concerning the latter possibility, we noted earlier that 1900–40 was not characterized by distinct Subsaharan rainfall epochs of the type that have occurred since, and Lough (1981, 1982, 1986) observed that the frequency of separate African (37°N–10°S) rainfall patterns identified by Nicholson (1980b) differed between the 1911–39 and 1948–72 periods.

Many of Lough's findings for 1948–72 were recently confirmed by Semazzi et al.'s (1988) similar (unrotated Empirical Orthogonal Function, EOF) Atlantic analyses for 1970–84, which were also largely confined to SST. These findings included identification of the mode of tropical Atlantic SST variation described above, confirmation of its existence (using the EOF time series) during some of the most deficient individual Subsaharan rainy seasons of the period (1972, 1973, 1984), and further documentation of its significant correlation with Subsaharan rainfall for an extended period as a whole. In addition, the unrotated EOF analysis of global SST by Parker et al. (1988) and Owen and Ward (1989) yielded a major mode of variation that contained a tropical Atlantic pattern that strongly resembled Lough's and which was also found to be "closely linked to the year to year (but not longer term decadal) variability" of Subsharan rainfall for a more recent (1940–80) but not earlier (1901–39) period. The Lough–Semazzi et al.–Parker et al.–Owen and Ward PC/EOF analyses have thus provided time series type confirmation of the earlier case study and composite analyses of Lamb (1976, 1978a,b) that utilized data from a limited number of extreme individual years. This use of complementary methods of analysis has strengthened the resulting knowledge base.

Further confirmation of the above results for the post-1970 period was sought by Lamb et al. (1986) and Lamb and Peppler (1989), when they performed case study analyses patterned after Lamb (1976, 1978a) for the four driest Subsaharan rainy seasons since 1941 (1972, 1977, 1983, 1984; Fig. 5.4) and the wettest rainy season since 1969 (1975; Fig. 5.4). This effort considered not just SST but, as for the earlier Lamb (1976, 1978a; present Figs. 5.12, 5.13) investigations, also involved the tropical Atlantic surface fields of pressure, wind, cloudiness, and rainfall. The opportunity to perform case study analyses for multiple years since 1970 was afforded by the data base for key individual years being much stronger than when we commenced our original work in the mid-1970s. This pursuit of multiple case studies was also motivated by recognition of the danger that the compositing approach used in Lamb (1976, 1978b; present Figs. 5.12, 5.13) and many other studies, including one discussed in the next section, could involve the combination of data from similar patterns of varying magnitude (and hence be dominated by particular years) or even from somewhat different patterns (e.g., Trenberth and Shea, 1987). The principal results obtained are summarized in Figs. 5.15 to 5.17.

The results for the extremely deficient Subsaharan rainy seasons of 1972, 1977, and 1984 (Figs. 5.15 and 5.16) essentially duplicate those yielded by the earlier drought-focused work (1968 case study, DRY composite; Fig. 5.12), particularly 1968. Their SST anomaly fields strongly feature the distinctive pattern described above and its associated southward displacement of the zone of maximum SST (Figs. 5.15 and 5.16, c). Those changes were accompanied by southward displacements (250–350 km relative to 1911–70 mean) of the near-equatorial pressure trough (Fig. 5.15 and 5.16, b), kinematic axis separating Northern and Southern Hemisphere trades (Figs. 5.15 and 5.16, a), and maxima of precipitation frequency and total cloudiness (Figs. 5.15 and 5.16, d). In addition, the North (South) Atlantic subtropical high again showed evidence of extending further (less) equatorward than normal (Figs. 5.15 and 5.16, b).

The foregoing common set of departure patterns is not nearly so characteristic of the other extremely deficient rainy season (1983) investigated by Lamb et al. (1986) and Lamb and Peppler (1989). The SST anomalies for that year (Fig. 5.16c) tend to be positive over most of the study domain, and the zone of maximum SST is expanded both northward and southward. Furthermore, while there is some evidence of a southward displacement of the 1983 near-equatorial pressure trough (Fig. 5.16b), this receives little support from the wind field anomalies (Fig. 5.16a) or rainfall fre-

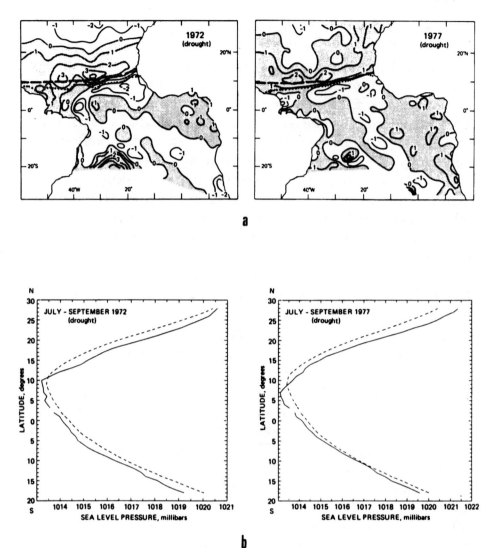

Fig. 5.15 Tropical Atlantic surface conditions during deficient Subsaharan
         rainy season cores (July–September) of 1972 and 1977. Taken from
         Lamb and Peppler (1989). a: Resultant wind fields for 1972 and 1977
         expressed as departures from 1911–70 average patterns. Solid lines
         are departure isotachs m/s, positive values shaded; dotted lines lo-
         cate discontinuity (kinematic axis) between northerly and southerly
         resultant wind directions for 1972 and 1977, with broken lines do-
         ing likewise for 60-year mean. b: Meridional transects of surface
         pressure for the north tropical Atlantic (30°–1°N; 10°–60°W) and
         south tropical Atlantic (4°N–20°S; 13°E–52°W). Solid line is indi-
         cated year, broken line is the 1911–70 mean.

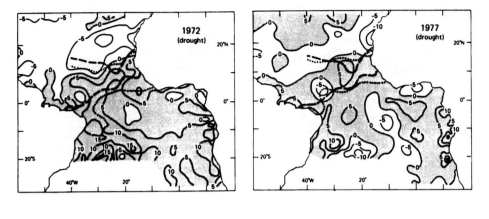

c

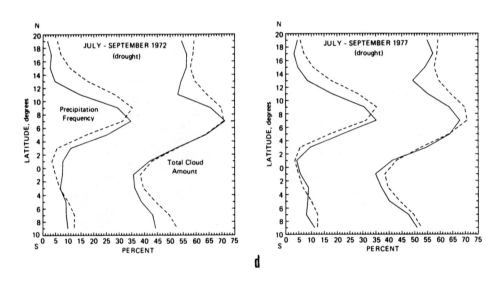

d

Fig 5.15 continued *c*: SST fields for 1972 and 1977 expressed as departure from the 1911–70 average patterns. Solid lines are departure isotherms (tenths of 1°C), positive values shaded; dotted lines enclose area of maximum sea surface temperature east of 40°W for 1968 and DRY (> 26.7°C), with broken lines doing likewise for 1911–70 mean. *d*: Meridional transects of rainfall frequency (left-hand profiles) and total cloudiness (right-hand profiles) for 10°–40°W. Solid line is indicated year; broken line is 1911–70 mean.

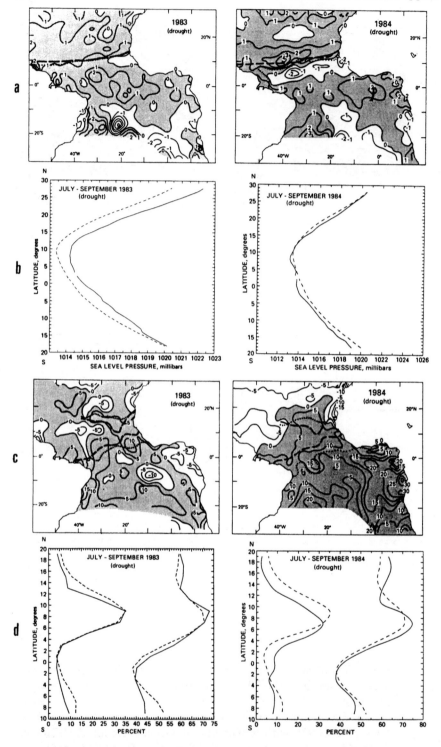

Fig. 5.16 Same as Fig. 5.15 except for deficient Subsaharan rainy seasons of
1983 and 1984. Taken from Lamb and Peppler (1989).

quency/total cloud amount results (Fig. 5.16d) obtained. Clearly, the tropical Atlantic surface atmospheric–oceanic conditions that prevailed during the very deficient 1983 Subsaharan rainy season were somewhat different from those that accompanied both the other recent years of near-comparable rainfall deficiency (1972, 1977, 1984; Figs. 5.15 and 5.16) and some years of more moderate drought (1968; years contained in DRY composite – 1942, 1949, 1970, 1971; Fig. 5.12). This finding provides a challenging situation with respect to our understanding of Subsaharan drought, particularly since 1983 was the year of a major ENSO event. This theme is developed further below.

The non-drought Subsaharan rainy season investigated by Lamb and Peppler (1989) for comparative purposes (1975) received rainfall that was close to the 1941–82 average, and the most abundant during 1970–88 (Fig. 5.4). Its tropical Atlantic surface atmospheric–oceanic conditions were also different from (but not opposite to) those of the very poor 1972, 1977, and 1984 rainy seasons. The 1975 SST anomaly pattern was rather fragmented and the zone of maximum SST was close to its average position (Fig. 5.17c). The near-equatorial pressure trough (Fig. 5.17b), kinematic axis (Fig. 5.17a), and rainfall frequency/total cloud maxima (Fig. 5.17d) for 1975 were also located in close to their 1911–70 mean latitudinal positions. These 1975 results thus do not duplicate those presented earlier for the slightly wetter 1967 rainy season (Fig. 5.13), which tended to be the opposite of the patterns identified here and previously as characteristic of most strongly deficient Subsaharan rainy seasons. The 1975 anomaly fields, particularly for SST and wind speed, are also somewhat different from those obtained for the above non-conforming drought year of 1983 (Figs. 5.16 and 5.17).

The principal results described above concerning the relationship between tropical Atlantic sea–air interaction and Subsaharan rainfall have been supported by several other recent empirical investigations that have used complementary approaches. This has further strengthened the resulting knowledge base. The studies involved include: (1) Hastenrath's (1984) identification of the differences in tropical Atlantic surface pressure, SST, surface wind (including its divergence), and cloud cover between composites for wet and dry Subsaharan rainy seasons; (2) Bah's (1987) correlation of Gulf of Guinea SST with Subsaharan rainfall; and (3)

Wolter's (1989) orthogonally rotated PC analysis of surface pressure and wind, cloud cover, and SST for the tropical Atlantic, eastern Pacific, and Indian Oceans, including both the spatial loading patterns obtained and the correlation of their associated PC score time series with Subsaharan rainfall.

Additional support for some of the major results described thus far in this section has emanated from Druyan's (1987) unique analysis of a 37-year simulation of global climate that was performed with the nine-level General Circulation Model (GCM) of the Goddard Institute for Space Studies (GISS). This simulation was originally undertaken as a "control" for a GCM experiment designed to evaluate the global climatic consequences of a doubled concentration of atmospheric greenhouse gases. The GISS GCM version used (Model II) included a sufficiently interactive ocean for interannual SST variations to occur in the simulation. The simulation was also characterized by considerable interannual variation of Subsaharan rainfall. For the purpose of "studying the synoptics of (Subsaharan) drought via a GCM," Druyan (1987) computed the average fields of several parameters (SST, sea level pressure, surface resultant wind, surface specific humidity, and others) for the Augusts of five very dry Subsaharan modeled-years and five very wet Subsaharan modeled-years, and also obtained the differences (DRY minus WET) between those average fields.

The results presented by Druyan (1987) were generally limited to Africa and the Atlantic Ocean. The DRY ensemble of modeled-Augusts had the following average characteristics for Subsaharan West Africa relative to its WET counterpart: weaker southwesterly monsoon flow that did not penetrate as far north, including along and off the Atlantic coast; lower surface specific humidity and total atmospheric water vapor; and implied weakening of the water vapor convergence. These results are in agreement with the empirically based findings of Lamb (1976, 1978a,b, 1983) and Newell and Kidson (1984) discussed earlier. However, the DRY minus WET tropical Atlantic SST pattern obtained by Druyan (1987) is only partially consistent (in the southeastern and northeastern tropical Atlantic) with the empirically based drought SST anomaly pattern presented in Figs. 5.12c, 5.14, 5.15c, and 5.16c (1984). Although the agreement elsewhere between the two patterns is rather poor, the SSTs in the two areas of consistency (particularly the southeastern tropical Atlantic) were considered

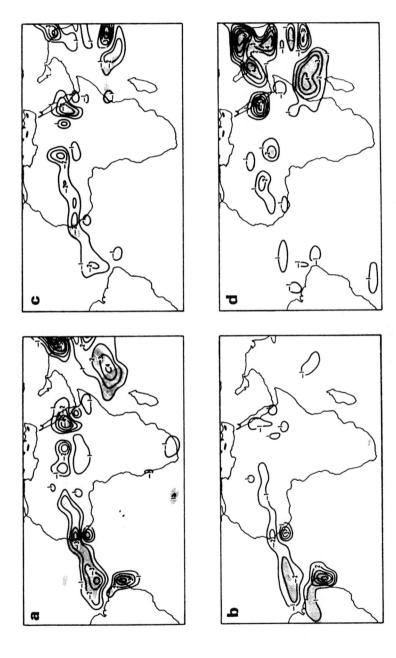

Fig. 5.19 Rainfall anomaly in mm/day (relative to control integration; see text) for Atmospheric General Circulation Model experiment with: (a) full global SST anomaly pattern in Fig. 5.18; (b) Atlantic portion of Fig. 5.18 pattern only; (c) Pacific portion of Fig. 5.18 pattern only; (d) Indian Ocean portion of Fig. 5.18 pattern only. See text for further details. Stippling indicates areas where the magnitude of the *t*-variate exceeds 2. Taken from Palmer (1986).

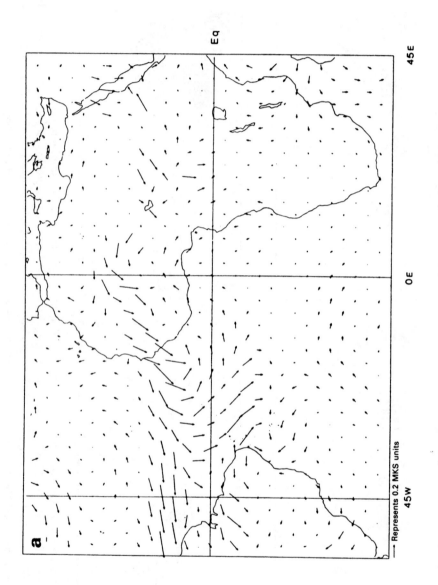

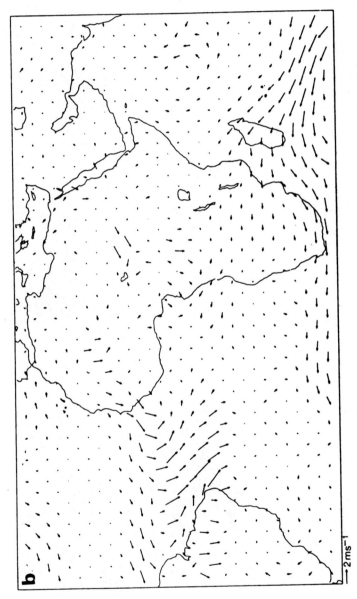

Fig. 5.20 Wind field and related anomalies at 950-mb level for UKMO GCM experiments described in text. Note scale beneath each panel. (a) 180-day mean steady moisture flux (anomaly minus control) corresponding to full global SST anomaly in Fig. 5.18 (see text). Taken from Folland et al. (1986). (b)180-day mean wind (anomaly minus control) corresponding to Atlantic SST anomaly in Fig. 5.18 and climatological average SST elsewhere (see text). Taken from Palmer (1986).

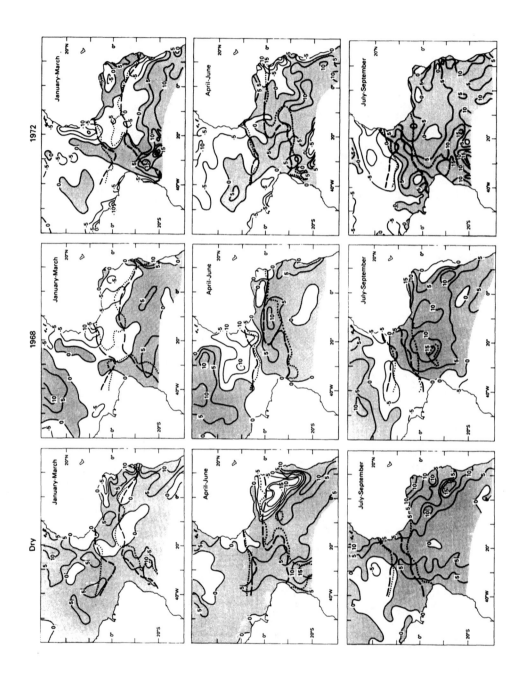

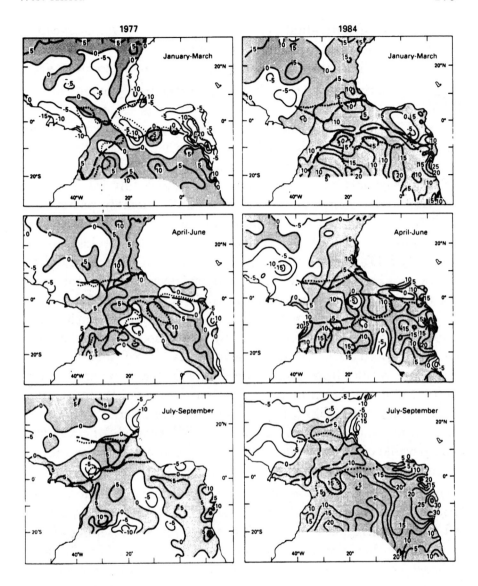

Fig. 5.21 Seasonal SST fields for 1968, DRY, 1972, 1977, and 1984 expressed
as departures from the 1911–70 average patterns. Solid lines are
departure isotherms (tenth of 1°C), positive values shaded; dotted
lines enclosed area of maximum SST east of 40°W for indicated years
and composite (>27.0°C for January–March; >27.2°C for April–
June; >26.7°C for July–September) with broken lines doing likewise
for 1911–70 mean. 1968 and DRY patterns are taken from Lamb
(1978a,b).

with the SST of individual oceans. That investigation suggested
the Atlantic and Indian Ocean components of Fig. 5.18 were more
important for Subsaharan rainfall than their Pacific counterpart.

The distinctive pattern in Fig. 5.18 prompted Folland et al.
(1986) and Palmer (1986) to conduct numerical experiments using the U.K. Meteorological Office (UKMO) 11-level Atmospheric
General Circulation Model (AGCM) to further investigate the influence on Subsaharan rainfall of both global and individual ocean
SST anomaly patterns. Separate 180-day fixed July integrations
were conducted using the entire global SST anomaly pattern in
Fig. 5.18 and each of its three individual ocean components. The
control integration involved was run with fixed climatological average July SSTs; the global anomaly integration added the entire
pattern in Fig. 5.18 to the climatological average pattern; each
"individual ocean" integration added only the portion of Fig. 5.18
for that ocean to the climatological average pattern, which it re-
tained elsewhere. The principal results obtained are summarized
in Figs. 5.19 and 5.20. The decrease in rainfall in Subsaharan West
Africa that results from SST anomalies is shown to be greater for
the total global pattern in Fig. 5.18 than for any of its individual
ocean components (Fig. 5.19). It was also considered by Palmer
(1986) that "the Atlantic and Pacific (SST anomaly) fields have
a comparable effect in reducing (that) rainfall whereas the Indian
Ocean field produces a slight enhancement."

Figure 5.20 reveals the following key tropical Atlantic–West
African atmospheric responses of the AGCM to *both* the global
and Atlantic SST anomalies in Fig. 5.18 – (1) a weakening of the
low-level southeast trade winds between approximately 5°N and
10°S, which was part of a reduction in strength of the southern
Hadley cell; and (2) a lessening and equatorward displacement of
the southwest monsoon flow into the western Subsaharan zone,
which in turn "reduced the supply of moisture for precipitation"
(Palmer, 1986). Note that these AGCM responses duplicate many
of the principal findings (surface wind, lower tropospheric water
vapor flux) of the empirical investigations of Lamb (1976, 1978a,
b, 1983) and Newell and Kidson (1984) for very deficient Subsaharan rainy seasons, and also for Druyan's (1987, 1988) GISS GCM-
simulated drought Augusts. Palmer (1986) suggests that the above
AGCM responses would likely have been intensified if the particularly pronounced 1984 tropical Atlantic SST anomaly pattern (see

Fig. 5.16c) had been used in his simulations in place of the much weaker departure pattern contained in Fig. 5.18. He further argues that the role of the "tropical Atlantic ... was probably paramount during the exceptional 1984 season." These views were supported by Owen and Folland's (1988) subsequent use of the 1984 and 1950 global SST patterns in similar UKMO AGCM experimentation. The SST anomaly difference pattern for these years possessed essentially the same global morphology as Fig. 5.18, but featured stronger tropical Atlantic anomalies; the 1950 Subsaharan rainy season was very wet (Fig. 5.4).

Palmer (1986) also suggested that the Pacific SST anomaly pattern in Fig. 5.18 contributes to the reduction of rainfall in Subsaharan West Africa via the generation and maintenance of anomalous upper tropospheric equatorial westerly winds that extend as far east as Africa. This anomaly feature, and its associated baroclinic vertical structure, apparently render the Subsaharan atmospheric environment less conducive to the development of the rain-producing West African Disturbance Lines ("Lignes de Grains") described earlier (Fig. 5.11b). This possible teleconnection mechanism receives some support from Semazzi et al.'s (1988, 1989) experimentation with the GCM of the Goddard Laboratory for Atmospheres (GLA). It is possible that this process contributed to the extreme Subsaharan rainfall deficiency of 1983, given the absence of the tropical Atlantic drought SST anomaly pattern noted earlier.

Prediction of Subsaharan rainfall from SST anomaly pattern

We established earlier that very deficient rainy seasons in Subsaharan West Africa tend to coincide with a distinctive SST anomaly pattern that spans at least the tropical Atlantic, and possibly much of the global ocean. The tropical Atlantic pattern contains anomalously cool (warm) water to the northwest (southeast) of a line linking southwestern West Africa and northeastern Brazil; the remainder of the global pattern features cool (warm) departures to the north (south) of approximately 5°N in the Pacific, and an Indian Ocean dominated by positive anomalies. Furthermore, empirical and modeling evidence presented earlier in the chapter suggested that both the tropical Atlantic SST anomaly pattern and the total global pattern of which it is sometimes part are accompanied by physically consistent southward displacements in the wind, surface pressure, and other meteorological fields over the tropical Atlantic and West Africa. These displacements, in turn, were noted to be consistent with the greatly reduced Subsaharan rainfall.

Lamb (1976, 1978a,b) found that the rainy season tropical Atlantic SST anomaly pattern in question tended to evolve or exist during preceding seasons for the 1968 drought case study and DRY composite discussed in the subsection on initial empirical investigations. This finding has also emerged from further analysis of the 1972, 1977, and 1984 drought case studies that were introduced in the subsection on subsequent empirical and numerical modeling investigations. Figure 5.21 presents the results obtained for each of these four case studies and the DRY composite. They particularly feature the northward expansion, during the seasons that precede a deficient Subsaharan rainy season, of an extensive area of positive SST anomalies that emanates from the southern tropical Atlantic.

The evolution of the 1968 and DRY SST departure patterns in Fig. 5.21 was seen by Lamb (1978b) to offer "encouragement that Subsaharan droughts may be predictable three to six months in advance." This view received additional empirical support from Lough's (1986) correlations of seasonal SST and surface pressure (SP) PC score time series for the tropical Atlantic with Subsaharan rainfall, which were discussed earlier, and also from the 1972, 1977, and 1984 case study results presented in Fig. 5.21. It is further suggested by several of the GCM studies that provided important evidence for the existence of a concurrent relationship between the aforementioned SST anomaly patterns and very deficient Subsaharan rainy seasons. The salient experimental and analytical features of this GCM work include: (1) Druyan's (1987) descriptive analysis of the month-to-month tropical Atlantic SST and SP evolution in his GISS GCM-generated ensembles of five DRY and five WET Subsaharan rainy seasons; (2) Druyan's (1988) formal lag-correlation analysis between the evolving modeled-SST in the southeastern tropical Atlantic and modeled Subsaharan rainfall for the above 10 modeled-years; (3) Druyan's (1988) addition of prescribed tropical Atlantic and global SST anomalies from March or May onward to a control simulation that originally yielded average Subsaharan modeled-rainfall; (4) Owen and Folland's (1988) use of evolving historical SST anomaly patterns (from late March onward) to drive their UKMO AGCM experimentation for the contrasting Subsaharan rainy seasons of 1984 (very dry) and 1950 (very wet); and (5) Owen and Ward's (1989) similar work that considered 1958 (very wet) and 1983 (very dry) as well as 1950 and 1984. See the preceding sections for further detail on all these investigations except Owen and Ward (1989).

It is therefore not surprising that attempts are now being made to predict the general quality of Subsaharan rainy seasons several months in advance from the evolution of the tropical Atlantic and global SST fields. While the UKMO effort has received the greatest exposure to date, work is proceeding at other institutions as well. According to Owen and Ward (1989), the UKMO endeavors are utilizing unrotated EOF counterparts to both: (1) the Subsaharan drought tropical Atlantic SST anomaly pattern identified in Figs. 5.12c, 5.14, 5.15c, and 5.16c (1984), which "is closely linked to year-to-year (Subsaharan rainfall) variability"; and (2) the Subsaharan drought global SST anomaly pattern shown in Fig. 5.18, which "is closely linked to the longer-term decadal (Subsaharan rainfall) variations." Details of the methods employed and results obtained appear in Parker et al. (1988) and Owen and Ward (1989). In general, three linear statistical techniques (discriminant analysis, multiple regression, single regression) are being used to predict Subsaharan rainfall from measures of the similarity between observed tropical Atlantic and global SST anomaly patterns for spring (e.g., March–April) and unrotated EOF SST representations (1) and (2) above. Clearly, this assumes that spring SST anomalies persist through July–September. Hindcast prediction equations that were used to obtain ranges or values of Subsaharan rainfall for 1946–86 (1901–45) were developed from historical SST and rainfall data and SST EOF loadings for 1901–45 (1946–86). However, this statistically desirable attempt to separate the periods used for equation development and their predictive deployment may not be entirely complete, since all SST EOF loadings utilized appear to be from an analysis for 1901–80. A verification of the results given by the multiple regression technique appears in Fig. 5.22. Parker et al. (1988) provide more comprehensive verification information concerning the discriminant analysis and single regression techniques. Overall, the skill of this hindcast prediction exercise for 1901–86 is perhaps best described as "mixed."

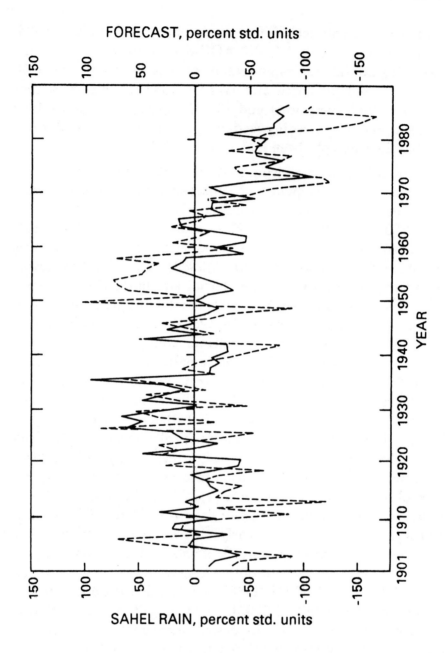

Fig. 5.22 Hindcast predictions of Subsaharan rainfall for each year from 1901–
86 (solid line) made from March–April global and tropical Atlantic
SST anomaly patterns using a multiple regression technique described in the text. Dashed line is observed Subsaharan rainfall.
Correlation between predicted and observed Subsaharan rainfall for
1901–45 is +0.52 and for 1946–86 is +0.68, both of which are significant at 95 percent level. Taken from Owen and Ward (1989).

Since 1986, the foregoing approaches have also provided the basis for the UKMO's annual operational issuance to the heads of African meteorological agencies of "a cautiously worded experimental forecast for Sahel rainfall for the (upcoming) summer" (Parker et al., 1988). The skill demonstrated by these predictions has also been somewhat varied, with the lack of spring–summer persistence of the global SST anomaly pattern having a particularly adverse impact in 1988 (Parker et al., 1988; Owen and Ward, 1989).

In summary, we agree with Parker et al. (1988) that the above UKMO effort constitutes "a useful start in ... forecasting Sahel rainfall," and also with Owen and Ward's (1989) view that "the work is still in its infancy. Much remains to be understood about the links between sea temperatures and rainfall in the Sahel." The challenge is indeed considerable.

Concluding remarks

Much of the rest of this book documents the pervasiveness of the ENSO phenomenon and attests to its wide-ranging and considerable impacts upon society. This results from it being such a dominant and conservative mode of atmosphere–ocean variation. In contrast, the present chapter has been concerned with two aspects of the regional climate of West Africa that are of great societal importance – the strongly seasonal rainfalls of Morocco and the Subsaharan zone – but which have been only very weakly related to ENSO for the past 50 years as a whole. The principal exception is that *some* extremely deficient Subsaharan rainy seasons have coincided with *some* ENSO events. As a result, ENSO does not in general provide a basis for the much needed skillful prediction of Moroccan and Subsaharan rainy season quality several months in advance.

Despite this situation, climate diagnostics and modeling research conducted during the past 15 years has succeeded in associating the interannual variations of Moroccan and Subsaharan rainfall with particular states of other components of the global atmosphere–ocean system. For both regions, the scientific endeavors were prompted by the dire societal consequences of recent periods of extended drought. They offer some hope for the development of teleconnection-based prediction schemes for Moroccan and Subsaharan rainfall of the type mentioned above. However, much work remains to be done.

The interannual variation of Moroccan (November–April) rainfall was found to be inversely related to the concurrent state of the NAO, with the relationship being relatively strong by the standards of recent research into the mechanisms of tropical and subtropical rainfall fluctuations. However, this relationship was shown to be only weakly foreshadowed by the evolution of the NAO during preceding months. This situation underlines the need for enhanced understanding of the NAO in general, and, more specifically, of the extent to which its winter state can be predicted from earlier observations of other components of the global climate system. Our present knowledge of these subjects is minimal. There is a clear need for a comprehensive investigation that exploits both empirical and numerical modeling approaches.

It was established that very deficient rainy seasons (July–September) in Subsaharan West Africa tend to coincide with a distinctive SST anomaly pattern that spans at least the tropical Atlantic, and possibly much of the global ocean. Furthermore, empirical and modeling evidence suggested that this rainy season SST anomaly pattern is accompanied by southward displacements in the wind, surface pressure, and other meteorological fields over the tropical Atlantic and West Africa. These displacements, in turn, were noted to be consistent with the greatly reduced Subsaharan rainfall. Because of this situation, attempts are now being made to predict the general quality of Subsaharan rainy seasons several months in advance from the evolution of the tropical Atlantic and global SST fields. The results of this effort have been varied.

While it is clear that a useful start has been made on an important problem, much also remains to be learned about the teleconnection relationship between SST anomalies and Subsaharan rainfall. Future empirical and numerical modeling research should seek further elucidation of the relative importance for Subsaharan rainfall of the tropical Atlantic and global SST anomaly patterns, and of the sea–air interaction and atmospheric processes involved in the teleconnection. In the latter regard, the role of the Subsaharan 700 mb easterly wind maximum for the zone's rainfall needs to be clarified, as do the mechanisms by which ENSO influences (or fails to influence) that rainfall. Progress would likely be accelerated by a greater diagnostic analysis of the computations made during GCM experiments than has occurred to date, particularly with respect to key links in the teleconnection chain.

Acknowledgments

The preparation of this review and the performance of the new research that it contains were supported by NOAA Grants NA86AA-D-AC065 and NA87AA-D-AC084 from the U.S. TOGA Program Office. We thank Stefan Hastenrath and Neville Nicholls for their formal reviews of the manuscript, Jeff Rogers for the provision of NAO data, and Linda Riggin and Rebecca Runge for their technical assistance. The helpful comments of J.A. Adedokun, A.T. Grove, O.J. Olaniran, Fredrick Semazzi, and Thompson Webb III were also much appreciated.

References

Adedoyin, J.A. (1989). Global-scale sea-surface temperature anomalies and rainfall characteristics in northern Nigeria. *International Journal of Climatology*, 9, 133–44.

Ångström, A. (1935). Teleconnections of climate changes in present time. *Geografiska Annaler*, 17, 242–58.

Bah, A. (1987). Towards the prediction of Sahelian rainfall from sea surface temperatures in the Gulf of Guinea. *Tellus*, 39A, 39–48.

Barnston, A.G. & Livezey, R.E. (1987). Classification, seasonality, and persistence of low-frequency atmospheric circulation patterns. *Monthly Weather Review*, 115, 1083–1126.

Bensari, A. & Benarafa, S. (1988). La pluie artificielle au Maroc: Project d'augmentationde l'emeigement d'hiver. In *Proceedings of Conference on Drought, Water Management and Food Production*, 139–48. Rabat: Kingdom of Morocco.

Bensari, A., Benarafa, S., Loukah, B., Benassi M., Mrabet, A., Mathews, D., Medina, J., Hartzell, C. & Deshler, T. (1989). Programme Al Ghait – Design and evaluation of a weather modification program in Morocco. Cloud Physics and Weather Modification Research Programme, WMP Report No. 12, 533–6. Geneva: World Meteorological Organization.

Brown, B.G. & Katz, R.W. (1989). Use of statistical methods in the search for teleconnections. *Proceedings of the Fourth International Meeting on Statistical Climatology*, March 27–31, 1989, Rotorua, N.Z., 87–92. Wellington: New Zealand Meteorological Service.

Burpee, R.W. (1972). The origin and structure of easterly waves in the lower troposphere of North Africa. *Journal of the Atmospheric Sciences*, 29, 77–90.

Byers, H.R. (1959). *General Meteorology* (Third Edition). New York: McGraw-Hill.

Charney, J.G. (1975). Dynamics of deserts and drought in the Sahel. *Quarterly Journal of the Royal Meteorological Society*, **101**, 193–202.

Charney, J.G., Quirk, W.J., Chow, S.-H. & Kornfield, J. (1977). A comparative study of the effects of albedo change on drought in semi-arid regions. *Journal of the Atmospheric Sciences*, **34**, 1366–85.

Dannmeyer, F. (1948). Zur Frage der Gegensätzlichkeit der kalten Winter in Grönland zu den warmen Wintern in Deutschland. *Polarforsch*, **2**, 29–30.

Davy, E.G. (1974). Drought in West Africa. *World Meteorological Organization Bulletin*, **23**, 18–23.

Defant, A. (1924). Die Schwankungen der atmosphärischen Zirkulation über dem Nordatlantischen Ozean im 25-jährigen Zeitraum 1881–1905. *Geografiska Annaler*, **6**, 13–41.

Dove, H. W. (1839). Über die geographische Verbreitung gleichartiger Witterungserscheinungen. *Abhandlungen Kön. Akademie Wissenschaften zu Berlin*, **1838**, 287–415.

Druyan, L.M. (1987). GCM studies of the African summer monsoon. *Climate Dynamics*, **2**, 117–26.

Druyan, L.M. (1988). Sea surface temperature - Sahel drought teleconnections in GCM simulations. In *Recent Climatic Change: A Regional Approach*, ed. S. Gregory, 154–65. London: Belhaven Press.

Druyan, L.M. (1989). Advances in the study of Subsaharan drought. *International Journal of Climatology*, **9**, 77–90.

Eldridge, R.H. (1957). A synoptic study of West African disturbance lines. *Quarterly Journal of the Royal Meteorological Society*, **83**, 303–14.

Folland, C.K., Palmer, T.N. & Parker, D.E. (1986). Sahel rainfall and world-wide sea temperatures. *Nature*, **320**, 602–7.

Griffiths, J. F., ed. (1972). *Climates of Africa*. World Survey of Climatology Volume 10. Amsterdam: Elsevier.

Grove, A.T. (1972). Climatic change in Africa in the last 20,000 years. In *Les Problémes de Développement du Sahara Septentrional*, **2**, Alger.

Grove, A.T. (1973) A note on the remarkably low rainfall of the Sudan zone in 1913. *Savanna*, **2**, 133–8.

Hamilton, R.A. & Archbold, J.W. (1945). Meteorology of Nigeria and adjacent territory. *Quarterly Journal of the Royal Meteorological Society*, **71**, 230–65.

Hann, J. (1890). Zur Witterungsgeschichte von Nord-Grönland, Westküste. *Meteorologische Zeitschrift*, **7**, 109–15.

Hare, F. K. (1977). Takoradi-Khartoum air route. *Climatic Change*, **1**, 157–72.

Hastenrath, S. (1978). On modes of tropical circulation and climate anomalies. *Journal of the Atmospheric Sciences*, **35**, 2222–31.

Hastenrath, S. (1984). Interannual variability and annual cycle: Mechanisms of circulation and climate in the tropical Atlantic sector. *Monthly Weather Review*, **112**, 1097–1107.

Hastenrath, S. (1985). *Climate and Circulation of the Tropics*. Dordrecht: D. Riedel Publishing Company.

Hastenrath, S. (1987). On the prediction of India monsoon rainfall anomalies. *Journal of Climate and Applied Meteorology*, **26**, 847–57.

Hastenrath, S. & Heller, L. (1977). Dynamics of climatic hazards in northeast Brazil. *Quarterly Journal of the Royal Meteorological Society*, **103**, 77–92.

Hastenrath, S., de Castro, L.-C. & Aceituno, P. (1987). The Southern Oscillation in the tropical Atlantic sector. *Beiträge zur Physik der Atmosphäre*, **60**, 447–63.

Jenkinson, A. (1973). A note on variations in May to September rainfall in West African marginal areas. In *Drought in Africa*, ed. D. Dalby & R. J. Harrison-Church, 27–8. London: School of Oriental and African Studies.

Kidson, J.W. (1977). African rainfall and its relation to the upper air circulation. *Quarterly Journal of the Royal Meteorological Society*, **103**, 441–56.

Kingdom of Morocco, 1988: *Proceedings of Conference on Drought, Water Management and Food Production*. Rabat: Kingdom of Morocco.

Klein, W.H. (1957). Principal tracks and mean frequencies of cyclones and anticyclones in the Northern Hemisphere. Research Paper No. 40, U.S. Weather Bureau, Washington, D.C.: U.S. Government Printing Office.

Kutzbach, J.E. & Otto-Bliesner, B.L. (1982). The sensitivity of the African-Asian monsoonal climate to orbital parameter changes for 9000 years B.P. in a low resolution general circulation model. *Journal of the Atmospheric Sciences*, **39**, 1177–88.

Kutzbach, J.E. & Street-Perrott, F.A. (1985). Milankovitch forcing of fluctuations in the levels of tropical lakes from 18 to 0 kyr B.P. *Nature*, **317**, 130–4.

Kutzbach, J.E. & Guetter, P.J. (1986). The influence of changing orbital parameters and surface boundary conditions on climate simulations for the past 18,000 years. *Journal of the Atmospheric Sciences*, **43**, 1726–59.

Lamb, P.J. (1976). Variations in General Circulation and Climate over the Tropical Atlantic and Africa: Weather Anomalies in the Subsaharan Region. Ph.D. Dissertation, Department of Meteorology, The University of Wisconsin, Madison.

Lamb, P.J. (1977). On the surface climatology of the tropical Atlantic. *Archiv für Meteorologie, Geophysik und Bioklimatologie*, **25B**, 21–31.

Lamb, P.J. (1978a). Case studies of tropical Atlantic surface circulation patterns during recent Sub-Saharan weather anomalies. *Monthly Weather Review*, **106**, 482–91.

Lamb, P.J. (1978b). Large-scale tropical Atlantic surface circulation patterns associated with Subsaharan weather anomalies. *Tellus*, **30**, 240–51.

Lamb, P.J. (1980). Sahelian drought. *New Zealand Journal of Geography*, **68**, 12–6.

Lamb, P.J. (1981). Do we know what we should be trying to forecast climatically? *Bulletin of the American Meteorological Society*, **62**, 1000–1.

Lamb, P.J. (1982a). Persistence of Subsaharan drought. *Nature*, **299**, 46–8.

Lamb, P.J. (1982b). Comments on "West African rainfall variations and tropical Atlantic sea surface temperatures." *Climate Monitor*, **11**, 46–9.

Lamb, P.J. (1983). West African water vapor variations between recent contrasting Subsaharan rainy seasons. *Tellus*, **35A**, 198–212.

Lamb, P.J. (1985a). Subsaharan drought. Paper prepared for the U.S. Department of State Conference on African Environmental Issues, Washington, D.C., October 1985. [Available from Illinois State Water Survey, 2204 Griffith Drive, Champaign, IL 61820.]

Lamb, P.J. (1985b). Rainfall in Subsaharan West Africa during 1941–83. *Zeitschrift für Gletscherkunde und Glazialgeologie*, **21**, 131–9.

Lamb, P.J. & Peppler, R.A. (1987). North Atlantic Oscillation: Concept and an application. *Bulletin of the American Meteorological Society*, **68**, 1218–25.

Lamb, P.J. & Peppler, R.A. (1988). Large-scale atmospheric features associated with drought in Morocco. In *Proceedings of Conference on Drought, Water Management and Food Production*, 103–27. Rabat: Kingdom of Morocco.

Lamb, P.J. & Peppler, R.A. (1989). Further case studies of tropical Atlantic surface circulation patterns associated with Subsaharan drought. *Proceedings of the Thirteenth Annual Climate Diagnostics Workshop*, 454–60. Washington, DC: U.S. Department of Commerce.

Lamb, P.J., Peppler, R.A. & Hastenrath, S. (1986). Interannual variability in the tropical Atlantic. *Nature*, **322**, 238–40.

Landsberg, H.E. (1973). An analysis of the annual rainfall at Dakar (Senegal) 1887–92. Institute for Fluid Dynamics and Applied Mathematics Paper, University of Maryland.

Landsberg, H.E. (1975). Sahel drought: Change of climate or part of climate? *Archiv für Meteorologie, Geophysik, und Bioklimatologie*, **23B**, 193–200.

Loewe, F. (1937). A period of warm winters in western Greenland and the temperature see-saw between western Greenland and Europe. *Quarterly Journal of the Royal Meteorological Society*, **63**, 365–71.

Loewe, F. (1966). The temperature see-saw between western Greenland and Europe. *Weather*, **21**, 241–6.

Lough, J.M. (1980). West African rainfall fluctuations and tropical Atlantic sea surface temperatures. *Climate Monitor*, **9**, 150–7.

Lough, J.M. (1981). Atlantic Sea Surface Temperatures and the Weather in Africa. Ph.D. Dissertation, School of Environmental Sciences, University of East Anglia, Norwich.

Lough, J.M. (1982). Response to comments by P. J. Lamb. *Climate Monitor*, **11**, 50–1.

Lough, J.M. (1986). Tropical Atlantic sea surface temperatures and rainfall variations in Subsaharan Africa. *Monthly Weather Review*, **114**, 561–70.

McBride, J.L. & Nicholls, N. (1983). Seasonal relationships between Australian rainfall and the Southern Oscillation. *Monthly Weather Review*, **111**, 1998–2004.

Meehl, G.A. & van Loon, H. (1979). The seesaw in winter temperatures between Greenland and Northern Europe. Part III: Teleconnections with lower latitudes. *Monthly Weather Review*, **107**, 1095–1106.

Meko, D.M. (1988). Temporal and spatial variation of drought in Morocco. In *Proceedings of Conference on Drought, Water Management and Food Production*, 55–82. Rabat: Kingdom of Morocco.

Moses, T., Kiladis, G.N., Diaz, H.F. & Barry, R.G. (1987). Characteristics and frequency reversals in mean sea level pressure in the North Atlantic sector and their relationship to long-term temperature trends. *Journal of Climatology*, **7**, 13–30.

Newell, R.E. & Kidson, J.W. (1984). African mean wind changes between Sahelian wet and dry periods. *Journal of Climatology*, **4**, 27–33.

Nicholls, N. (1981). Air–sea interaction and the possibility of long-range weather prediction in the Indonesian Archipelago. *Monthly Weather Review*, **109**, 2435–43.

Nicholls, N. (1985). Impact of the Southern Oscillation on Australian crops. *Journal of Climatology*, **5**, 553–60.

Nicholls, N. (1989). Sea surface temperatures and Australian winter rainfall. *Journal of Climate*, **2**, 965–73.

Nicholson, S.E. (1978). Climatic variations in the Sahel and other African regions during the past five centuries. *Journal of Arid Environments*, **1**, 3–34.

Nicholson, S.E. (1979). Revised rainfall series for the West African subtropics. *Monthly Weather Review*, **107**, 620–3.

Nicholson, S.E. (1980a). Subsaharan climates in historic times. In *The Sahara and the Nile*, ed. M.A.J. Williams & H. Faure, 173–200. Rotterdam: Balkhema.

Nicholson, S.E. (1980b). The nature of rainfall fluctuations in subtropical West Africa. *Monthly Weather Review*, **108**, 473–87.

Owen, J.A. & Folland, C.K. (1988). Modelling the influence of sea-surface temperatures on tropical rainfall. In *Recent Climatic Change: A Regional Approach*, ed. S. Gregory, 141–53. London: Belhaven Press.

Owen, J.A. & Ward, M.N. (1989). Forecasting Sahel rainfall. *Weather*, **44**, 57–64.

Palmer, T.N. (1986). Influence of the Atlantic, Pacific, and Indian Oceans on Sahel rainfall. *Nature*, **322**, 251–3.

Palmer, T.N. & Zhaobo, S. (1985). A modelling and observational study of the relationship between sea surface temperature in the northwest Atlantic and the atmospheric general circulation. *Quarterly Journal of the Royal Meteorological Society*, **111**, 947–75.

Parker, D.E., Folland, C.K. & Ward, M.N. (1988). Sea surface temperature anomaly patterns and prediction of seasonal rainfall in the Sahel region of Africa. In *Recent Climatic Change: A Regional Approach*, ed. S. Gregory, 166–78. London: Belhaven Press.

Quenouille, M.H. (1952). *Associated Measurements*. London: Butterworths.

Ramage, C.S. (1974). Structure of an oceanic near-equatorial trough deduced from research aircraft traverses. *Monthly Weather Review*, **102**, 754–9.

Rogers, J.C. (1984). The association between the North Atlantic Oscillation and the Southern Oscillation in the Northern Hemisphere. *Monthly Weather Review*, **112**, 1999–2015.

Rogers, J.C. (1985). Atmospheric circulation changes associated with the warming over the North Atlantic in the 1920s. *Journal of Climate and Applied Meteorology*, **24**, 1303–10.

Rogers, J.C. & van Loon, H. (1979). The seesaw in winter temperatures between Greenland and Northern Europe. Part II: Some oceanic and atmospheric effects in middle and high latitudes. *Monthly Weather Review*, **107**, 509–19.

Ropelewski, C.F. & Jones, P.D. (1987). An extension of the Tahiti-Darwin Southern Oscillation index. *Monthly Weather Review*, **115**, 2161–5.

Rowntree, P.R. (1976). Response of the atmosphere to a tropical Atlantic ocean temperature anomaly. *Quarterly Journal of the Royal Meteorological Society*, **102**, 607–25.

Seddon, D. (1986). Bread riots in north Africa: Economic policy and social unrest in Tunisia and Morocco. In *World Recession and the Food Crisis in Africa*, ed. P. Lawrence, 177–92. Boulder, CO: Westview Press.

Semazzi, F.H.M., Mehta, V. & Sud, Y.C. (1988). An investigation of the relationship between Sub-Saharan rainfall and global sea surface temperatures. *Atmosphere–Ocean*, **26**, 118–38.

Semazzi, F.H.M., Mehta, V. & Sud, Y.C. (1989). Reply to "Comments on 'An investigation of the relationship between Sub-Saharan Rainfall and Global Sea Surface Temperatures'." *Atmosphere–Ocean*, **27**, 601–5.

Stockton, C.W. (1988a). Preface. In *Proceedings of Conference on Drought, Water Management and Food Production*, 15–6. Rabat: Kingdom of Morocco.

Stockton, C.W. (1988b). Current research progress toward understanding drought. In *Proceedings of Conference on Drought, Water Management and Food Production*, 21–35. Rabat: Kingdom of Morocco.

Stockton, C.W., and others (1985). Long-term Reconstruction of Drought in Morocco: A Project Commissioned by His Majesty King Hassan II of Morocco. Tucson: AZ: C.W. Stockton and Associates. [Available from C.W. Stockton and Associates, Inc., 10451 E. Plumeria Road, Tucson, Arizona 85749].

Street-Perrott, F.A. & Grove, A.T. (1979). Global maps of lake level fluctuations since 30,000 B.P. *Quaternary Research*, **12**, 83–118.

Street-Perrott, F.A. & Roberts, N. (1983). Fluctuations in closed-basin lakes as an indicator of past atmospheric circulation patterns. In *Variations in the Global Water Budget*, ed. F.A. Street-Perrott, M.A. Beran, & R.A.S. Ratcliffe, 331–45. Dordrecht: Reidel.

Sud, Y.C. & Fennessy, M. (1982). A study of the influence of surface albedo on July circulation in semi-arid regions using the GLAS GCM. *Journal of Climatology*, **2**, 105–25.

Sud, Y.C. & Fennessy, M. (1984). A numerical study of the influence of evaporation in semi-arid regions on the July circulation using the GLAS GCM. *Journal of Climatology*, **4**, 383–98.

Swearingen, W.D. (1987). *Moroccan Mirages: Agrarian Dreams and Deceptions, 1912–1986*. Princeton: Princeton University Press.

Trenberth, K.E., and & Shea, D.J. (1987). On the evolution of the Southern Oscillation. *Monthly Weather Review*, **115**, 3078–96.

van Loon, H. & Rogers, J.C. (1978). The seesaw in winter temperatures between Greenland and Northern Europe. Part I: General description. *Monthly Weather Review*, **106**, 296–310.

Walker, G.T. (1924). Correlations in seasonal variations of weather, IX. *Memoirs of the Indian Meteorological Department*, **24**, 275–332.

Walker, G.T. & Bliss, E.W. (1932). World Weather V. *Memoirs of the Royal Meteorological Society*, **44**, 53–84.

Walker, J. & Rowntree, P.R. (1977). The effect of soil moisture on circulation and rainfall in a tropical model. *Quarterly Journal of the Royal Meteorological Society*, **103**, 29–46.

Weare, B.C. (1977). Empirical orthogonal analysis of Atlantic surface temperatures. *Quarterly Journal of the Royal Meteorological Society*, **103**, 467–78.

Whittaker, L.M. & Horn, L.H. (1984). Northern Hemisphere extratropical cyclone activity for four mid-season months. *Journal of Climatology*, **4**, 297–310.

Wolter, K. (1989). Modes of tropical circulation, Southern Oscillation, and Sahel rainfall anomalies. *Journal of Climate*, **1**, 149–72.

Wu, M.-C. & Hastenrath, S. (1986). On the interannual variability of the Indian monsoon and the Southern Oscillation. *Archiv für Meteorologie, Geophysik und Bioklimatologie*, **B36**, 239–61.

6

The Asian snow cover–monsoon–ENSO connection

T. P. Barnett

Climate Research Division, Scripps Institution of Oceanography
La Jolla, CA 92093

L. Dümenil, U. Schlese and E. Roeckner

Meteorologisches Institut, University of Hamburg
Hamburg, Germany

M. Latif

Max-Planck-Institut für Meteorologie
Hamburg, Germany

Introduction

It has been speculated for over a century that the varying extent and thickness of the Eurasian snow cover exerted some degree of control over both regional and, perhaps, global climate change. Blanford (1884) was one of the first to suggest the summer monsoon over India and Burma might be influenced by the spring snow cover on the Himalayas. He further guessed that such a relation, if thermally driven, might involve "major portions of the Asiatic continent rather than merely a relatively small portion of its mountain axis." These suggestions, or variants thereof, have been repeated and partially confirmed empirically by other investigators in the intervening period, e.g., Hahn and Shukla (1976), Dey and Bhanu Kumar (1982, 1983) and Dickson (1984) to name a few.

The role of the monsoon in larger scale global climate variations has only begun to be studied. What evidence exists seems to suggest that there is a reasonably strong relation between El Niño/Southern Oscillation (ENSO) events and monsoon variations, with the warm phase of these events being typified by lower than normal precipitation in Southeast Asia. Such results were found empirically by Weare (1979), Angell (1981) and Mooley and

Parthasarathy (1983), among others. All suggest that positive equatorial sea surface temperature (SST) anomalies in the central and eastern Pacific tend to follow a poor (summer) monsoon over Asia. Tanaka (1980) came to a similar conclusion regarding the interrelation of ENSO events, the Tropical Easterly Jet, and the monsoon. In contrast, Rasmusson and Carpenter (1983) use a composite analysis to suggest that the changes in the sea surface temperature off the coast of South America, which they found normally to peak in April and May, precede the monsoon season by a month or two. The sign of the correlation they discuss is as found by the previous authors. A timing of maximum SST off South America is subjective, poorly defined, and not well related to the seasonal cycle in these composites so lead/lag relations are hard to discern. A rigorous study of the lead/lag relations between monsoon rainfall and eastern Pacific SST using 80 years of data clearly shows SST changes to be coincident with or lag summer monsoon rainfall (e.g., Khandekar and Neralla, 1984). Elliott and Angel (1987) show the same lag relationship and also suggest it includes the Southern Oscillation as well (cf. Wu and Hastenrath, 1986). In summary, there is an apparent link between the warm extremes of ENSO events and poor summer monsoons with many quantitative analyses suggesting monsoon variations occur with or foreshadow some, but not all, of the main ENSO signals.

In this chapter we combine the above two hypotheses. We investigate first the hypothesis that snow-induced changes in the monsoon are significant. The second, or subsequent, interaction, i.e., that changes in the monsoon can effect the SST (sea surface temperature) in the tropical Pacific and hence induce ENSO events is investigated next. Both studies are carried out with sophisticated numerical models but the results compare favorably with observations where the latter are available. The following section investigates the effect of anomalous snow depth on atmospheric properties on both a regional and a global scale. It will be shown that the snow mass (depth) appears to be the key climatological variable in effecting an atmospheric response. The physics associated with this complex interaction are discussed next. A numerical demonstration is then offered to show that snow-induced monsoon variations create perturbations in the Pacific tradewind that could serve as a trigger for El Niño events. The present report is an abbreviated version of the results described in Barnett et al.

(1989) and the interested reader may refer to that reference for a more detailed description of the numerical experiment.

Atmospheric model performance

Model

The general circulation model used in this set of low resolution climate integrations is a derivative of the ECMWF (European Centre for Medium-Range Weather Forecasts) spectral medium-range forecasting model (Fischer, 1987). T21 denotes the spectral model version with triangular truncation of the spherical harmonic representation of model fields at wavenumber 21. Non-linear terms and physical processes are evaluated at grid points of an almost regular 'Gaussian' grid providing a resolution of 5.625° in latitude and longitude. The T21 model uses 16 levels combining a sigma-coordinate near the surface with a pressure-coordinate in the stratosphere (Simmons and Burridge, 1981). The uppermost level is at 25 hPa.

At land points, surface temperature as the lower boundary condition is a prognostic variable derived from a surface heat balance equation and diffusion through the soil. The diffusion of heat and moisture in the ground is based on values representative of three layers in the soil. For the lowest layer in the ground, climatological temperature and moisture content are prescribed by updating the fields with climatology every fourth day during a seasonal cycle integration. The soil characteristics are taken as equal for all land areas and the albedo of the surface is prescribed according to geographic conditions over the globe except that it is a function of local snow conditions. An albedo of 0.60 is taken for full snow coverage although eventually a more sophisticated representation may be required (Robock, 1983).

Atmospheric forcing of the surface temperature (T_S) is given by the heat budget at the surface due to the sum of solar and thermal radiative fluxes, sensible and latent heat fluxes. Additionally, T_S is modified by the melting of snow. In the equation of surface wetness (the amount of water present in the top soil layer), precipitation, surface moisture fluxes in snow free areas or snow melt water, and diffusion in the soil are considered. At sea-points, sea surface temperatures are prescribed according to climatology. They are updated every fourth day. The model distinguishes between open water and sea-ice. The underlying sea-ice mask is defined if the climatological sea surface temperatures are below $-2°C$.

Monsoon simulation

All in all, the model does a surprisingly good job of reproducing the main features of the monsoon and other important Asian climate fields in averages obtained from a 10-year control run. For instance, the distribution of model-simulated precipitation in June is typical of other summer monsoon months and is compared with observations of Jaeger (1976) in Fig. 6.1. The largest area of intense precipitation occurs in both model and observations over the Bay of Bengal–Burma region with secondary maxima along the equator between New Guinea and the dateline, and in northeast China. Other maxima on the west coast of India in the vicinity of the Ghat mountains and over the Philippines are not well simulated, but these features are likely related to orography which is smaller than the model's spatial resolution so the failure is not surprising. The observed 10 cm contour of precipitation in the western Pacific is well reproduced by the model. The same cannot be said for the 10 cm contour in the western Indian Ocean, an area where the data is sparse. However, the overall impression is that this relatively coarse resolution model is doing a rather good quantitative job of reproducing the large-scale space and time distribution of precipitation over the region of interest.

The distribution and depths of snow in the control simulation are compared in Fig. 6.2 with observations of the snow depth field over Eurasia for the critical months of April and May. The observations, courtesy of the U.S. Air Force ETAC, were derived largely from station data and are therefore least reliable in the central and southeastern parts of Asia. It is clear that the snow depth produced by the model are reasonably close to the observations over most of Asia, although there is a distinct tendency for the model to have more snow than observed. The model snow line for selected months was also compared with the ETAC and Soviet climatologies and with the monthly snowlines position obtained from satellite data (cf. Matson and Wiesnet, 1981). We found that the model snowline was generally located in a more southerly

JUNE PRECIPITATION (cm)

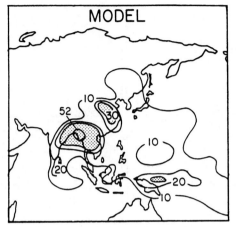

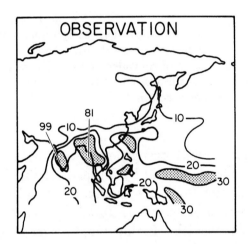

Fig. 6.1 Comparison between the June precipitation climatology from the model and from the observations of Jaeger (1976). Regions with more than 30 cm precipitation during the month are stippled (units are centimeters). Regions with less than 10 cm precipitation are omitted.

position than given by the satellite or station data. The results suggested the model's snow line retreats with about a one month lag relative to the observed snow line. In summary, the model's ability to simulate the amount and distribution of snow over Eurasia, while in need of some improvement, is better than we would have guessed and seems adequate for the numerical experiments to be reported below.

The seasonal cycle wind field developments are similar to those found in the observations. For instance, the onset of the monsoon

SNOW DEPTH (cm H₂O)

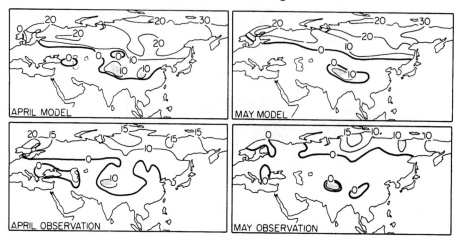

Fig. 6.2 The comparison of the snow distribution and depth given by the model's climatology with observations. The contours estimate snow depth in centimeters of equivalent water. The heavy solid line indicates the approximate position of the snow line. The stippled areas represent large snow accumulations associated with the Himalayas, etc.

finds the development and strengthening of a low latitude easterly jet at 200 mb just as described by Tanaka (1980). The maximum speed of the model jet (–27 m/s) compares fortuitously well with the long term mean values given by Tanaka (–24 m/s). The surface (10 m) wind field over the Indian Ocean agrees well with long term means from ship observations. In fact, a detailed comparison between the near-surface winds from T21 and ship observations shows that the model does a surprisingly good job of reproducing the observations throughout the tropics, although it does have several important shortcomings (Graham et al., 1988).

In summary, the T21 model does a creditable, qualitative job of reproducing key climatic features associated with the Asian monsoon. Many of the fields (e.g., winds, precipitation) are also accounted for with good quantitative accuracy. The snow field simulations in the model are realistic with respect to depth but the snow line tends to retreat somewhat more slowly than observed. Considering the model's strengths and weaknesses, it appears quite capable of providing a reliable base for monsoon perturbation experiments.

The snow–monsoon connection

Experimental setup

The basic idea to be tested was that variations in the snow *depth* and its subsequent melting, evaporation, and influence on the soil hydrology would cause an atmospheric response which could effect monsoon development. Our major modeling problem was to alter snow depth without dramatically affecting the energy balance in the model. The results expected a priori were that a deeper than normal Eurasian snow cover would lead to a "poor" monsoon while a shallower than normal snow cover would lead to a "good" monsoon. By "poor" monsoon we mean a reduction in precipitation over Southeast Asia, increased surface pressure, reduction in both the surface winds and Tropical Easterly Jet, and colder than normal atmospheric temperature over Asia (cf. Hastenrath, 1985; Fein and Stevens, 1987). A "good" monsoon is taken to be the converse of a "poor" monsoon.

In this experimental series we wished to change the snow depth but leave the model otherwise free to carry through its hydrological cycle. This was a difficult task since we are dealing with an interactive variable, not a passive one (like in SST sensitivity experiments). The way we accomplished this was to simply double or halve the snowfall rate given by the model's condensation schemes without implications for the atmospheric heat and moisture balance at the time it falls. At the surface, snow depth is accumulated at twice (or one half) the normal rate and therefore a larger (smaller) amount of snow is available for melting and evaporation processes at the later stages of the simulation. This approach is reasonably consistent from an energetics point of view. However, this procedure is not without its drawbacks. For instance, over how large an area should snowfall be increased/decreased? We applied the altered snowfall rates applied only to the Eurasian continent since the satellite-observed snowline variations suggest a continental scale is appropriate to the snow depth anomaly field. In fact, this may be an over estimation but there is little evidence to guide us otherwise. All other aspects of the hydrological cycle were handled as in the control runs, as was the specification of all other seasonal forcing terms.

Four sets of integrations were initialized with 1 January data from four different years of the control run. We designated the group of four doubled snowfall rate experiments by "D" and the group of four halved-rate experiments by "H". This provided four realizations of D and four of H conditions; an adequate sample for the nonparametric significance tests employed here, i.e., the PPP test described by Preisendorfer and Barnett (1983). Results from these groups of simulations were averaged together and compared with the control. Thus the results below will represent, say, the average of the four D experiments, referred to simply as D, minus the mean control; the difference typically being normalized by the standard deviation of the control run.

Perhaps the most critical aspect of the experiments is the size of the induced snow anomalies. The variability between the D (H) experiments and the control should be comparable to the variations observed in nature. We investigated the relative size of the model anomalies in four ways. The result of these studies showed that the relative size of the perturbations during the D and H experiments relative to the control was comparable to those observed in nature. Since we are comparing differences between perturbation runs and the control, our results should be relevant to the real world.

Results

Regional response   The numerical simulations with varied snow depth over Eurasia produced large, significant changes in the atmospheric variables normally associated with the Asian summer monsoon. In general, *a heavier than normal snow pack over Eurasia (D experiments) led to a weaker than normal, or "poor," monsoon.* A lighter than normal snow pack led to a stronger than normal, or "good" monsoon. The following paragraphs give a brief descriptive account of the evidence that led to these conclusions and, hence, verification of the first link in the snow–monsoon–El Niño connection.

In the discussion, it is useful to compare composited observed anomaly fields with the model simulations. One set of composites came from the 14-year snow depth data at the 18 Asian stations. The two years with the deepest spring snow pack (1979 and 1983) and the two years with the shallowest spring snow pack (1978 and

1984) were used to develop composites for the D and H experiments, respectively. We would have preferred to use more than two years in the composites but the shortness of the data set and the desire to study its extremes left us little choice. Thus, while many of the comparisons between model and data to be shown are favorable, one must remember the small number of realizations involved.

It was also convenient to develop composites based on the strength of the Southeast Asian monsoon. The first EOF of 45 precipitation stations scattered across India and Southeast Asia had components of like sign and so represents a signal in the precipitation field that is coherent over the entire region. The three years between 1950 and 1980 with the largest/smallest values of the associated principal component were designated good/poor monsoon years. Good monsoon years were 1956, 1961 and 1970 while poor monsoon years were taken to be 1965, 1968 and 1972. These definitions are in reasonable, though not exact, accord with the definitions of Parthasarathy and Mooley (1978), Mooley et al. (1986) and Khandekar and Neralla (1984) for the Indian monsoon.

Surface temperature   The normalized difference (Δ) between the average of the four D/H experiments and the control (C) run mean (e.g., $(D - C)/\sigma$) are shown in Fig. 6.3 for the land's surface temperature in June. Values of Δ greater/smaller than $+2/-2$ are common over much of Asia for the H/D experiments for the March–August time frame. The largest signals are seen at the surface in the D experiment with the Δ exceeding 2 over much of Asia. These values correspond roughly to a June surface temperature averaged over Asia that is 3.2°C colder during the D experiment than C. Note the region of large negative anomaly lies below (south) of a region of little or no anomaly in May and June. The latter regions are still snow covered in both the D and C simulations. The large negative temperatures are associated with regions of either rapid snow melt or areas that have recently become snow free (see below). The large negative anomaly retreats northward with the season until it covers only the northern fringe of Siberia in August (not shown). The H experiments have positive surface temperature anomalies; a result of the early snow melt associated with less snow depth and subsequent heating of the land by solar radiation. In this group of experiments, June temperatures aver-

aged over Asia for the four experiments were 1.1°C warmer than
in C. The spatial distribution of temperature anomalies and their
temporal evolution in H is similar to those in D (but of opposite
sign) if one allows a one month lag, i.e., the June temperature
anomaly field in H resembles that for July in the D experiment.
The response in the H experiments closely resembles, but is weaker
than, that described by Kutzbach (1981) for the monsoon circulation in the Holocene; a time of reduced snow cover over Asia.
By the end of August, the temperatures have largely returned to
normal in the H experiments.

Fig. 6.3 Normalized difference (Δ) fields between the D and H experiments
       and the control run for surface temperature for the months of May
       through July. The regions of stippling or hatching indicate Δ values
       less than −2.0 or greater than 2.0, respectively.

Tropospheric temperatures The entire atmospheric column over much of Eurasia reflects the patterns seen at the surface in the D and H experiments (cf. Fig. 6.4). The regional Δ values are highly significant. Thus a cold atmospheric column over Eurasia goes with heavy snow cover. The magnitude of the meridional temperature gradient in May/June is of critical importance to the establishment of the monsoon. During typical D and H experiments there is little difference in the 200 mb temperature, approximately 219–220°K in June, over the Arabian Sea/Indian Ocean. However, over the Middle East and Central Asia the net difference between the two experiments is 4–6°C (in June). Thus, during H, the meridional gradient is positive with temperature values over the warmest regions of the Asian heartland (229°K) approximately 9°C higher than over the ocean (220°K). During D, the warmest values over Asia are of order 224°K and the meridional gradient is reduced in strength by a factor of two. The same behavior is seen at all levels between the surface and 200 mb. Since it is the strength of these meridional gradients that helps initiate and sustain the monsoon, it is clear the D experiments represent a "poor" monsoon situation and the H experiments the opposite.

The model temperature response at 850 mb is compared against observed temperature, approximated by the (1000–700) mb thickness field over Eurasia (Fig. 6.4). The light snow composite shows a large region of higher than normal temperature over central and eastern Asia with lower than normal temperatures over Europe, the Middle East and western tropical Pacific. The corresponding results from the H experiments (Fig. 6.4, top) show a remarkable similarity in spatial distribution. Even the magnitude of the largest anomalies, approximately +2°C, are in accord. Comparison of the thickness composites for heavy snow condition with the D experiments shows lower than normal temperatures over central Asia, again a favorable comparison. However, where the D experiments give an anomaly that is coherent over all of Eurasia, the thickness fields mostly show a dipole response: cold over Asia and warm over Europe. The failure of the model to reproduce this dipole structure may well indicate the zonal extent of our snowfall perturbation is too large. Or it may indicate we have no really large heavy snowfall realizations in the snow data set. It is therefore with caution that we conclude that the model, to first order, compares favorably with observations.

JUNE TEMPERATURE 850 mb

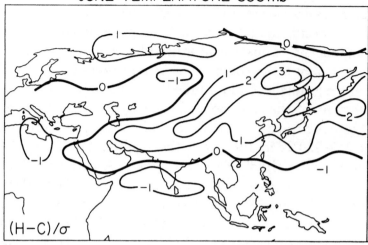

JUNE (1000-700)mb THICKNESS COMPOSITE

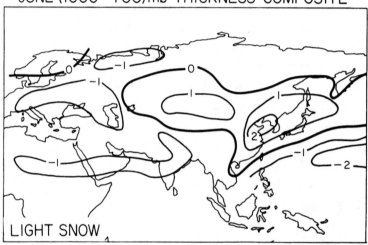

Fig. 6.4 Top Same as Fig. 6.3 except for the temperature at 850 mb during the month of June. Bottom: Observed (1000–700) mb thickness
anomalies for composited light snow conditions. Units are in standard deviations with 1 $\sigma \approx 2°$C.

<u>Wind fields</u>   The Δ values for the zonal surface winds for the
D experiment show a decrease over the Arabian Sea and Indian
Ocean for the May–August period. These anomalies are of order 1–2 m/s or roughly 10–30 percent of the mean. The decrease
in the eastward-directed wind component over the land mass of
southern Asia is as large or larger. These results suggest a reduction in the surface convergence and are expected during a poor

PRECIPITATION : AVE EXPERIMENT-CONTROL (cm)

Fig. 6.6 Difference in precipitation between the average of the four D/H runs and control for the months shown. Units are cm.

gave the results in Fig. 6.7. The observed precipitation anomaly has the correct sense as predicted by the model. However, the magnitude is generally a factor of two to three less than the model values but again, the composite consists of only two realizations. A separate investigation showed the heavy/light snow composite years were also poor/good monsoon rainfall years according to the all-Southeast Asia precipitation index. Both results support the model result that heavy Asian snow cover precedes a poor monsoon.

Soil hydrology   A crucial item in the experiment will turn out to be the differential behavior of the soil moisture during the D and H runs. The surface soil moisture content (Fig. 6.8) illustrates well the time lag between comparable states of the D and H experiments. In the light snowfall case, most of the snow is gone over central Asia by the end of May and the soil moisture is being evaporated. Thus the soil moisture content is less than normal over most of Eurasia (March–July), except in areas where snow is

OBSERVED COMPOSITE RAINFALL : MAY – JULY

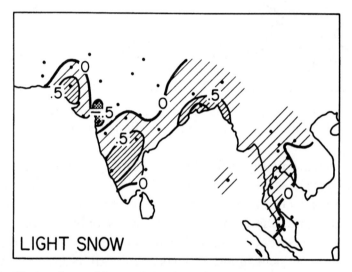

Fig. 6.7 Observed monthly precipitation anomalies for heavy and light snow
year composites. Units are in standard deviations.

still melting and/or on the ground. By June the large standardized
anomalies are gone in central Asia. Inspection of the actual values
shows this result is due to a nearly complete utilization of the surface moisture in these dry regions of Asia. In the D experiment, by
contrast, the standardized anomalies are largely positive (April–
August), reflecting (1) the increased moisture available through
prolonged, excess snow melt and (2) the protective effect of the
snow cover that lasted later into the seasonal cycle than usual.

ZONAL AVERAGE U–COMPONENT (m/s)

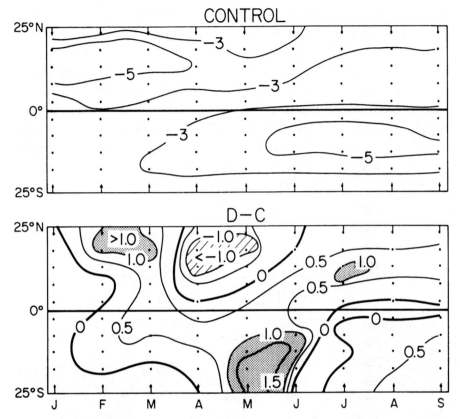

Fig. 6.9 Zonal wind component (m/s) averaged along latitude lines from 106°E to 152°W for the months of January through September. The upper panel shows the average value for the 10 year control run. The lower panel shows the result of subtracting the control values from those obtained during a typical D experiment.

responsible for the regional response described under "Results" in the previous section, and then the causes of the remote responses.

Regional physics

The physical processes that conspire in the model to control the land–ocean temperature contrast, and hence the summer monsoons, work in stages depending on the state of the snow pack and phase of the seasonal cycle. Let us concentrate initially on the physics operative in the D experiments:

(1) When excess snow is present the dominant terms in the surface heat balance are short wave (solar) radiation (S) and latent heat flux (Q_L). The increased albedo of the snow reduces the S received by the land mass. But the sublimation of the snow is generally less than evaporation from the soil, while the ground and air are also colder and surface wind reduced so Q_L is reduced. These two terms largely cancel each other, leaving generally a small positive (land warming) residual. The smaller sensible heat flux (Q_S) and long wave radiation IR anomalies are positive owing essentially to the colder surface temperature and also represent anomalous warming of the land mass. These results are illustrated in Fig. 6.10, which also shows the spatial complexity of the flux fields.

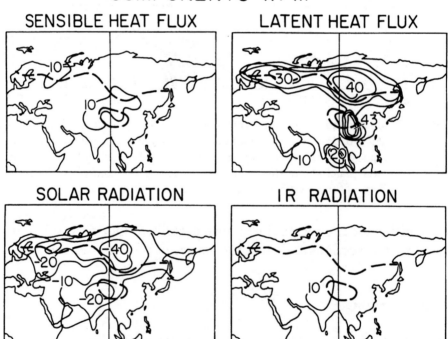

Fig. 6.10 Components of the surface heat budget for the averaged D experiments minus control for June (W/m²). The dashed line is the approximate position of the snowline in the model. Positive/negative values represent fluxes into/out of the earth/ocean surface.

a near-simultaneous relation between variations in Indian monsoon rainfall and ENSO.

However, the work reported in the prior sections failed to establish conclusive evidence for a monsoon–ENSO relationship for the following reasons:

(1) The climate variations associated with snow/monsoon perturbation have a time scale of three to four months; far shorter than the ENSO cycle. In the observational study of Barnett (1985a,b) the large-scale SLP anomalies developed over Asia in the spring just as simulated here. However, they then expand and intensify during the early summer, eventually covering much of the eastern hemisphere. Our simulations did not reproduce this latter effect.

(2) The large-scale tropical responses away from the monsoon are most apparent at height but less strongly evident at the surface and then for only several months.

It is possible the results produced in our simulations could play a critical *triggering* role in ENSO, a role which we would have missed given our experimental setup. Consider the following scenario: The perturbations in the Pacific wind stress field suggested by the T21 model could trigger an equatorial ocean warming that would then force a lower frequency ENSO event in the atmosphere. Although the stress fields differed from the control with high significance in only two months, they appear to have had a low frequency variation, or bias, to which the ocean, acting as an integrator, would be sensitive. Such a signal would escape our significance tests in the presence of high frequency noise. Since SSTs were specified in our snow/monsoon simulations, the T21 model would have no chance to develop a prolonged ENSO event. The above idea was tested in the two ways described below.

Ocean model experiment

A high resolution, sophisticated model of the equatorial Pacific Ocean (Latif et al., 1985; Latif, 1987) was brought to steady state using the annual cycle stress fields from the observations of Goldenberg and O'Brien (1981). The ocean model was then driven with surface stress anomalies derived from two D experiments beginning 1 January and extending through October. The resulting development of the equatorial Pacific SST field is given in the form of a Hovmöller diagram (Fig. 6.13, upper). The reader is here cautioned that the following results should be considered both preliminary and at best qualitative in nature for reasons that will be stated shortly.

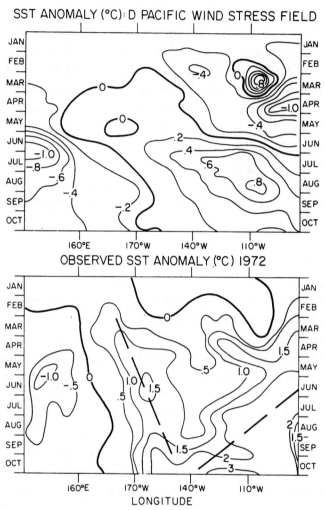

Fig. 6.13 Upper: Longitude–time diagram showing the evolution of SST anomalies within ± 5° of the equator in a model of the tropical Pacific Ocean driven by the average wind stress fields obtained from two D experiments. The ocean model was integrated with wind data beginning 1 January. Anomalies (°C) are relative to values obtained by driving the ocean model with T21 control run wind stresses. Lower: Similar display for SST anomalies (°C) observed in 1972. The dashed lines suggest the sense of zonal displacement of SST anomalies.

The ocean model developed SST anomalies that showed a characteristic eastward propagation. The largest elements of the signal are seen to occur nearly simultaneously with the large May/June stress anomaly in the western Pacific (Rossby wave upwelling) and several months later (the Kelvin wave transit time) in the eastern Pacific. The maximum magnitude of warming in the eastern Pacific peaked briefly at 0.8°C. The short-lived perturbations in that area in March are due to local wind stress anomalies; apparently transients. In summary, the stress field associated with the D experiments produced a mini-El Niño, a result in concert with the empirical studies noted above.

The eastward translation of the SST anomalies shown in Fig. 6.13 is not characteristic of a "typical" El Niño, (e.g., Barnett, 1977; Rasmusson and Carpenter, 1982). However, some El Niños do demonstrate this feature to some extent, e.g., the 1972 event (Fig. 6.13, lower), the 1982–83 event, and the 1986 event (Fig. 6.14, lower) which showed eastward anomaly motion in the western and central Pacific. Clearly the strongest anomaly signal goes westward and that's what has dominated prior analyses. Note that 1972 was a poor monsoon year and the positive SST anomalies first appear in both the western Pacific and coast of South America simultaneously in the early spring. If one imagines "classes" of El Niño events each triggered by different physical processes then the above results may belong to one such class. Interestingly, the observed SST anomaly field evolution during other poor monsoon years (not shown) demonstrate features common to Fig. 6.13 (upper). But they contain large, fundamental differences also.

Based on the above result one might guess that our experimental procedure (specified SST) precluded the simulation of an ENSO event. However, the snow/monsoon perturbation *by itself* is apparently not adequate to sustain a full blown ENSO for the following reasons:

(1) The D experiment stress anomalies imposed on the ocean model are roughly two times smaller than observations suggest are required for a moderate equatorial warming. It is known, however, that the T21 underestimates the observed stress field over the ocean (Fischer, 1987).

(2) The large-scale response of the ocean model is generally small, typically less than 0.5°C over the time and space scales associated with ENSO. Further, the corresponding ocean model

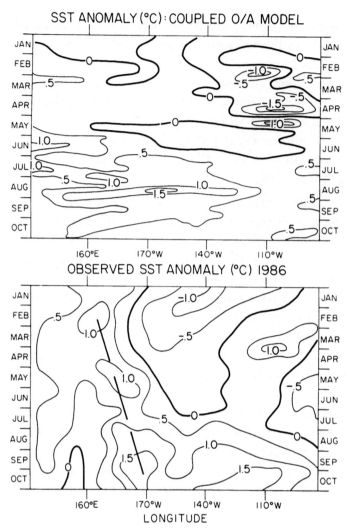

Fig. 6.14 Upper: same as Fig. 6.13 but for a coupled ocean–atmosphere model
with doubled snowfall over Eurasia as the only perturbation. Note
the amplified response of the ocean over that for the uncoupled run
(Fig. 6.13, upper). Lower: same as Fig. 6.13 but for observed SST
anomalies during 1986.

sea level responses are also modest with maximum values of
5 cm.

(3) As noted above, the time scale of the snow/monsoon perturbation is 3–4 months while the ENSO time scale is of order
1–2 years. Thus, Asian climate anomalies can, at most, trigger ENSO events.

All in all, the snow cover/monsoon signal has all the necessary
characteristics to trigger the Pacific portion of an ENSO event; but

of ETAC kindly provided the observed snow data and that data played a critical role in ascertaining the realism of the snow cover perturbation developed in the model.

References

Angell, J.K. (1981). Comparison of variations in atmospheric quantities with sea surface temperature variations in the equatorial eastern Pacific, *Monthly Weather Review*, **109**, 230–43.

Barnett, T.P. (1977). An attempt to verify some theories of El Niño, *Journal of Physical Oceanography*, **7**, 633–47.

Barnett, T.P. (1985a). Variations in near-global sea level pressure, *Journal of the Atmospheric Sciences*, **42**, 478–501.

Barnett, T.P. (1985b). Three-dimensional structure of low frequency pressure variations in the tropical atmosphere, *Journal of the Atmospheric Sciences*, **42**, 2798–2803.

Barnett, T.P., Dümenil, L., Schlese, U., Roeckner, E. & Latif, M. (1989). The effect of Eurasian snow cover on regional and global variations, *Journal of the Atmospheric Sciences*, **46**, 661–85.

Blanford, H.F. (1884). On the connexion of Himalayan snowfall and seasons of drought in India, *Proceedings of the Royal Society of London*, **37**, 3–22.

Dey, B. & Bhanu Kumar, O.S.R.U. (1982). An apparent relationship between Eurasian snow cover and the advanced period of the Indian summer monsoon, *Journal of Applied Meteorology*, **21**, 1929–32.

Dey, B. & Bhanu Kumar, O.S.R.U. (1983). Himalayan winter snow cover area and summer monsoon rainfall over India, *Journal of Geophysical Research*, **88**, 5471–4.

Dickson, R.R. (1984). Eurasian snow cover versus Indian monsoon rainfall – An extension of the Hahn–Shukla results, *Journal of Climate and Applied Meteorology*, **23**, 171–3.

Elliott, W. and Angel, J. (1987). The relation between the Indian monsoon rainfall, the Southern Oscillation, and hemispheric air and sea temperature: 1884–1984, *Journal of Climate and Applied Meteorology*, **26**, 943–8.

Fein. J. and Stevens, P. (1987). *Monsoons*. New York: John Wiley and Sons.

Fischer, G., ed. (1987). *Large-scale Atmospheric Modelling*. Rep. No. 1, Hamburg: Meteorological Institute of the University of Hamburg.

Flohn, H. (1957). Large-scale aspects of the "summer monsoon" in south and east Asia, *Journal of the Meteorological Society of Japan* (75th annual volume), 180–6.

Goldenberg, S. and O'Brien, J. (1981). Time and space variability of the tropical Pacific wind stress, *Monthly Weather Review*, **109**, 1190–1207.

Graham, N.E., Barnett, T.P., Chervin, R.M., Schlesinger, M.E. & Schlese, U. (1989). Comparisons of GCM and observed surface wind fields over the tropical Indian and Pacific Oceans, *Journal of the Atmospheric Sciences*, **46**, 760–88.

Hahn, D.J. and Shukla, J. (1976). An apparent relationship between Eurasian snow cover and Indian monsoon rainfall, *Journal of the Atmospheric Sciences*, **33**, 2461–2.

He, H., McGinnis, J., Song, Z. & Yanai, M. (1987). On the onset of Asian summer monsoon in 1979 and the effect of the Tibetan plateau, *Monthly Weather Review*, **115** (9–10), 1966–95.

Jaeger, L. (1976). Monatskarten des Niederschlags für die ganze Erde, *Berichte des Deutschen Wetterdienstes*, Band 18, No. 139.

Hastenrath, S. (1985). *Climate and Circulation of the Tropics.* Boston: D. Reidel Publishing Co.

Khandekar, M.L. & Neralla, V.R. (1984). On the relationship between the sea surface temperatures in the equatorial Pacific and the Indian monsoon rainfall, *Geophysical Research Letters*, **11**, 1137–40.

Kutzbach, J. (1981). Monsoon climate of the early Holocene: Climate experiment with the earth's orbital parameters for 9000 years ago, *Science*, **214**, 59–61.

Latif, M. (1987). Tropical ocean circulation experiments, *Journal of Physical Oceanography*, **17**(2), 246–63.

Latif, M., Biercamp, J. & von Storch, H. (1988). The response of a coupled ocean–atmosphere general circulation model to wind bursts, *Journal of the Atmospheric Sciences*, **45**(6), 964–79.

Latif, M., Maier-Reimer, E. & Olbers, D.J. (1985). Climate variability studies with a primitive equation model of the equatorial Pacific, in *Coupled Ocean–Atmosphere Models*, ed. J.C.J. Nihoul. Amsterdam, Netherlands: Elsevier Science Publishers.

Matson, M. & Wiesnet, D. (1981). New data base for climate studies, *Nature*, **288**, 451–6.

Meehl, G.A. & Washington, W.M. (1988). A comparison of soil moisture sensitivity in two global climate models, *Journal of the Atmospheric Sciences*, **45**(9), 1476–92.

Mooley, D. & Parthasarathy, B. (1983). Variability of the Indian summer monsoon and tropical circulation features, *Monthly Weather Review*, **111**, 967–78.

Mooley, D., Parthasarathy, B. & Pant, G.B. (1986). Relationship between Indian summer monsoon rainfall and location of the ridge at the 500-mb level along 75°E, *Journal of Climate and Applied Meteorology*, 25, 633–40.

Parthasarathy, B. & Mooley, D.A. (1978). Some features of a long homogeneous series of Indian summer monsoon rainfall, *Monthly Weather Review*, **106**, 771–81.

Preisendorfer, R. & Barnett, T.P. (1983). Numerical model-reality intercomparison tests using small-sample statistics, *Journal of the Atmospheric Sciences*, 40, 1884–96.

Rasmusson, E. and Carpenter, T. (1982). Variations in tropical sea surface temperature and surface winds associated with the Southern Oscillation/El Niño. *Monthly Weather Review*, 110, 354–84.

Rasmusson, E. and Carpenter, T. (1983). The relationship between eastern equatorial sea surface temperatures and rainfall over India and Sri Lanka. *Monthly Weather Review*, 111, 517–27.

Robock, A. (1983). Ice and snow feedbacks and the latitudinal and seasonal distribution of climate sensitivity. *Journal of the Atmospheric Sciences*, **40**, 986–97.

Simmons, A.J. & Burridge, R. (1981). A energy and angular-momentum considering finite difference scheme, hybrid coordinates, and medium-range weather prediction. ECMWF Tech. Rep. No. 28.

Tanaka, N. (1980). Role of the circulation of 150 mb level in the winter in summer monsoon in the Asian and Australian regions, *Climatological Notes*, **26**. Tsukuba, Japan: Inst. of Geoscience, University of Tsukuba.

Weare, B. (1979). Statistical study of the relationship between ocean surface temperatures and the Indian monsoon, *Journal of the Atmospheric Sciences*, **36**, 2279–91.

Webster, P., Chou, L. & Lau, K.-M. (1977). Mechanisms effecting the state, evolution, and transition of the planetary scale monsoon, *Pageoph*, **115**, 1463–90.

Wu, M.-C. & Hastenrath, S. (1986). On the interannual variability of the Indian monsoon and the Southern Oscillation, *Archiv für Meteorologie, Geophysik, und Bioklimatologie Serie B*, **36**, 239–61.

Yeh, T.-C., Wetherald, R. & Manabe, S. (1983). A model study of the short term climatic and hydrological effects of sub-snow cover removal, *Monthly Weather Review*, **111**, 1013–24.

7

Teleconnections in global rainfall anomalies: Seasonal to inter-decadal time scales

K.-M. LAU

Laboratory for Atmospheres
NASA/Goddard Space Flight Center
Greenbelt MD 20771

P.J. SHEU

Applied Research Corporation
Landover MD 20785

Introduction

For centuries, the occurrences of large-scale floods and droughts have had devastating socio-economic impacts on a large segment of the world population. During the 1982–83 El Niño–Southern Oscillation (ENSO), for example, extreme drought conditions were experienced over Indonesia, northern Australia and Brazil while northern Peru, the western United States and southern China were stricken with extensive flooding. These extreme conditions lasted for one to two seasons and normal conditions resumed following the end of the ENSO. Yet, it is known that droughts over East Africa, for example, often persist for several years and even decades, and are not necessarily related to the ENSO (Nicholson, 1983). Thus, anomalies in large-scale precipitation appear to possess different modes of variability. It is important to know, for socio-economic reasons, if the severe drought/flood conditions over geographically separated regions occur independently and appear simultaneously only by chance or if they are the manifestation of a global scale coherent shift in rainfall patterns that is likely to recur in the future.

Study of global rainfall variability is also essential for scientific reasons. It is now recognized that the heat of condensation released in tropical precipitation accounts for a large part of intraseasonal to interannual variability of the general circulation and the overall climate of the earth. Variation of tropical rainfall is closely related to large-scale sea surface temperature (SST) anomalies. Scores of recent general circulation experiments have produced global rainfall anomaly patterns associated with sea surface temperature anomalies (e.g., Lau, 1985; Blackmon et al., 1983; and others). Because of the lack of documentation of global rainfall anomaly patterns, these model patterns can only be compared to relative distributions in proxy rainfall quantities such as outgoing longwave radiation (OLR) anomalies (e.g., Lau and Chan, 1983a, b).

While present rainfall station coverage in the tropics is obviously inadequate for detailed comparison of model rainfall to observations, long-term station rainfall data still provide the best global scale precipitation information that can be used for an absolute comparison between model and observation on a coarse spatial scale.

Up to the present, most observational studies of rainfall anomalies have separately focused only on the regional monsoon areas of India, Australia, Brazil and Africa (e.g., Rasmusson and Carpenter, 1983; Shukla and Paolino, 1983; Bhalme and Jadhav, 1984; McBride and Nicholls, 1983; Chu, 1983; and Nicholson, 1983). However, possible interrelationships between rainfall anomalies over these areas and other regions are not well documented. Some examples of previous studies on global rainfall inter-relationships are as follow. Kidson (1975) studied tropical rainfall variation in relation to surface temperature and pressure and found strong correlation with the Southern Oscillation. However, the rainfall data he used were only from a 10-year period, so that results may be subject to sampling errors and inter-decadal variations cannot be studied. Stoeckenius (1981) examined areal averages of tropical rainfall anomalies and found some teleconnection in rainfall associated with the fluctuation of the Walker circulation. Fleer (1981) studied the correlation of rainfall between tropical stations and found a significant inverse relationship between the rainfall anomaly over the maritime continent and the dateline in the time scale of ENSO. Recently, Ropelewski and Halpert (1988) showed the patterns of worldwide drought and flood conditions in relation to ENSO. Lau and Sheu (1988) showed that a strong biennial signal can be found in global rainfall and suggested a close link between a biennial oscillation and ENSO.

In this chapter, we expand on the above studies by enlarging both the spatial and temporal domain of the rainfall record, focusing on the global teleconnections of regional rainfall anomalies and possible secular changes in large-scale rainfall distribution. Results from a global analysis of rainfall climatology and anomalies in both the tropics and the extratropics for the last 80 years will be presented. The outline of this chapter is as follows. We describe the data and analysis procedures used. Then the climatological seasonal variation is presented as a background. The main results are presented in the next sections and a general discussion and concluding remarks are given in the final section of the chapter.

Data and analysis procedure

The data used in this study are obtained from the NCAR monthly world surface climatological record. Some of the data record date back to the mid-1700s. For this study, we use only data from 1901–1980. Monthly rainfall totals for about 4000 stations all over the world were extracted. Only stations within the global band between 50°N and 50°S are used. The data are selected in two categories. The first category consists of station records with missing data (not exceeding 10 percent of the total) from the period 1901–80. These stations are then grouped within 5° × 5° latitude–longitude boxes. Only one station with the least missing data within the group will be chosen to represent the group. The second category includes remote stations not covered by the first category, provided they show missing data of not more than 40 percent. This covers some island stations in the equatorial central and western Pacific and northern China. The latter are checked against raw data obtained independently from China and are found to be in good agreement. The two categories were then merged to form one final data set.

The above procedure is necessary to reject bad data and to optimize the spatial and temporal resolution of the data. As a result, 155 station records were selected. Of these, about half show 5 percent or less missing data. Less than one quarter of the stations belong to the second category. The monthly climatology is computed based on the above-mentioned processed data. The

anomalous rainfall is then defined with respect to this climatology. In the analyses of the anomaly, missing data are replaced
with the value zero. A rainfall index is defined at each station by
normalizing the yearly anomaly by its standard deviation. The
main results of this work are based on empirical orthogonal function (EOF) analyses of the rainfall index for anomalous annual or
seasonal rainfall totals.

Seasonal variation

Harmonic analysis is applied to the 80-year monthly precipitation climatology. Figure 7.1 shows the distribution of the zeroth
harmonic or the mean rainfall. The geographic locations of the
individual stations are indicated by squares in Fig. 7.2. Rainfall amounts of over 100 mm/month are found over an extensive
area in the equatorial western Pacific. Maximum rainfalls of over
300 mm/month are found near the equator and 160°E. This rainy
region extends southeastward into the Southern Hemisphere subtropics where it can be identified with the well-known presence
of the South Pacific Convergence Zone (SPCZ). In the Northern
Hemisphere the rain band covers the southeast coast of China,
Japan, the Indian subcontinent, and the Bay of Bengal. Another
major rain area is found over northern South America and the Atlantic ITCZ and western Africa. Large scale semi-arid or desert-
like regions with mean rainfall less than 25 mm/month are found
over an extensive area stretching from northern Africa, through
the Middle East to western Asia. Desert-like conditions are also
found over extreme western central Australia and Peru.

Figure 7.2 shows the harmonic dial for the first (annual) harmonic component. The length of the line segment pointing away
from the station location indicates the amplitude relative to the
zeroth harmonic. The phase is represented by the direction of the
line segment according to that of a clock. A line segment pointing
directly north indicates maximum rainfall occurs on 1 December.
Unless otherwise specified, the four seasons are defined with respect to the Northern Hemisphere. The major summer monsoon
rainfall region of India and East Asia is readily identified with
maximum in rainfall in late July or August. The East Asian monsoon regime appears to extend well into the extratropics to 40°N
near northern Korea. A somewhat distinct seasonal regime can be

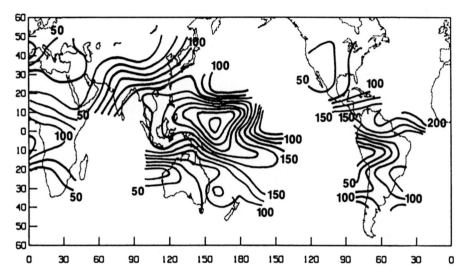

Fig. 7.1 Distribution of monthly mean (zeroth harmonic) of global rainfall in mm/month.

found over the southeast coast of China and Japan. Here, maximum rainfall tends to occur in late June or early July, about a month before the peak monsoon activity over India. This may be identified with the pre-monsoon period of the Mei-yu (Plume rain) or the Baiu over China and Japan, respectively. Summer monsoon-like rainfall is also found over central America and the west coast of Africa.

The winter monsoon areas are characterized by maximum rainfall occurring in late January or February. Here, three major regions are identified. These are the maritime continent of Indonesia/Australia, the land mass of South Africa and of South America between 50° and 25°S. The harmonics dials over India/East Asia, Indo-China and the maritime continent/New Guinea indicate maximum rainfall occurring in July-August, September-October and January-February, respectively. Winter monsoon-like characteristics are also found over some island stations between 0° and 20°S in the region of the SPCZ where the rainfall maxima tends to occur in late February to early March. The above phase variation suggests a migration southeastward of the maximum rainfall zone along Indo-China from India/East Asia to the maritime continent during the monsoon transition season. This migration has been identified by Lau and Chan (1983b) from outgoing longwave radiation.

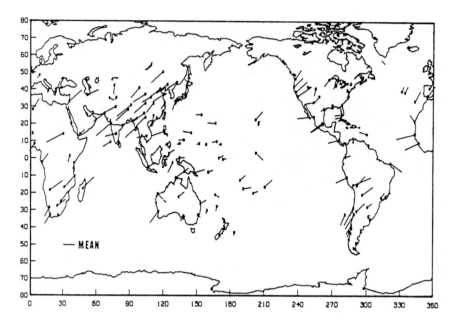

Fig. 7.2 Harmonic dials showing amplitude and phase of the annual (first
harmonic) variation of global rainfall. Amplitude is normalized with
respect to the zeroth harmonics.

The amplitude of the annual harmonic shows strong latitude
dependence in addition to the zonal asymmetry due to the land–
sea distribution. Near the equator, the small relative amplitude
of the annual harmonic indicates the lack of a pronounced rainy
season. However, as is apparent from Fig. 7.1, the equatorial western Pacific is the rainiest area in the world with mean precipitation of over 300 mm/month. In the Southern Hemisphere, along
30°S, over the southern tips of Africa, Australia and Chile, the
rainfall regime shifts abruptly from Southern Hemisphere summer to Southern Hemisphere winter maximum. The locations of
the latter correspond to the centers of the Southern Hemisphere
winter storm tracks. For the Northern Hemisphere extratropics,
the zonality is less marked because of the presence of the major
continents. Winter rainfall maxima are found over the northwest
United States and the west coast of Canada. Over the east coast of
North America, the seasonality in rainfall is not very pronounced.

The amplitude of the semi-annual harmonic (not shown) is in
general small compared with the first harmonic. Because rainfall
over monsoon areas generally occurs in one season, i.e., no negative rainfall during other seasons, the second harmonic over these

regions is phase-locked to the first harmonic and does not necessarily represent a real double maximum in rainfall (cf. Hsu and
Wallace, 1976). Only stations very close to the equator show some
real amplitude in the semi-annual variation but they are not our
main concern in this chapter.

Standard deviation

The standard deviation (SD) of the monthly anomalies are shown
in Fig. 7.3. The regions of large monthly variability (SD > 100
mm/month) approximately coincides with the regions of large
monthly means shown in Fig. 7.1. The center of maximum variability is found over the Bay of Bengal and Indo-China. Large
SD is also found over the maritime continent and the equatorial
central Pacific. Other regions of large SD are found over Central
America and northeast South America. The ratio of the monthly
standard deviation to the mean (or the coefficient of variability)
expressed in percentages is shown in Fig. 7.4. In general, a large
coefficient would imply a high degree of variability due to either
large fluctuations in the monthly means or a small climatological
mean. Thus the desert-like belt stretching from northern Africa
through the Arabian Peninsula to western Asia shows large values
(>100) because of the persistent dry condition over these regions.
Other regions of high variability are found over the central Pacific, southern Africa, western Australia and the west coast of the
United States and Mexico. In contrast, due to the large mean,
rainfall over the western equatorial Pacific and East Asia is relatively less variable.

   The space/time variations of the above global rainfall variability
will be studied using EOF analysis. In the following only the first
three EOF eigenvectors will be discussed. The first eigenvector
(EOF1) is highly significant and the second (EOF2) and the third
(EOF3) are only marginally significant by the test of North et al.
(1982).

Global rainfall anomalies associated with ENSO

The first eigenvector (EOF1) explains 11.4 percent of the total
variance. Its time coefficient (T1) is shown in Fig. 7.5. The time

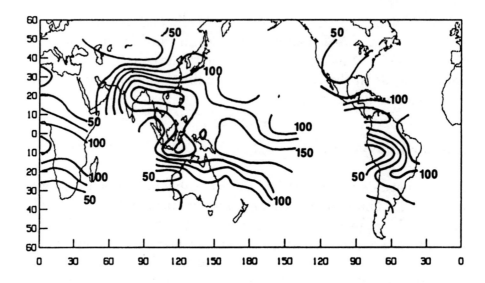

Fig. 7.3 Global distribution of standard deviation of monthly departures of rainfall in mm/month.

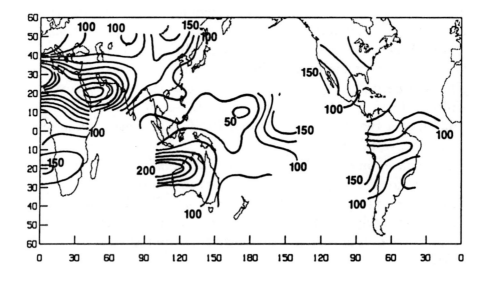

Fig. 7.4 Global distribution of the ratio (percent) of standard deviation of monthly departures to monthly mean rainfall.

series does not shown any obvious periodicity. Large amplitude fluctuations occur at irregular intervals from two to 10 years. To establish a possible relationship between this temporal mode of variation and that of ENSO, we compare T1 with the years of so-called warm events (WE) and cold events (CE). These years are defined based on sea surface temperature (SST) variation and a variety of other meteorological parameters over the tropical Pacific. The WEs generally coincide with other definitions of ENSO (e.g. Rasmusson and Carpenter, 1982; Quinn, 1979; van Loon, 1984). Tables 7.1 and 7.2 show, respectively, the years of the WEs and CEs. The WEs (CEs) are marked by black (hatched) bars in Fig. 7.5. In spite of some ambiguity in the definition of WEs (cf. Rasmusson and Carpenter, 1982), there is a remarkable correspondence between the occurrence of a WE and the fluctuation of T1. Out of the 18 WEs, 15 cases (except 1939, 1951, 1953) show negative deviations in T1. Figure 7.5 also shows the years of large negative swings in T1. They are defined by the amplitude of a negative deviation exceeding unity and are marked by the letter N in Fig. 7.5. Each of the 10 large negative events coincides with a major WE. In addition, there appears to be a higher frequency of occurrence of these strong events during the first two decades of the century. This feature will be further discussed in the section on inter-decadal variations.

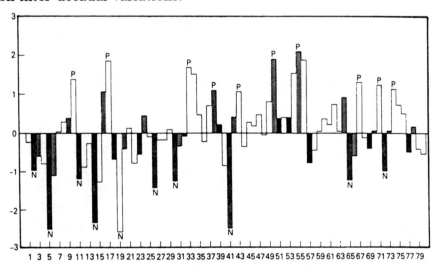

Fig. 7.5 Time coefficient of EOF1 in relative unit. Solid and hatched bars denote years of WEs and CEs respectively.

Table 7.1 Precipitation anomaly (mm/day) over the maritime continent and
the equatorial central Pacific for warm events. The values shown
are largest negative anomalies among the three stations, Meanado
(2.5°N, 124.5°E), Manokwari (0.1°N, 134.3°E) and Ambon (2.7°S,
128.2°E) for the maritime continent, and the largest positive anomalies among the three stations, Ocean (0.10 N, 169.5°E), Abiang
(0.7°S, 173.1°E) and Christmas (3°N, 157.5°W) for the equatorial
central Pacific. Also shown is the strength of the dipole anomaly.

Year	Maritime continent A	Central Pacific B	Dipole anomaly $(A - B)/2$
1902	−3.65	—	—
1905	−3.04	6.47	4.77
1911	−2.27	4.96	3.62
1914	−4.36	5.76	5.06
1918	−1.98	1.91	1.95
1923	−2.88	1.91	2.36
1925	−0.50	1.78	1.14
1930	−2.55	3.82	3.19
1932	−2.00	0.01	1.01
1939	−0.00	0.30	0.15
1941	−2.88	5.12	4.00
1951	−1.21	1.17	1.19
1953	−0.53	4.21	2.37
1957	−0.46	3.33	1.90
1965	−5.50	2.99	4.25
1969	−1.68	5.56	3.62
1972	—	3.77	—
1976	−1.76	0.64	1.20
Average	−2.19	3.16	2.66

There is a similar correspondence between positive swings in T1
and the occurrences of CEs. Positive deviations are shown in 11
out of the 16 CEs observed during the data period. Three out
of the 10 large positive deviation in T1 (marked by the letter P)
are CEs. Six of the large positive swings occur one year after a
CE. The one-year lag may be due in part to the ambiguity in the
definition of the CEs because such events tend to last over one
year and generally overlap two consecutive years. If this is taken
into account, the occurrences of CEs and large positive deviations
may still be considered highly correlated because nine out of 10

Table 7.2 Same as in Table 7.1 except for the cold events. The values shown
are the maximum anomaly of the three stations over the maritime
continent and the minimum anomaly of the three stations over the
equatorial central Pacific.

Year	Maritime continent A	Central Pacific B	Dipole anomaly (A − B)/2
1903	0.98	—	—
1906	1.19	−1.94	−1.57
1908	1.31	−1.24	−1.28
1916	8.52	−4.10	−6.31
1920	0.46	—	—
1924	5.56	−1.40	−3.48
1931	2.22	−2.01	−2.12
1938	3.14	−4.16	−3.65
1942	—	−1.15	—
1949	5.80	−1.93	−3.87
1954	5.41	−3.12	−4.27
1964	−0.92	−1.50	−0.29
1966	−1.77	0.08	0.93
1970	−1.07	−2.09	−0.51
1973	—	−2.11	—
1978	2.17	−0.85	−1.51
Average	2.76	−1.92	−2.34

large positive deviations either coincide with or occur one year
after a CE. In general, while there appears to be a correspondence
between the occurrences of both WEs and CEs with respect to
the fluctuation of T1, a better relationship seems to hold between
WEs and negative deviations in T1.

Figure 7.6 shows the scatter diagram of T1 with the monthly
surface pressure variation over Darwin, the latter being a measure of the strength of the Southern Oscillation. An inverse relationship can be seen with higher-than-normal pressure at Darwin
corresponding to large negative deviations of T1. The correlation between the two time series is 0.65 significant at the 99 percent confidence level. Thus the global rainfall pattern depicted by
EOF1 is closely related to the Southern Oscillation.

Figure 7.7 shows the spatial distribution of EOF1. Since each
EOF is determined to within a phase of 0 and π, for convenience

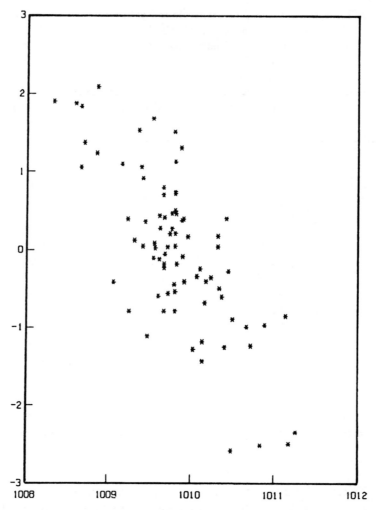

Fig. 7.6 Scatter diagram of time series of EOF1 and annual surface pressure
(mb) over Darwin.

we refer to the negative (positive) anomalies in Fig. 7.7 as flood
(drought) conditions with the understanding that these conditions
can be reversed depending on the phase of the principal component
time series. The major flood and drought regions as depicted by
EOF1 are denoted by the letters F and D as follows:

(1) *Flood regions*

F1. A wedge-shaped region between 10°N and 10°S over the equatorial central Pacific near the date line.

F2. Northern Mexico and southern United States, from California
to Florida.

F3. Central Chile and Argentina.

F4. Coastal region of south and southeastern China and Japan.

F5. Eastern Africa and the Middle East.

F6. Atlantic coast of western Europe.

(2) *Drought regions*

D1. A horseshoe pattern stretching from the South Pacific Convergence Zone (SPCZ), northeastern Australia, the maritime continent and the Caroline Islands engulfing F1.

D2. The monsoon region of northwestern India and northern China.

D3. Venezuela, Colombia and the Amazon.

D4. Southern Africa and Madagascar.

D5. Northern Africa, the Sahel and Sudan.

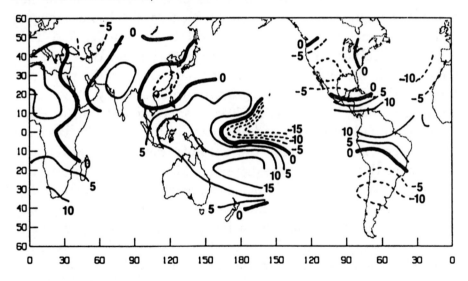

Fig. 7.7 Spatial distribution of EOF1 of global precipitation in relative units.

The strongest signals are found in F1 and D1 which coincide with the locations of the east–west dipole in anomalous convection found during ENSO (Lau and Chan, 1983a). The anomalies over the regions from F2 to F3 and from D2 to D3 are quite strong. For regions F4-6 and D4-5, the signals are much weaker. The above pattern is consistent with the ENSO global precipitation anomalies observed during the 1982–83 ENSO (Gill and Rasmusson, 1983) and the composite rainfall pattern discussed by Ropelewski and Halpert (1988). One important point to note is that large-scale drought and flood conditions do not occur alone but in a compensatory way (i.e., the sum of the station loading for the EOF is small compared with the sum of its absolute deviation). This implies that the above rainfall anomalies are probably due to a shift in the general circulation of the atmosphere rather than due to a uniform increase or decrease in total energy input such as resulting from a change in solar insolation.

The statistical significance of the above rainfall pattern is ascertained by repeating the above analysis with the rainfall data divided into two 40-year segments. Results (not shown) indicate that the main features described above remain essentially unchanged using either data segment. This implies that the above statistics are fairly robust throughout the 80-year period.

Since a basic requirement for the validity of tropical SST experiments with GCM is the ability to produced realistic rainfall anomalies, a quantity useful in such context is the absolute magnitude of the precipitation dipole, F1 and D1 shown in Fig. 7.7. This is given in Tables 7.1 and 7.2, respectively, for the WEs and the CEs. For all the WEs, the dipole anomaly exhibits the same polarity, with the strength varying from 1 to 5 mm/day. In general, strong CEs (e.g., 1905, 1914, 1941) correspond to a strong development of rainfall anomalies at both poles of the anomaly. Because there is no "negative rain," the absolute rainfall deviations are larger over the wet than the dry pole. Maximum increase in the rainfall rate over the central Pacific can reach to about 6.5 mm/day and decrease over the maritime continent to over 4 mm/day consistent with that obtained from recent GCM results (e.g., Lau, 1985). These values should be used as a standard reference for other GCM simulations. For the CEs, the dipole anomalies are formed with the negative polarity. The average of the anomalies over the two regions and that of the dipole is significantly less than those for the WEs. In particular, the dipole is much less well developed for CEs during the 1960s than for WEs during the same period.

Inter-decadal variations

Figure 7.8 shows the time series for EOF2 which explains 7.4 percent of the total variance. The year-to-year fluctuation is quite small. A decreasing trend is observed starting from around 1940.

During the period 1901–40, the time series appears flat and the deviations remain largely positive. The negative trend seems to have reach bottom in 1979 with an indication of an upward trend after that time. Fletcher (1982) suggested the presence of an ultra-long period of about 120 years in the tropical ocean–atmosphere system in addition to the ENSO time scale. Figure 7.9 shows the departures of the seasonal mean from a 120-year climatology of the surface easterly wind speed and sea surface temperature between 30°N and 30°S over the entire Pacific. It is noted from Fig. 7.9b that the surface wind over the Pacific shows an increasingly west-erly trend during the the early 1900s and remains westerly for the 4 decades from 1920 to 1960. A decreasing trend begins around the 1940s, resulting in more easterly wind from the early 1960s to 1980. The above transitions coincide very well with T2 (Fig. 7.9a) during the overlapping period, suggesting that the long-term vari-ation between rainfall and surface wind may be correlated. Fig-ure 7.9c shows the SST anomaly with the scales inverted to allow for easier comparison. It is interesting to note that SST over the Pacific also shows a similar variation with below normal SST dur-ing the period 1900–1940 and above normal SST thereafter. The transition from below to above normal SST in the 1940s appears to lead the surface wind and the precipitation by about 10 years. The entire 120-year time series of wind and SST indicates that the trend from 1900 to 1980 may be part of an ultralong time scale oscillation with period in the order of 100 years.

The structure of EOF2 is shown in Fig. 7.10. The largest co-herent signal is found over the monsoon complex of the eastern hemisphere. A zero contour runs northwest to southeast from Southeast Asia to New Guinea and northeastern Australia sep-arating the negative anomalies over India, western Asia, north-eastern China and northern Australia from the positive anomalies over the South China Sea, the Phillipines, the Solomon Islands and extreme southwestern Australia.

Parts of Central and South America (except Peru and southern Chile) also show negative anomalies. The whole African continent and the Middle East show positive deviations. The actual sign of the anomalies depends on the product of the eigenvector and its time coefficient. Figs. 7.8 and 7.10 indicate that rainfall was reduced during the early part of the century and enhanced dur-ing the period 1960–80 over areas with negative EOF loadings,

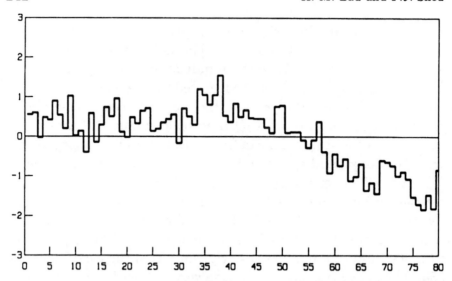

Fig. 7.8 Time series of EOF2 of global precipitation in number of standard
    deviations from the mean.

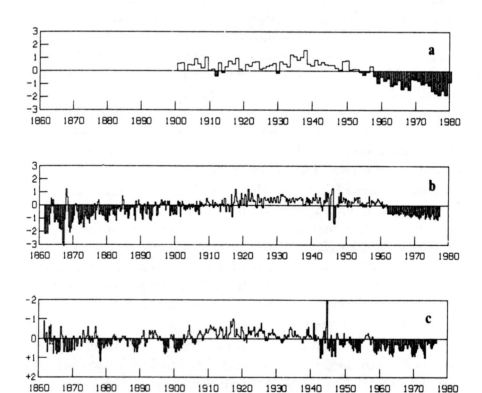

Fig. 7.9 Time series of (a) time coefficient of EOF2, (b) anomalous surface
    easterlies (m/s) and (c) anomalous SST (inverted ordinate) over the
    entire tropical Pacific 30°S–30°N (°C). Both (b) and (c) are adopted
    from Fletcher (1982).

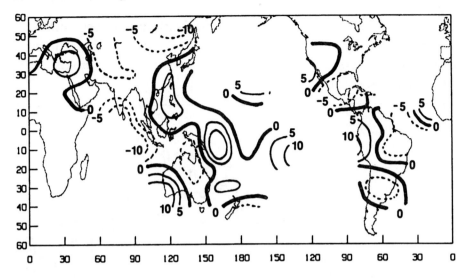

Fig. 7.10   Structure of EOF2 for global precipitation in relative units.

in particular, the western and northern portion of monsoon Asia, and northern Australia (Fig. 7.10). On the other hand, rainfall over the regions stretching from Southeast Asia to New Guinea, the SPCZ and Africa and the Middle East varied with the opposite sense as the above. This pattern represents a westward shift of the major monsoon rainy region shown in Fig. 7.3. This pattern is consistent with the changes in the trade wind shown in Fig. 7.9 for the following simple reason. A strong easterly trade wind during 1960–80 would imply a strong setup by the tropical ocean–atmosphere system, leading to surface convergence further west of the Pacific trade wind region and leaving the SPCZ and New Guinea deficient in rainfall. The stronger easterlies during the past two decades are also consistent with the presence of a stronger Hadley-type overturning commensurate with a warmer Pacific ocean.

A second inter-decadal variation with period about 60 years is found in the EOF3 time coefficient, T3. This EOF explains 6.8 percent of the total variance. The time coefficient (Figure 7.11) shows an increasing trend around the 1930s and a decreasing trend in the 1960s with relatively small year-to-year variability. Figure 7.12 shows the spatial structure of this EOF. The center of the negative anomalies appear to shift eastward in approximate quadrature with that for EOF2. As a result, the subtropical western and the Borneo–Sumatra regions are dominated by negative

anomalies. This negative anomaly extends further across northern
Australia to the region of the SPCZ. The anomalies over India and
northwestern Asia appear to be in phase with the anomaly over
the equatorial central Pacific.

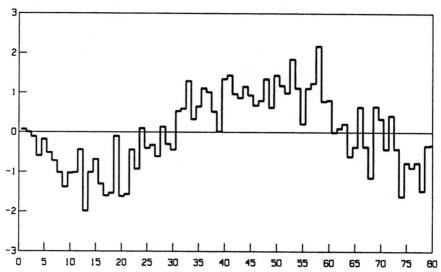

Fig. 7.11 Time series of EOF3 of global precipitation in number of standard
deviations from the mean.

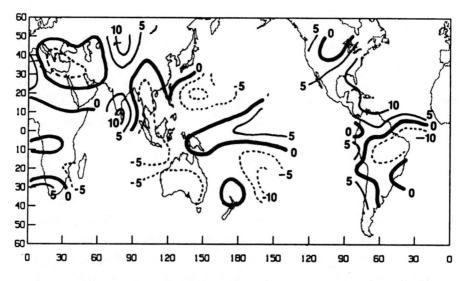

Fig. 7.12   Structure of EOF3 of global precipitation in relative units.

Another region with large organized anomalies is found along
the North Atlantic coast and the Caribbean. Here all stations,

including some mid-Atlantic island stations, all show large positive EOF loadings, suggesting that the anomaly is likely to cover the entire North Atlantic including the Atlantic ITCZ and central America. A negative anomaly is also found over the Middle East. Elsewhere, the pattern is not very well organized. The temporal variation in this mode implies a trend opposite in sense to those suggested by EOF2 over the major monsoon regions during the last 20 years.

As an independent check on the space/time variation of EOF3, Fig. 7.13 shows the drought-area index (DAI) over India as defined by Bhalme and Mooley (1980). This index is expressed as a percentage of area of the Indian subcontinent under moderate to severe drought conditions. Years with DAI of 25 or higher were defined as large-scale drought years. It is noted that there appears to be a higher tendency for the occurrence of large-scale drought over India during 1900–20 and after the 1960s. Drought occurrence is minimal from the 1920s to the 1950s. This variation compares remarkably well with that of T3. The sign of the anomaly over India in EOF3 is consistent with the trend described by the DAI. This means that the occurrence of severe drought over India is only part of a global change in rainfall pattern. Further evidence of the global nature of the long-term variation depicted by T3 is shown in Fig. 7.14, which depicts temporal variation of the Northern Hemisphere annual mean surface temperature anomalies (Jones et al., 1982). The three distinct climatic regimes representing an interdecadal alternation between cold and warm climates is remarkably similar to the variation of T3.

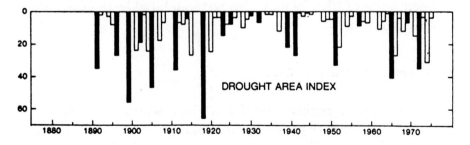

Fig. 7.13 Time series showing the Drought Area Index (DAI) over India (after Bhalme and Mooley, 1980). See text for explanation.

The variation suggests an increase (decrease) in rainfall over the North Atlantic and the Indian/East Asia monsoon region during a

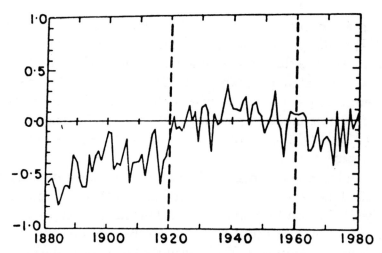

Fig. 7.14 Northern Hemisphere annual mean surface temperature (after Jones et al., 1982). Vertical dotted lines delineate the different climatic regimes.

warm (cold) climate. The association between the Indian monsoon and Northern Hemisphere surface temperature is consistent with that suggested by Verma et al. (1984, private communication) based on Indian subdivision rainfall.

We recall that there appears to be a stronger tendency for occurrence of large negative swings in T1 during the early part of the twentieth century (1900–20) than the middle of the century (1930–60). As discussed in an earlier section and found by many other studies (e.g., Rasmusson and Carpenter, 1982), drought conditions over India are associated with WEs or ENSOs (which coincides with large negative swings in T1). Thus, it is inferred that the occurrence of ENSOs may also be modulated in the same time scale as shown in T3.

Time series reconstruction

It is important to realize that for a given station or region the actual rainfall trend depends approximately on a linear combination of the dominant EOF modes. For example, for Indian rainfall during the last two decades (1960–80), EOF1 suggests no trends, EOF2 indicates an increasing trend and EOF3 a decreasing trend. To determine if the trends depicted by T2 and/or T3 are real, we have reconstructed the rainfall index using just the first three EOFs and compare them to the actual time series. Figure 7.15

shows the reconstructed and the actual rainfall index at selected stations. These stations are chosen to be at the center of, or close to, maxima in EOF2 or EOF3.

The actual rainfall index at Colombo (8°N, 80°E) shows below normal rainfall between 1900–20 and normal or slightly above normal rainfall from 1950 to 1970. This is well represented by the reconstruction where the long-term trend is accentuated. Apparently, T3 contributes much to the relatively dry condition during the early part of the century and a compensation between the two trends occurs between 1960 and 1980, resulting in a less well defined trend in the actual time series during this period. The magnitude of the combined inter-decadal trend throughout the entire period is estimated to be on the order of 3 mm/day, somewhat less but comparable to the interannual variation.

For Perth (31°S, 116°E), a drying trend beginning near the mid-1940s and extending to the 1970s is obvious in the actual rainfall index. This is also closely approximated in the reconstructed index. It is clear that long-term rainfall variation at Perth is strongly affected by the EOF2 mode. The ENSO signal is also quite strong for this station.

Similar trends in rainfall are found in the actual time series for Manila (16°N, 121°E), the Solomon Islands (8°S, 160°E), the Azores (39.5°N, 28.7°W), and other stations (not shown) and they are all fairly well approximated by the reconstructed time series. Thus the long-term variation in rainfall for these widely separated stations are likely to be real and intercorrelated, suggesting that coherent global scale changes in precipitation occur on the inter-decadal time scale. The magnitude of this global trend is in general less than, but comparable to, the year-to-year changes.

Seasonality

As we have shown in the beginning of this chapter, rainfall amounts show pronounced seasonality except in the equatorial region. It is well known that large-scale anomalies observed during ENSO are strongly dependent on the seasons. It is therefore important to know if there are any significant seasonal changes associated with the EOF patterns shown in the preceding analyses. We have repeated the analyses in the previous sections by

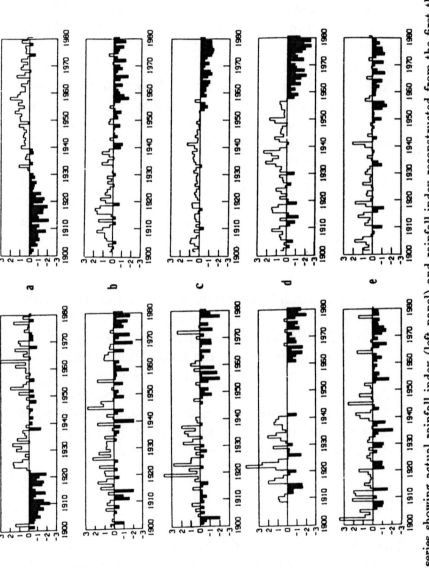

Fig. 7.15 Time series showing actual rainfall index (left panel) and rainfall index reconstructed from the first three EOFs (right panel) for (a) Colombo (8°N, 80°W), (b) Perth (31°S, 116°E), (c) Manila (16°N, 121°E), (d) the Solomon Islands (8°S, 160°E) and (e) the Azores (39.5°N, 28.7°W).

using seasonally averaged data to obtain EOF patterns for winter (DJF), summer (JJA), spring (MAM), and fall (SON). As the seasonal stratification is found to have little impact on the interdecadal time scale, only the EOFs for the interannual variation are shown (Fig. 7.16). The DJF pattern closely resembles the annual pattern shown in Fig. 7.7. In addition to the dipole features over the western Pacific, several features become more prominent during the winter. The negative anomaly over the southwestern United States is very intense and appears to be coupled with a large positive anomaly over northern South America. The orientation of this anomaly pair is along the prevailing upper level jetstream over North America during winter. A negative anomaly is found off the coast of East Asia and southern Japan, coinciding with the location of cold surges from the main continent (Lau et al., 1983). Both the North American and the East Asian anomaly are found to be strongest on the east coast of a major continent, suggesting that the above rainfall anomalies may be related to winter-time storms through some large-scale setup by stationary planetary waves.

The summer pattern is quite different. Here, the western Pacific is dominated by a north–south oriented pattern centered at 120°E. An east–west oriented but weaker dipole is found between India/the Arabian Sea and Indo-China/the South China Sea. The Indian anomaly extends northeastward in a narrow belt through central China to northern Japan. This belt is probably associated with the Mei-yu rainbelt over China during summer. It is noticed that the relative positions of the primary centers of action remain essentially the same as those for winter, i.e., the horseshoe shape of positive anomalies surrounding a negative anomaly can still be discerned. The anomalies over North and Central America appear to be displaced northward with the the northern anomaly much diminished. This is consistent with the poleward migration and reduction of the North America jetstream during the summer which may lead to a reduction in downstream teleconnection from tropics to midlatitudes (Lau and Lim, 1984). A new feature that emerges during JJA is a strong negative anomaly over southern Chile and Argentina. Lau and Chan (1983a) have suggested a connection between the SPCZ and rainfall over Argentina based on outgoing longwave radiation.

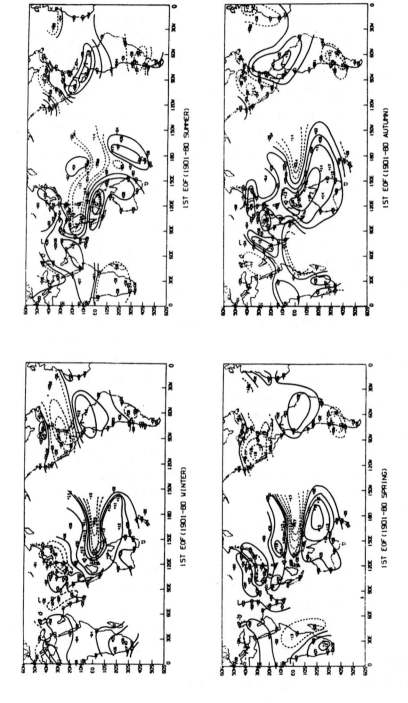

Fig. 7.16 Structure of most dominant global precipitation EOF for (a) winter, (b) summer, (c) spring and (d) autumn.

The EOF for spring and autumn are similar to that for winter. The buildup of the horseshoe pattern over the Indonesia/western Pacific and the anomalies over North and Central America are already apparent in autumn. During the spring, the western Pacific anomaly splits into a pattern somewhat symmetric about the equator with the equatorial anomaly sandwiched between two elongated subtropical anomalies. There appears to be strong continuity in the EOF patterns between autumn and winter, indicating a somewhat smooth transition between these seasons. In contrast, the key summer EOF features appear to have a distinct geographical location compared with those of spring and autumn.

The time coefficients for the seasonal EOFs are shown in Fig. 7.17. The years of WEs and CEs as defined earlier are marked by black and hatched bars respectively. For the winter pattern, the marked bars denote the year following either WEs or CEs. With the exception for spring, the EOF time series for the other seasons show very good correspondence between the occurrence of WEs and CEs. More than three-quarters of the WEs (CEs) coincide with negative (positive) deviations in the respective time series. The weaker correspondence between the spring time series and the WEs or CEs suggests that these events are not quite locked in phase with the global precipitation pattern. This is consistent with the observation that a large positive SST anomaly over the central Pacific associated with ENSO events generally develops during the early part of the year, becomes fully established during the summer of the warm year and lasts until the following winter. In other words, the spring rainfall pattern is more variable with respect to the occurrence of WEs or CEs because the global control associated with these events may not yet reach full intensity during their early stage of development.

Summary and discussion

A study of global rainfall variability for seasonal to inter-decadal time scales has been carried out. Given the limitations of historical rainfall data, we have shown that some valuable scientific insight can be gained from analyses of such a data set. Harmonic analysis clearly delineates the magnitude and phase of the seasonal cycle in precipitation over key monsoon regions in the tropics, desert regions in the subtropics and winter storm tracks in the extratropics.

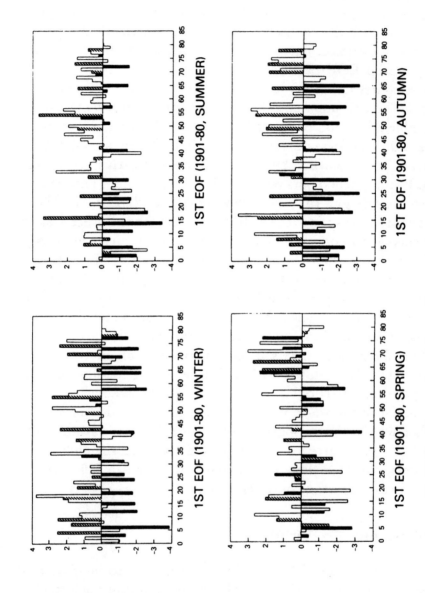

Fig. 7.17 Time series corresponding to the seasonal EOFs in Fig. 7.16.

Results suggest that the transition from the summer to the winter monsoon may take place continuously, following the migration of the main precipitation zone from the Bay of Bengal along the Indo-China subcontinent to northern Australia/New Guinea.

For anomalies of annual precipitation amounts, three major global modes are detected. The first is an interannual mode (EOF1), which is closely related to ENSO and is strongly phase-locked to the season cycle. The second (EOF2) and third (EOF3) are inter-decadal modes. EOF2 is a secular trend which is manifested in a major shift in tropical precipitation pattern associated with a westward migration of the Indian/East Asian monsoon between the early and the latter part of the twentieth century. EOF3 is an oscillation with a period of the order of 60 years. This mode affects precipitation over the major monsoon areas as well as that over the North Atlantic and appears to modulate the interannual variability associated with EOF1. In addition, the temporal variation of EOF3 appears to be consistent with the long-term variation of drought over India and Northern Hemisphere surface temperature.

Undoubtedly, for both the interannual and the inter-decadal modes, the ocean plays an essential role. Because of its close association with ENSO, the physical processes responsible for the spatial structure of EOF1 is likely to arise from the zonal asymmetry of the Walker-type atmospheric circulation associated with the east–west contrast in SST over the tropical Pacific. The episodic and irregular occurrence of extreme events in this mode suggests that it may arise stochastically by triggering inherent instability in the coupled ocean–atmosphere system in the tropics (Lau, 1985). The time scale (of one to two years' duration) of individual events is likely to be governed by the dynamics of the interaction between the atmosphere and the upper ocean in the equatorial regions (Lau, 1981; Philander et al., 1984; and many others).

On the other hand, for EOF2 and EOF3, the long time scales are more likely to be associated with the deeper oceans. In particular for EOF2, the rainfall variability appears to be correlated with an overall cooling or warming of the entire tropical Pacific, whose time scale is consistent with those occurring as a result of the meridional thermocline overturning involving the deep ocean in the extratropics. The observed long-term variation of the Pacific trade wind is consistent with the above SST trend. Therefore, for the inter-decadal modes, it is likely that the ocean and atmosphere may be interacting via Hadley-type (in its classical sense) circulations which involve north–south temperature contrasts. Needless to say, this has to be substantiated by further investigations.

At the time of the final draft of this chapter, the authors became aware of recent studies by Folland and Owen (1988) showing long-term variation of annual rainfall anomaly over the Sahel region. A comparison of their Sahel rainfall time series suggests that there are strong resemblances between the long-term Sahel rainfall and the principal component of EOF3, in particular the decline of the Sahel rainfall starting from the 1960s to the 1980s. However, central African rainfall stations over the Sahel are not well sampled in the present analysis. While the EOF3 pattern shows a weak drying trend over central Africa from 1960 to 1980, it is not due to Sahel rainfall. This then raises an interesting question: is the Sahel rainfall trend only a regional manifestation of a global signal, or is such a trend unique to the Sahel? The present analysis tends to support the former hypothesis. Much more work is needed to confirm or refute this hypothesis. No matter what the outcome is, it appears that in order to better understand the mechanisms for the regional occurrence of extensive drought and flood it is necessary to adopt a global view. While, undoubtedly, the often-cited regional anthropogenic factors (e.g., overgrazing, deforestation, etc.) may play an important role in shaping the detailed distribution of deserts vs. vegetated land, these factors are likely to contribute to trends in drought/flood conditions but not necessarily to the global oscillatory behavior. Thus, to understand the variability of, for example, African drought and India/East Asia monsoon, the anthropogenic factors have to be separated from the naturally occurring processes, as shown in the interannual and inter-decadal modes suggested in this study.

References

Bhalme, H.N. & Jadhav, S.K. (1984). The Southern Oscillation and its relations to the monsoon rainfall. *Journal of Climate*, **4**, 509–24.

Bhalme, H.N. & Mooley, D.A. (1980). Large scale droughts/floods and monsoon circulation. *Monthly Weather Review*, **108**, 1197–1211.

Blackmon, M., Geisler, J.E. & Pitcher, E.J. (1983). A general circulation model study of the January climate anomaly patterns associated with interannual variation of Pacific sea surface temperature. *Journal of the Atmospheric Sciences*, **40**, 1410–25.

Chu, P.-S. (1983). Diagnostic studies of rainfall anomalies in northeast Brazil. *Monthly Weather Review*, **111**, 1655–64.

Fleer, H.E. (1981). Teleconnection in rainfall anomalies in the tropics and subtropics. In *Monsoon Dynamics*, ed. J. Lighthill & R. Pierce. Cambridge, U.K.: Cambridge University Press.

Fletcher, J. (1982). The difference between the Southern Oscillation and the El Niño. *Proceedings of the Seventh Annual Climate Diagnostic Workshop*. NTIS, PB83-208033. Washington, DC: U.S. Department of Commerce.

Folland, C.K. & Owen, J.A. (1988). GCM simulation and prediction of Sahel rainfall using global and regional sea surface temperatures. In *Report of a Workshop at the European Centre for Medium-Range Weather Forecasting*. WMO Publication No. 254, 102–14. Geneva, Switzerland: WMO.

Gill, A. & Rasmusson, E.M. (1983). The 1982/83 climate anomaly in the equatorial Pacific. *Nature*, **306**, 229–34.

Hsu, C.-P. & Wallace, J.M. (1976). The global distribution of the annual and the semiannual cycles in precipitation. *Monthly Weather Review*, **104**, 1093–1101.

Jones, P.D., Wigley, T.M.L. & Kelly, P.M. (1982). Variation in surface air temperatures: Part 1. Northern hemisphere, 1881–1980. *Monthly Weather Review*, **110**, 59–70.

Kidson, J.W. (1975). Tropical eigenvector analysis and the Southern Oscillation. *Monthly Weather Review*, **103**, 187–96.

Lau, K.M. (1981). Oscillations in a simple equatorial climate system. *Journal of the Atmospheric Sciences*, **38**, 248–61.

Lau, K.M. & Chan, P.H. (1983a). Short-term climate variability and atmospheric teleconnection from satellite derived outgoing longwave radiation. Part I: simultaneous relationships. *Journal of the Atmospheric Sciences*, **40**, 2735–50.

Lau, K.M. & Chan, P.H. (1983b). Short-term climate variability and atmospheric teleconnection from satellite derived outgoing longwave radiation. Part II: lagged correlations. *Journal of the Atmospheric Sciences*, **40**, 2752–67.

Lau, K.M., Chang, C.P. & Chan, P.H. (1983). Short-term planetary scale interactions over the tropics and midlatitudes. Part II. Winter-MONEX period. *Monthly Weather Review*, **111**, 1372–88.

Lau, K.M. & Lim, H. (1984). On the dynamics of equatorial forcing of climate teleconnections. *Journal of the Atmospheric Sciences*, **41**, 161–76.

Lau K.M. & Sheu, P.J. (1988). Annual cycle, quasi-biennial oscillation and Southern Oscillation in global precipitation. *Journal of Geophysical Research*, **93**, 10975–88.

Lau, K.M. (1985). Elements of a stochastic-dynamical theory of the long-term variability of the El Niño/Southern Oscillation. *Journal of the Atmospheric Sciences*, **42**, 1552–8.

Lau, N.C. (1985). Modeling the seasonal dependence of the atmospheric response to observed El Niños in 1962-76. *Monthly Weather Review*, **113**, 1970–96.

McBride, J.L. & Nicholls, N. (1983). Seasonal relationships between Australian rainfall and the Southern Oscillation. *Monthly Weather Review*, **111**, 1998–2004.

Nicholson, S.E. (1983). Sub-Saharan rainfall in the year 1976-1980: Evidence of a continued drought. *Monthly Weather Review*, **111**, 1646–54.

North, G.R., Bell, T.L., Cahalan R.F. & Moeng, F.J. (1982). Sampling errors in the estimation of empirical orthogonal functions. *Monthly Weather Review*, **110**, 699–706.

Philander, S.G.H. (1985). El Niño and La Niña. *Journal of the Atmospheric Sciences*, **42**, 2652–62.

Philander, S.G.H., Yamagata T. & Pacanowski, C. (1984). Unstable air–sea interactions in the tropics. *Journal of the Atmospheric Sciences*, **41**, 604–13.

Quinn, W.H. (1979). Monitoring and predicting short-term climate changes in the south Pacific Ocean. *Proceedings of the International Conference on Marine Science and Technology*, Part 1. Valparaiso, Chile: Catholic University of Valparaiso, 26–30.

Rasmusson, E.M. & Carpenter, T.H. (1982). Variations in tropical sea surface temperature and surface wind fields associated with the Southern Oscillation/El Niño. *Monthly Weather Review*, **110**, 354–84.

Rasmusson, E.M. & Carpenter, T.H. (1983). The relationship between eastern equatorial Pacific seas surface temperature and rainfall over India and Sri Lanka. *Monthly Weather Review*, **111**, 517–28.

Ropelewski, C. & Halpert, J. (1988). Global and regional precipitation pattern associated with the El Niño southern Oscillation. *Monthly Weather Review*, **115**, 1606–26.

Shukla, J. & Paolino, D. (1983). The Southern Oscillation and the long-range forecasting of monsoon rainfall over India. *Monthly Weather Review*, **111**, 1830–7.

Stoeckenius, T. (1981). Interannual variations of tropical precipitation pattern. *Monthly Weather Review*, **109**, 1233–47.

van Loon, H. (1984). The Southern Oscillation. Part III: Associations with the trades and with the trough in the westerlies of the South Pacific Ocean. *Monthly Weather Review*, **112**, 947–54.

8

El Niño and QBO influences on tropical cyclone activity

WILLIAM M. GRAY and JOHN D. SHEAFFER

Colorado State University
Fort Collins, CO 80523

Introduction

The formation and intensification of tropical cyclones are governed by many factors, the interactions and relative importance of which are still not fully understood. It is clear, however, that optimum conditions for the development of tropical cyclones include an elusive and rather easily disrupted combination of atmospheric circulation features. The peculiar combination of proximity of the Northwest Atlantic to the east Pacific El Niño area and to the atmospheric circulation anomalies associated with El Niño is such that the upper–tropospheric climate in the Atlantic–Caribbean basin is strongly affected by major El Niño events.

Six seasonal climatological factors have been identified (Gray, 1979) as important to the seasonal formation frequency of tropical storms. These six factors can be subdivided into two circulation parameters, a single geographic parameter and three thermodynamic parameters. The geographic factor is the Coriolis parameter (f). Because geostrophic considerations dictate that pressure gradients near the equator must be weak, intensification of tropical vortices must be displaced at least 4–5° from the equator. It is observed that the frequency of tropical cyclogenesis increases linearly from about 5° to 15° latitude. The three thermodynamic parameters governing tropical cyclogenesis include ocean thermal energy (E), the vertical gradient of equivalent potential temperature (Θ_e) and relative humidity (RH) in the middle troposphere. A warm (greater than 26°C) ocean and deep thermocline provide a vast reservoir of thermal energy for maintaining developing storms. A reasonably strong decrease (10 K or more) of equivalent potential temperature between the ocean surface and 500 mb plus relative humidity values of 60 percent or more between 500–700 mb are critical factors for the occurrence of deep convection, and hence, for vertical coupling of upper level outflow and lower tropospheric inflow circulations. The two larger-scale atmospheric circulation features whose day to day variations are important to the formation of tropical storms include areas of broad-scale, low level cyclonic vorticity (ζ_r) and the presence of minimal vertical wind shear (S_Z) in the troposphere.

Combinations of these six parameters can be used to define a dynamic potential ($f\zeta_r/S_Z$) and a thermal potential ($E\Delta\Theta_e RH$). The product of these two potential terms specifies a "Seasonal Genesis Parameter" or SGP where SGP = ($f\zeta_r/S_Z$)($E\Delta\Theta_e RH$). The spatial distribution of mean long-term values of SGP provide excellent estimates of the relative seasonal cyclone frequency for nearly all global cyclone basins (see Gray, 1975, 1979).

It is through the vertical wind shear component of the SGP relationship that El Niño events act to diminish the frequency of tropical cyclones in the Atlantic–Caribbean area. Tropical cyclones are relatively infrequent in areas where vertical wind shear is large. In the following section, we elaborate on this point and review evidence indicating that anomalous vertical shear conditions appear over the tropical Atlantic on seasonal time scales during El Niño events and disrupt the formation of tropical cyclones. Subsequently, extensive statistical data are presented which compare the frequency of Atlantic hurricanes for El Niño versus non-El Niño seasons. The next section provides a summary of current knowledge regarding associations between El Niño events and activity in other tropical cyclone basins around the globe. The chapter concludes with a brief description of the modulation of tropical cyclone frequency and intensity by the stratospheric Quasi-Biennial Oscillation, or QBO.

Large scale features linking circulation in the tropical Atlantic to El Niño

Persistent favorable flow patterns must occur during seasons with relatively frequent development of tropical cyclones. The more favorable the seasonal background environment is, the greater the probability that individual cloud clusters will develop into cyclones

rather than dissipate or remain as traveling depressions and disturbances (Gray, 1988a). Figure 8.1 shows an analysis of composited
rawinsonde data for areas around developing tropical disturbances
which eventually became tropical storms within the Caribbean
basin. This figure illustrates the anticyclonic north–south zonal
wind shear in the upper troposphere which is typically associated
with the development and maintenance of hurricanes.

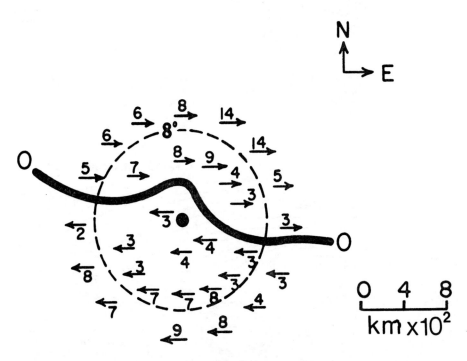

Fig. 8.1 Composite of 200 mb zonal winds (in m/s) about the center point
(large dot) of Caribbean basin tropical weather systems in an early
stage of cyclone development (adapted from Gray, 1968).

For a large number of hurricanes to form and be maintained
through an active hurricane season, it is important that seasonal
upper tropospheric (200 mb) winds in the latitude belt of 5–15°N
not blow strongly from the west, especially in the normally active region just to the east of the Antilles. The upper panel of
Fig. 8.2 shows typical 200 mb conditions in this region favorable
for August-September hurricane activity whereas the lower panel
shows typical unfavorable conditions. Hurricane activity is increased by any process which enhances seasonally averaged upper
tropospheric wind patterns similar to those in the top of Fig. 8.2.

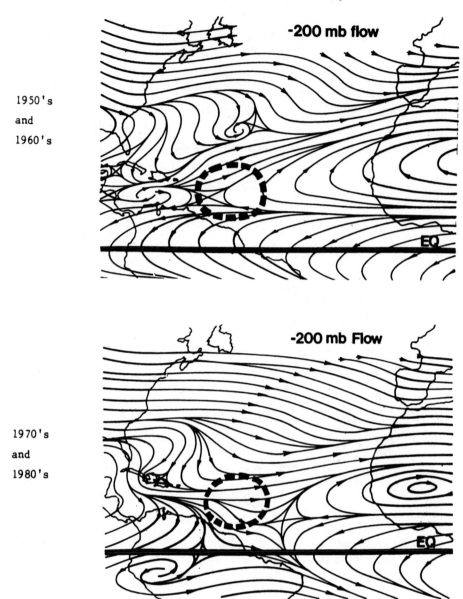

Fig. 8.2 Typical favorable upper tropospheric (200 mb) late summer wind patterns during the 1950s and 1960s (top) in comparison with typical less
favorable wind patterns during the 1970s and 1980s (bottom). The
encircled areas are likely locations for developing tropical disturbances
to transform themselves into hurricanes. A comparison of the encircled areas shows the greater westerly winds of the more recent period
(compare also with Fig. 8.9).

Thermodynamic and lower tropospheric wind conditions generally have a smaller influence on seasonal variability of Atlantic cyclone activity.

Studies of satellite imagery and other data show that warm water in the eastern tropical Pacific during El Niño years is associated with a broad area of increased deep cumulus convection. This enhanced convection contributes to anomalous upper tropospheric (200 mb) outflow patterns. Other factors being equal, these outflow patterns appear to cause enhanced westerly winds over the Caribbean and western equatorial Atlantic regions, and thus create upper air conditions which differ from non-El Niño years. A conceptual representation of this process is shown in Fig. 8.3.

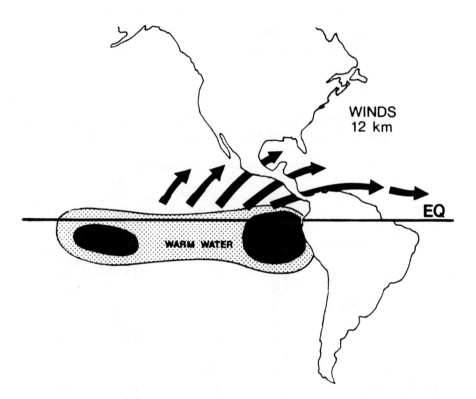

Fig. 8.3 Illustration of anomalously warm ocean surface temperatures in the eastern equatorial Pacific which are associated with El Niño events. Dark regions have the greatest warm anomalies. Arrows show the resulting anomalous upper tropospheric westerly winds which result from the enhanced deep cumulus convection occurring in the equatorial Pacific as a result of these warm water events.

Figure 8.4 shows the average August–September zonal wind differences observed at upper levels during six moderate or strong El Niño events, in comparison with non-El Niño years since 1957 for the Caribbean basin stations of Curacao and Barbados. Notice that the largest differences occurred in the upper troposphere at or near the 200 mb level. Similar average zonal wind differences at 200 mb for various periods within the hurricane season at these and other Caribbean basin stations are shown in Table 8.1. In general, upper tropospheric westerly winds are 2–7 m/s above average during El Niño years. A Wilcoxon two-sample rank test of the premise that average regional 200 mb winds are the same for both El Niño and non-El Niño years for the August–September period is rejected at the 0.1 percent level for both of these periods. Surprisingly, El Niño events have very little influence on regional sea level pressure or the summertime–early autumn rainfall in the Caribbean basin (see Tables 8.2 and 8.3). Mean El Niño year changes of 200 mb zonal winds over stations in the southeast U.S. and the Bahamas (not shown) are also much smaller than over the Caribbean basin stations.

Table 8.1 Caribbean basin 200 mb multimonth average zonal winds (m/s) by station for the last five El Niño years and 21 other non-El Niño years.

Station	El Niño		Non-El Niño	
	Aug–Sept mean	June–Oct mean	Aug–Sept mean	June–Oct mean
Swan Island	4.0	5.3	1.2	2.5
Santo Domingo	4.3	5.8	2.1	3.9
San Andras Island	4.0	3.0	−1.8	−0.7
Seawell Airport, Barbados	5.0	7.0	2.1	4.6
Raizet, Guadeloupe	5.0	7.8	2.1	4.8
Juliana, St. Maarten	3.8	6.6	2.5	4.3
Kingston, Jamaica	4.3	6.0	1.8	3.5
Grand Cayman, B.W.I.	2.4	4.8	0.1	2.5
Plesman, Curaçao	7.5	8.0	1.7	3.6
San Juan, Puerto Rico	4.8	6.6	2.5	4.3

Evidently, these anomalous lower latitude (5–15°) winds are the primary El Niño factor in the Atlantic tropical cyclone suppres-

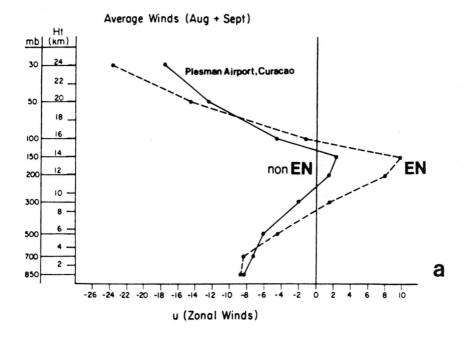

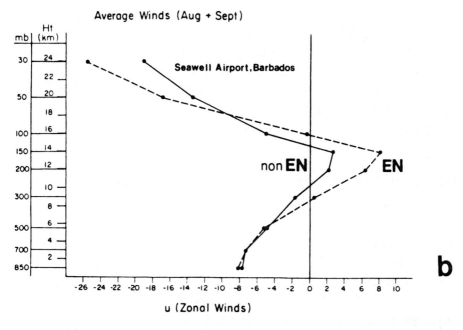

Fig. 8.4 (a) Average vertical profile of zonal wind during August and September at Curaçao (12°N, 69°W) for the last 5 El Niño years (1957, 1965, 1972, 1976 and 1982) (denoted EN) and for 18 other non-El Niño years (non EN). (b) As in Fig. 8.4a but for Seawell Airport, Barbados (13.5°N, 60.0°W).

Table 8.2 Average percentage departure from normal precipitation for 15 West
Indies region stations for each summer month of the last six strong
and moderate El Niño events.

Year	June	July	Aug.	Sept.	Oct.	Avg.
1953	−23	−3	−10	+20	−3	−4
1957	−3	+17	+20	−13	+17	+8
1965	−11	−8	−6	+11	−10	−5
1972	−12	+4	−15	−19	−12	−11
1976	−5	−17	−9	−11	+2	−8
1982	−13	−23	−22	−7	−8	−13
Avg.	−11	−5	−7	−3	−2	−5

sion mechanism. In El Niño years, an anticyclonic north–south
wind shear anomaly (or $\partial u / \partial y$) at 200 mb tends to make upper
tropospheric relative vorticity less negative. In addition to this
tendency toward diminished regional scale anticyclonic shear in
the upper troposphere, the increased westerly wind component at
200 mb during El Niño years is important as a factor in increased
vertical wind shear. As shown in Fig. 8.5, an appreciable westerly
wind anomaly in the upper-troposphere will disrupt the vertical
coupling between areas of low level inflow and upper level outflow, and thereby diminish the prospects for further development
of incipient storms.

Observational evidence for associations between
El Niño and seasonal Atlantic hurricane activity

The effects of El Niño on Atlantic tropical cyclone activity can
be isolated by considering the 18 moderate and strong El Niño
events which have occurred since 1900. These events, as reported
by Quinn et al. (1978, 1987) and updated here by the authors
through 1988, are listed in Table 8.4. If we can accept these 18
periods as significant El Niño events, then we can compare the
frequency of hurricanes and related occurrences during these El
Niño years to the frequency of such events occurring during the
other 71 non-El Niño years of this century.

Table 8.3 Sea level pressure (in mb, −1000) occurring in various months at Caribbean Basin stations during El Niño and non-El Niño years between 1950–82 and differences between these pressures.

Station		El Niño			Non-El Niño	
	May	Aug–Sep Avg.	Jun–Oct Avg.	May	Aug–Sep Avg.	Jun–Oct Avg.
Cayenne French Guiana	12.6	12.9	12.8	12.5	12.8	12.9
Jacksonville Florida	16.6	17.1	17.2	16.8	16.8	17.1
Maracay Venezuela	12.6	12.2	12.6	12.7	13.7	13.6
Merida Mexico	11.8	13.0	13.3	11.7	12.5	13.1
Nassau Bahamas	16.1	15.3	16.1	16.8	16.0	16.3
Plesman Curaçao	11.7	11.6	11.8	11.6	11.3	11.4
San Juan Puerto Rico	15.9	15.3	15.6	15.7	15.1	15.5
Seawell Barbados	13.9	13.7	14.1	14.2	13.4	13.7
Swan Island	12.0	13.0	13.0	13.4	12.9	12.7
Raizet Guadeloupe	15.2	14.3	14.8	14.8	14.1	14.5

A simple but striking illustration of the effects of El Niño can be seen in Fig. 8.6, which shows the tracks of hurricane intensity tropical cyclones in moderate and strong El Niño years in comparison with tracks for the years immediately prior to and immediately after El Niño episodes. Few hurricanes cross the Caribbean–West Indies region in an east-to-west direction during the El Niño years. By contrast, in the non-El Niño years, total hurricane frequency and the frequency of tracks crossing the Caribbean, especially the southern part, are much greater. El Niño-year suppression of westerly tracking cyclones in the Atlantic equatorwards of 20°N is no-

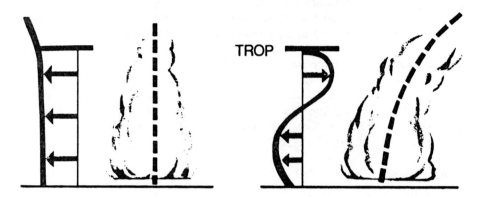

Fig. 8.5 Conceptual illustration of the type of tropospheric vertical wind shear
which is (left diagram) and is not (right diagram) conductive to hurricane formation. Excessive westerly wind shear with height causes
the upper portions of tropical disturbances to be blown off. This
shearing effect inhibits hurricane formation while the unsheared conditions in the left diagram do not. TROP signifies the top of the
troposphere, approximately 16 km.

Table 8.4 El Niño years since 1900 by intensity as determined by Quinn et al.
(1978) and updated by the authors through 1988.

Strong	Moderate	Weak	Very Weak
1983	1987	1969	1975
1982	1986	1951	1963
1972	1976	1943	1948
1957	1965	1932	1946
1941	1953	1923	
1925	1939	1917	
1918	1929		
1911	1914		
	1905		
	1902		

ticeably larger than the overall difference in hurricane activity between El Niño and non-El Niño years.

Figure 8.7 shows a plot of the seasonal number of hurricane
days for the years of 1900–88. (A hurricane day is any day when
a tropical cyclone was considered to have a maximum sustained
wind in excess of 34 m/s.) A Wilcoxon two-sample rank test of

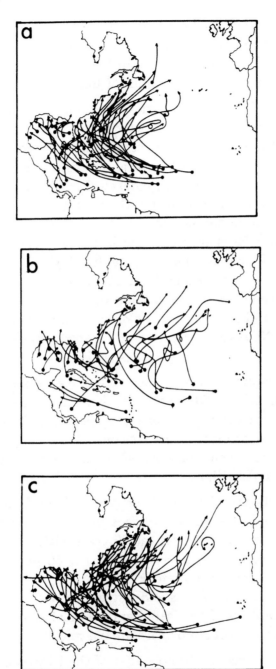

Fig. 8.6 (a) Fourteen years of trajectories or "tracks" of hurricane intensity storms occurring during the year immediately before each of 14 El Niño years between 1900 and 1976. (b) Fourteen years of hurricane intensity storm tracks during 14 El Niño years between 1900 and 1976. (c) Hurricane intensity tracks for the years immediately after each of these 14 El Niño years.

the null hypothesis of no distinction between hurricane activity for 18 El Niño and 71 non-El Niño years in Fig. 8.7 is significant at the 0.1 percent level for the seasonal frequency of hurricanes, for hurricanes plus tropical storms and for hurricane days as well.

The tendency for reduced hurricane activity in El Niño years is further illustrated in Table 8.5, which lists, in decreasing order, the number of hurricane days occurring in each year since 1900. Inspection of this table reveals several interesting features. Specifically, 15 of the 18 strong and moderate El Niño years fall in the second half (right-hand column) of the rankings in this table, and 11 of these 18 fall in the last fourth (lower portion of column 2). None of the first 23 years (with the largest number of hurricane days) is an El Niño year. Further, the five highest hurricane day totals for El Niño years range from 15 to 27. The corresponding five values for the highest non-El Niño years range between 46 and 57. The mean and median number of hurricane days in moderate and strong El Niño years was 10.8 and 8 respectively, versus 23.3 and 22 during non-El Niño years, respectively.

A simple ranking of the number of hurricanes per year yields similar results. Of the 29 years with three hurricanes or less, 13 (or 45 percent) were moderate or strong El Niño years. Of the 56 seasons with four or more hurricanes only five (or 8 percent) were El Niño years. The mean numbers of hurricanes per season during El Niño and non-El Niño years are 3.0 and 5.4, respectively.

Similar results are also found for seasonal totals of both hurricanes and tropical storms (cyclones with maximum sustained winds greater than 22 m/s). Of the 22 years with a total of five or fewer tropical storms and hurricanes, 11 (or 50 percent) were El Niño years. By contrast, only four of 55 years (or 7 percent) with seven or more tropical storms and hurricanes were El Niño years. The average numbers of hurricanes and tropical storms per season for El Niño and non-El Niño years are 5.4 and 9.1, respectively.

Finally, it is interesting to note that of the 59 major hurricanes striking coastal areas of the United States during the period of 1900 to 1988 (Hebert and Taylor, 1978 and updated through 1988 by the authors), only 4 occurred during the 18 strong or moderate El Niño years. During the other 71 non-El Niño years from 1900 to 1988, there were 55 major hurricane strikes on the U.S. coast. Hence, the frequency of major hurricanes per El Niño is 0.22 per

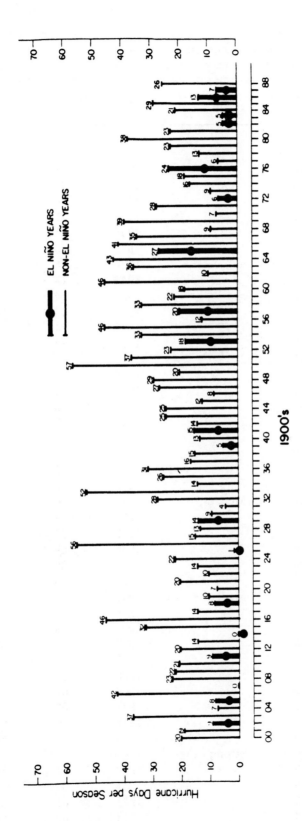

Fig. 8.7 The number of hurricane days (figure at tops of vertical lines) in El Niño and non-El Niño years by year from 1900 to 1988. El Niño years are represented by heavy lines.

Table 8.5 Ranking of Atlantic tropical cyclone seasons from 1900 to 1988 by
number of hurricane days. Indication of a moderate or strong El
Niño event for each year is given on the right of each column.

Year	Hurricane days	El Niño	Year	Hurricane days	El Niño
1950	57		1901	19	
1926	56		1975	18	
1933	52		1960	18	
1961	46		1953	18	Moderate
1955	46		1974	16	
1916	46		1937	16	
1964	43		1941	15	Strong
1906	42		1938	15	
1966	41		1927	15	
1969	39		1942	14	
1980	38		1934	14	
1951	37		1929	14	Moderate
1903	37		1923	14	
1963	36		1917	14	
1967	35		1913	14	
1958	33		1986	13	Moderate
1954	33		1978	13	
1915	32		1940	13	
1936	31		1928	13	
1985	29		1956	12	
1948	29		1945	12	
1971	28		1962	10	
1932	28		1922	10	
1965	27	Moderate	1919	10	
1947	27		1973	09	
1988	26		1968	09	
1935	26		1930	09	
1944	25		1911	09	Strong
1943	25		1902	09	Moderate
1976	24	Moderate	1946	08	
1981	23		1918	08	Strong
1979	23		1905	08	Moderate
1952	23		1987	07	Moderate
1908	23		1970	07	
1959	22		1920	07	
1924	22		1904	07	
1909	22		1977	06	
1984	21		1972	06	Strong
1910	21		1939	05	Moderate
1957	20	Strong	1983	05	Strong
1949	20		1982	05	Strong
1921	20		1931	04	
1912	20		1925	01	Strong
1900	20		1914	00	Moderate
			1907	00	

Table 8.6 Comparison of average values of hurricane number and Hurricane Destruction Potential (HDP) for El Niños of various intensity during this century. HDP as defined in the text, is expressed in units of 10^4 kt^2. Values in parentheses represent the ratio of the HDP values for each El Niño class to the value for all non-El Niño years.

	8 Strong EN seasons	8 Moderate EN seasons	10 Weak EN seasons	Average for Non-El Niño years
Average No. of Hurricanes Per Season (n)	2.8 (0.55)	3.3 (0.65)	5.6 (1.12)	5.0 (1.00)
Mean HDP Per Hurricane (10^4 kt^2)	8.7 (0.69)	11.7 (0.92)	13.7 (1.08)	12.7 (1.00)
Average Total Seasonal (10^4 kt^2)	24.0 (0.38)	38.1 (0.60)	76.8 (1.21)	63.1 (1.00)

year whereas that of major hurricanes per non-El Niño is 0.77 per year, a ratio of more than 3 to 1.

The potential wind and storm surge destruction of a hurricane can be approximated by the square of the storm's maximum wind speed (Gray, 1988b). This potential for damage from hurricane winds and storm surge can be termed Hurricane Destruction Potential or HDP. We define this as follows:

$$HDP = \sum (V_{max})^2$$

HDP values are computed only for conditions wherein maximum wind (V_{max}) values greater than 33.5 m/s (65 kt) are observed for each six-hour period of a hurricane existence and values are totaled for each storm. Comparative HDP statistics for El Niño versus non-El Niño years are summarized in Table 8.6. The values shown in parentheses in this table represent the ratio of total HDP for the specified El Niño intensity class relative to the average for non-El Niño years. Note that total seasonal HDP for the eight strong El Niño seasons is less than 40 percent of the mean non-El Niño value.

Hurricane activity as measured by the standard of HDP has been very low since 1970. The mean HDP for the 23-year period of 1947–69 was more than twice that of the 18-year period of 1970–87 (see Fig. 8.8). Although the average number of hurricanes per season was only 31 percent higher during the period of 1947–69 as compared with 1970–87, the Hurricane Destructive Potential was more than twice as great during the earlier period. Similarly, although named storm activity was only 19 percent higher for the earlier period, the seasonal average of total potential destruction from hurricanes with maximum winds greater than 51 m/s (100 kt) was nearly three times greater in 1947–69 as compared with 1970–87.

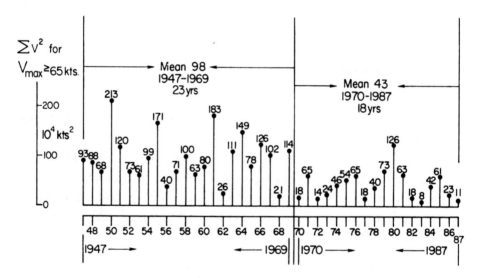

Fig. 8.8 Yearly variations of Hurricane Destruction Potential (HDP) for 1948–87. HDP is defined as the sum of all hurricane maximum wind speeds squared for V_{max} (greater than 33.5 m/s (65 kt) for each 6-hour observing period throughout the hurricane season (units 10^4 kt^2).

Although the apparent long-term trend in HDP shown in Fig. 8.8 has numerous parallels in other long-term meteorological time series, an important association for the purpose of this discussion is shown in Fig. 8.9 (see also Fig. 8.2). Note that the difference between upper tropospheric winds over the lower Caribbean for the diminished HDP period of 1970–87 versus the high HDP period of 1948–69 shows a vertical shear anomaly similar to that for El Niño versus non-El Niño conditions which were shown in Fig. 8.5.

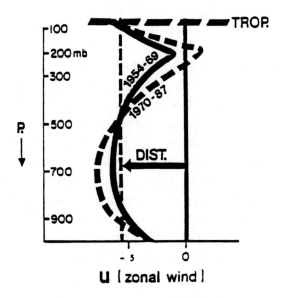

Fig. 8.9 Vertical profile of lower Caribbean basin zonal winds for the August-September period of 1954–69 vs. the period of 1970–87. The arrow labeled DIST. represents the typical westward velocity of a tropical disturbance in m/s. The area represented by these profiles is shown encircled in Fig. 8.2.

El Niño modification of tropical cyclone activity in other basins

The Atlantic–Caribbean basin is directly downwind from the area of major east Pacific El Niño SST warming. In addition, many of the factors required for the genesis of tropical cyclones tend to be transient or marginally developed in much of the Atlantic–Caribbean region. For these reasons, the net frequency of cyclone formation in the Atlantic is, in general, more sensitive to teleconnected El Niño influences than are other, more remote or more robust tropical cyclone areas. The west Atlantic basin is unique in being so influenced by El Niño events.

The large reduction in the total seasonal frequency of tropical cyclones which occurs during El Niño years in the northwest

Atlantic does not appear to be as prominent in other tropical cyclone basins. At present, most of these other areas are less well studied and limited meteorological data in some areas (particularly the northeast Pacific) restricts the extent to which physical explanations can be advanced to account for apparent El Niño-linked differences. Nevertheless, new findings by Chan (1985), Dong (1988) and Collimore (1990a) show that statistically significant differences do occur in the total frequency of tropical cyclones, especially in the northwest Pacific. Dong (1988) has shown El Niño modulation of cyclone frequency in both the western north Pacific and the Australian region when: (1) El Niño periods were specified as peaks in 12-monthly running means of sea surface temperatures in the equatorial east Pacific; (2) cyclone data were carefully stratified by the locations where they developed; and (3) the strength of the El Niño events was considered. In this way, Dong was able to show 40 percent reductions in cyclone frequency in the core of the northwest Pacific basin during El Niño years.

Nicholls (1979) reported that negative (high-pressure anomalies at Darwin) values of the Southern Oscillation Index (SOI) cause a significant reduction in early season Australian region tropical cyclone activity and a modest lowering of the whole Australian area seasonal cyclone activity. In addition to being the first to show evidence of an association between El Niño and tropical cyclone frequency, Nicholls' 1979 and subsequent results (Nicholls, 1984, 1985) also illustrate basic differences between the nature of El Niño effects in the northeast Australia region versus those in the northwest Atlantic. Whereas the previously noted vertical shear mechanism appears to account for tropical cyclone–El Niño teleconnections in the latter area, large differences both in regional sea surface temperatures and surface pressure may be related to El Niño linked variations in cyclone frequency in the Australia region. Cool sea surface temperature anomalies and relatively high barometric pressure throughout the region accompanying El Niño events may combine to diminish the frequency of Australia area cyclones with the opposite conditions enhancing cyclone frequency during non-El Niño periods. Gray (1984b) described a similar association wherein regional scale surface high pressure anomalies are associated with diminished cyclone activity in the tropical Atlantic. However, whereas high pressure anomalies in Northeast Australia are a fundamental component of El Niño, anomalies of regional scale pressure in the Atlantic–Caribbean basin are largely independent of El Niño.

Recent work by Hastings (1990) clearly demonstrates an eastward displacement of the primary center of Australian tropical cyclone activity, from the northeast coast to near the dateline during El Niño years. Hastings' results also show a later than normal start of the tropical cyclone season and suggest an overall decrease of total activity during El Niños. Other recent work by Landsea and Gray (1989) and Hastenrath and Wendland (1979) indicates that the incidence of strong cyclones (i.e., wind speeds exceeding 51 m/s) in the northeast Pacific basin increases by more than 50 percent during El Niño years and is a factor of two greater than during distinct cold water (high pressure) events in the eastern Pacific. However, when all northeast Pacific cyclones (including tropical storms) are compared, no seasonal El Niño/non-El Niño differences were observed.

Collectively, these plus other recent studies might be summarized as follows: It appears that El Niño events are associated with an increased frequency of intense tropical cyclones in the northeastern Pacific basin which tend to track somewhat farther westward than normal. El Niño events also cause southwest Pacific storms to form farther eastward than normal. In the year following an El Niño year, the seasonal storm activity in the northwest Pacific Ocean and Australia sometimes commences at a later date. Tropical cyclone and hurricane activity during the intense 1982–83 El Niño illustrates these features. Many more tropical cyclones (eight systems) formed east of 180° in the South Pacific in 1982–83 than in any previous year on record. However, because tropical cyclone activity in the south Indian Ocean and Australian region was below normal during the 1982–83 hurricane season, the Southern Hemisphere overall had a near normal frequency of total tropical cyclone activity during this season.

It appears that, in general, El Niño modulation of tropical cyclone frequency and intensity in non-Atlantic Ocean basins may be less than in the Atlantic. However, strong displacements of tropical cyclone activity occur in the northwest and (especially) in the southwest Pacific basins. Indeed, following Dong (1988) and Revell and Goulter (1986a,b), the total frequency of tropical cyclones in the whole South Pacific area is little affected by El Niño. There is, however, a considerable eastward displacement of the center of South Pacific activity during El Niños, with a similar but less prominent displacement occurring in the northwest Pacific (Hastings, 1990; Dong, 1988; Collimore, 1990a; Chan, 1985).

Tropical cyclone modulation by the QBO

In addition to El Niño, there are other global-scale meteorological factors whose slowly varying properties are also strongly reflected in both seasonal frequency and intensity of hurricanes. One of these global factors is the Quasi-Biennial Oscillation or QBO of stratospheric zonal wind anomalies near the equator. Though presently thought to be largely independent of El Niño, the QBO also influences seasonal hurricane activity and complicates the task of isolating variability due to El Niño events. The basic characteristics of the QBO are illustrated in Fig. 8.10 and the directional mode of the QBO at 50 mb during September for the last 40 years is shown in Fig. 8.11. Atlantic hurricane formation is inhibited when stratospheric QBO winds just above the tropopause are in the easterly anomaly mode (i.e., stronger than normal easterly winds) as shown in Fig. 8.12. Presently, we believe that the circulation configuration shown in Fig. 8.12 causes convective clouds and related effects in the core area of incipient cyclones to be strongly sheared off to the west. In the contrasting (westerly) QBO condition when lower stratospheric zonal winds blow only lightly from the east, vertical wind shear conditions tend to remain weak and inner-core deep convection and related effects are not strongly sheared in the lower stratosphere. The latter, westerly mode conditions, are more favorable for hurricane formation, especially for the development of very intense hurricanes.

The association between the frequency of more intense hurricanes (maximum winds greater than 51 m/s) and the QBO at 50 mb is summarized in Table 8.7. Note that more than twice as many intense hurricanes occur during the westerly phase of the QBO. Of the 101 most intense hurricanes in the 39-year period of 1950–88, 59 (3.5 per year) occurred in seasons when QBO winds were in the westerly mode versus 29 (1.7 per year) in the seasons when they were in the easterly mode. Intermediate (transitional) QBO winds were present for the other 13 (2.2 per year) seasons. The average number of hurricane days during west mode years is also about twice that of east mode years (Gray, 1984a).

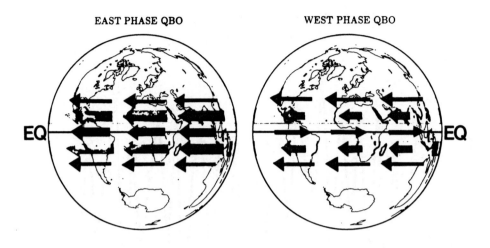

Fig. 8.10 Illustration of the two basic wind conditions for the stratospheric Quasi-Biennial Oscillation (QBO) which occur over the tropics at 50 mb (or 20 km altitude) during the summer seasons of both hemispheres. The left diagram shows conditions during the easterly phase when moderate easterly winds occur on the equator and strong easterly winds occur at 10°N. During the westerly phase (right diagram) stratospheric winds on the equator are from the west and weak easterlies occur at 10°N.

Recently Collimore (1990a) has reported evidence of a sharp reduction in the frequency of intense tropical cyclones in the west Pacific region during the east phase of the QBO. An analysis of all west Pacific cyclones for the past 30–40 years, shown in Table 8.8, suggests that a modest reduction of the frequency of west Pacific tropical cyclones may occur during the west phase of the QBO, especially in the Australian region. This result, which is based on QBO winds observed at Truk (7.5°N, 152°E), is opposite to what is observed in the Atlantic (i.e., Table 8.7). However, when we restrict the west Pacific QBO analysis to only the more intense cyclones, a very different association emerges, as shown in Table 8.9. Note that, in general, northwest Pacific cyclones tend to be considerably stronger than those in the Australia region. For this reason it was necessary to lower the wind speed criteria for intense storms in the latter area. The frequency of intense storms, as defined in Table 8.9, is nearly a factor of two greater in both the northwest and southwest Pacific during the QBO west phase.

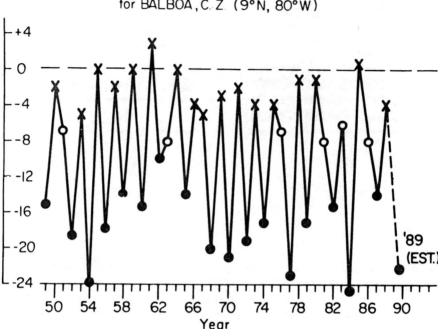

Fig. 8.11 Average September zonal winds (m/s) at 50 mb (20 km) for Balboa,
C.Z. (9°N, 80°W). Heavy dots represent seasons of strong easterly
stratospheric winds (easterly phase), X symbols represent seasons of
weak easterly stratospheric winds (westerly anomaly phase) and open
circles indicate seasons with intermediate easterly winds.

Also, the ratios of the frequency values for the west phase to east
phase are 5.0 and 3.3 for the most intense storm categories in
the northwest Pacific and for the Pacific portion of the Australian
region (i.e., east of 145°E), respectively.

Clearly, the stronger lower stratosphere–upper troposphere
shear of the QBO east phase appears to be a significant factor
governing the development of intense tropical cyclones. The quasiperiodicity of the QBO also allows for reasonably reliable estimates
of the lower stratospheric zonal wind shear in tropical areas well
before the start of the hurricane season. Primarily because of its
comparatively long record, QBO wind data from the measuring
station at Balboa (9°N, 80°W) in the Canal Zone are useful for
this purpose in the Atlantic basin. This station is also fairly representative of the western Atlantic latitude belt between 8° and

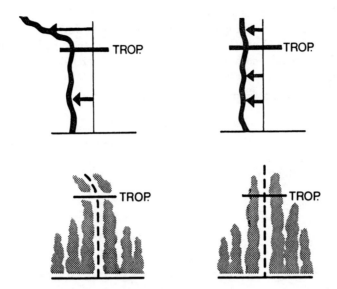

Fig. 8.12 Illustration of the type of vertical wind shear and cloudiness which extend through the tropopause (TROP.) into the tropical stratosphere
in association with developing tropical cyclones. Typical west phase
QBO conditions (diagram on the right) are more conducive to the
formation of intense hurricanes than are the east phase conditions
shown on the left.

Table 8.7 QBO stratification of the 101 most intense Atlantic hurricanes (maximum winds greater than 51 m/s) during the period of 1949–88 based
on September average 50 mb (20 km) winds at Balboa, C.Z. (9°N,
80°W).

Category of easterly wind	Intense hurricanes	Frequency (intense hurricanes per year)
West Phase (17 years)	59	3.5
East Phase (17 years)	29	1.7
Intermediate (6 years)	13	2.2

15°N where most of the intense Atlantic hurricanes form. Hence,
because of the QBO–hurricane association, it is possible to offer
useful probabilistic estimates on the potential for frequent, intense
and long-lived hurricanes during the hurricane season (see Gray,
1984a,b, 1989a,b).

Table 8.8 Total frequency of summer and autumn tropical cyclones of all inten-
         sity classes occurring in the west Pacific area during easterly versus
         westerly QBO phase. The QBO phase was determined from monthly
         50 mb wind anomalies at Truk (7.5°N,152°E) during each season
         (from Collimore, 1990b).

	QBO West phase	QBO East phase	Ratio West phase to East phase
Northwest Pacific, 0-20°N			
June–August	7.8	8.2	0.95
September–November	10.3	10.3	1.00
Australia Region, 0–20°S			
January–March	9.3	11.2	0.83
TOTAL	27.4	29.7	0.92

Conclusions

On the evidence presented above there appears to be little doubt
that seasonal Atlantic hurricane activity during El Niño years is
suppressed in comparison with hurricane activity occurring during non-El Niño years. The role of the El Niño in suppressing
seasonal Atlantic hurricane activity consists primarily of the dis-
ruptive shearing effects of abnormally strong upper tropospheric
westerly winds on developing tropical cloud clusters in the equatorial west Atlantic and Caribbean basin. The El Niño influence on
cyclone frequency is generally less prominent, but still detectable
in many of the other tropical basins. At present, it is believed that
broadscale variations of tropical sea surface temperatures are the
primary El Niño factors affecting tropical cyclone formation and
distribution in these other areas.

   This chapter has illustrated the nature and implications of a
teleconnected El Niño circulation component in the regional problem of seasonal hurricane variability. The linkage of El Niño with
seasonal hurricane activity opens up a new dimension for understanding west Atlantic hurricane variability, especially when the
effects of regional surface pressure anomalies, the QBO, and El
Niño signals are combined. It is interesting to note the excep-
tionally small amount of hurricane activity that occurred during
the years of 1972 and 1983 when an easterly stratospheric QBO

Table 8.9 Intense cyclones stratified by the phase of the 50 mb QBO winds at Truk. Cyclone data for the northwest Pacific and Australia regions (roughly 90°E to 180°E) start in 1952 and 1958, respectively (from Collimore, 1990b).

Cyclones with maximum winds exceeding:	Total number occurring during:			Ratio of West to East phase
	QBO West phase	QBO East phase	Intermediate QBO period	
Northwest Pacific, 0–20°N				
160 knots	15	3	4	5.0
150 knots	31	13	6	2.4
140 knots	45	20	8	2.3
130 knots	63	35	10	1.8
Entire Australia Region, 0–20°S				
100 knots	17	9	0	1.9
80 knots	40	23	2	1.7
65 knots	87	50	8	1.7
Australia Region 0–20°S, East of 145°E				
80 knots	23	7	0	3.3
65 knots	52	21	3	2.5
Australia Region 0–20°S, West of 145°E				
80 knots	17	16	2	1.1
65 knots	35	29	5	1.2

regime occurred in conjunction with a strong El Niño event. Only six hurricane days occurred in 1972 and only five in 1983, whereas 25 hurricane days is the long term average. Similarly, 1987 experienced a moderate El Niño event with easterly QBO winds and only seven hurricane days were observed.

In view of the growing awareness of significant biennial variability of a number of tropospheric phenomena (Angell et al., 1969; Brier, 1978; Trenberth, 1980; Rasmusson et al., 1981; Labitzke and van Loon, 1988, 1989), it should be expected that a QBO-seasonal tropical cyclone activity relationship might be superimposed on the El Niño signal. What is surprising is the large amount of total seasonal hurricane variability associated with these oscillations for the Atlantic region and for intense tropical cyclones, in both the Atlantic and the west Pacific regions.

The El Niño and QBO influences on Atlantic storm frequency are more pronounced than in other ocean basins because the Atlantic is a less robust region for hurricane activity. The typical type of tropical storm development which occurs in association with a monsoon trough does not typically occur in the Atlantic, where hurricane activity can vary from zero (as in 1907 and 1914) or one (as in 1906, 1919, and 1925) to 10 (as in 1933 and 1969) or 11 (as in 1916 and 1950). Variability of total cyclone frequency on this scale further demonstrates that the Atlantic region is more sensitive to modulation by large-scale general circulation influences than other basins and will likely be the place most influenced by both short- and longer-term alterations of the general circulation.

Acknowledgements

Portions of the research described here were supported by the National Science Foundation.

References

Angell, J., Korshover, J. & Cotten, G. (1969). Quasi-biennial variations in the centers of action. *Monthly Weather Review*, **97**, 867–72.

Brier, G.W. (1978). The quasi-biennial oscillation and feedback processes in the atmosphere–ocean–earth system. *Monthly Weather Review*, **106**, 938–46.

Chan, J.C.L. (1985). Tropical cyclone activity in the Northwest Pacific in relation to the El Niño/Southern Oscillation phenomenon. *Monthly Weather Review*, **113**, 599–606.

Collimore, C. (1990a). Long term fluctuations of the west Pacific tropical circulation and how they affect east Pacific SSTs. Department of Atmospheric Sciences Paper. Ft. Collins, CO: Colorado State University.

Collimore, C. (1990b). Interannual variability of west Pacific tropical cyclones. Department of Atmospheric Sciences Paper. Ft Collins, CO: Colorado State University.

Dong, K. (1988). El Niño and tropical cyclone frequency in the Australian region and the northwest Pacific. *Australian Meteorological Magazine*, **36**, 219–26.

Gray, W.M. (1968). Global view of the origin of tropical disturbances and storms. *Monthly Weather Review*, **96**, 55–73.

Gray, W.M. (1975). Tropical cyclone genesis. Department of Atmospheric Sciences Paper No. 234. Ft. Collins, CO: Colorado State University.

Gray, W.M. (1979). Hurricanes: Their formation, structure and likely role in the tropical circulation. Supplement to *Meteorology over the Tropical Oceans*, ed. D.B. Shaw. Bracknell, Berkshire: Royal Meteorological Society.

Gray, W.M. (1984a). Atlantic seasonal hurricane frequency, Part I: El Niño and 30 mb quasi-biennial oscillation influences. *Monthly Weather Review*, **112**, 1649–68.

Gray, W.M. (1984b). Atlantic seasonal hurricane frequency, Part II: Forecasting its variability. *Monthly Weather Review*, **112**, 1669–83.

Gray, W.M. (1988a). Environmental influences on tropical cyclones. *Australian Meteorological Magazine*, **36**, 3, 127–39.

Gray, W.M. (1988b). Forecast of Atlantic seasonal hurricane activity for 1988. *Southern Building* (July–August), 6–23.

Gray, W.M. (1989a). Background information for assessment of expected Atlantic hurricane activity for 1989. Talk presented to the 11th Annual Hurricane Conference, 7 April 1989, Miami, FL.

Gray, W.M. (1989b). Forecast of Atlantic seasonal hurricane activity for 1989 (as of 26 May 1989). Department of Atmospheric Sciences Report. Ft. Collins, CO: Colorado State University.

Hastenrath, S. & Wendland, W.M. (1979). On the secular variation of storms in the North Atlantic and Eastern Pacific oceans. *Tellus*, **31**, 28–38.

Hastings, P.A. (1990). Southern Oscillation influence on tropical cyclone activity in the Australian/South-West Pacific region. *International Journal of Climatology*, **3**, 291–8.

Hebert, P.J. & Taylor, G. (1978). The deadliest, costliest and most intense United States hurricanes of this century. NOAA Technical Memorandum NSW-NHC-7. Miami, FL: National Hurricane Center.

Labitzke, K. & van Loon, H. (1988). Associations between the 11-year solar cycle, the QBO, and the atmosphere, Part I: The troposphere and the stratosphere in the Northern Hemisphere in winter. *Journal of Atmospheric and Terrestrial Physics*, **50**, 197–206.

Labitzke, K. & van Loon, H. (1989). Associations between the 11-year solar cycle, the QBO, and the atmosphere, Part III: Aspects of the association. *Journal of Climate*, **2**, 554–65.

Landsea, C.W. & Gray, W.M. (1989). Eastern north Pacific tropical cyclone climatology – low frequency variations. Presented at the Second International Workshop on Tropical Cyclones, 25 November–7 December 1989, Manila, Philippines.

Nicholls, N. (1979). A possible method for predicting seasonal tropical cyclone activity in the Australian region. *Monthly Weather Review*, **107**, 1221–4.

Nicholls, N. (1984). The Southern Oscillation, sea-surface temperature, and interannual fluctuations in Australian tropical cyclone activity. *Journal of Climatology*, **4**, 661–70.

Nicholls, N. (1985). Predictability of interannual variations of Australian seasonal tropical cyclone activity. *Monthly Weather Review*, **113**, 1143–9.

Quinn, W.H., Zopf, D.O., Short, K.S. & Kuo Yang, R.T.W. (1978). Historical trends and statistics of the Southern Oscillation, El Niño, and Indonesian droughts. *Fishery Bulletin*, **76**, 663–78.

Quinn, W.H., Neal, V.T. & de Mayolo, S.E.A. (1987). El Niño occurrences over the past four and a half centuries. *Journal of Geophysical Research*, **92**, c13:14, 449–61.

Rasmusson, E.M., Arkin, R.A., Chen, W.Y. & Jalickee, J.B. (1981). Biennial variations in surface temperature over the United States as revealed by singular decomposition. *Monthly Weather Review*, **109**, 587–98.

Revell, C.G. & Goulter, S.W. (1986a). Lagged relations between the Southern Oscillation and numbers of tropical cyclones in the south Pacific region. *Monthly Weather Review*, **114**, 2669–70.

Revell, C.G. & Goulter, S.W. (1986b). South Pacific tropical cyclones and the Southern Oscillation. *Monthly Weather Review*, **114**, 1138–45.

Trenberth, K.E. (1980). Atmospheric quasi-biennial oscillations. *Monthly Weather Review*, **108**, 1370–7.

9

The rudimentary theory of atmospheric teleconnections associated with ENSO

JOSEPH J. TRIBBIA

National Center for Atmospheric Research*
Boulder, CO 80307

Introduction

It is almost tautological to state that the primary quality of atmospheric teleconnections which must be theoretically explained is their global extent. That is, the mechanism through which a local disturbance may be extended throughout the troposphere must be explained by theory and corroborated by experiments. With regard to atmospheric teleconnections associated with the El Niño–Southern Oscillation, this requires that the local SST anomalies in the equatorial Pacific be extended by some mechanism vertically to the troposphere and horizontally to extra tropical latitudes.

To illustrate the nature of the teleconnective signal that must be explained, the composite 500 mb anomaly of the Northern Hemispheric geopotential height for an equatorial warm ocean event is shown in Fig. 9.1 (van Loon, 1989, personal communication). The hemispheric extent and large scale of the anomaly pattern is evident, which is to be contrasted with the local equatorially trapped scale of the sea surface temperature anomalies of the composite warm phase (Shea, 1986). This disparity in scale between forcing and response (SST and anomaly patterns) is evident by comparing Fig. 9.1 with Fig. 9.2 from Rasmusson and Carpenter (1982). The ultimate goal of this chapter will be to explain why the atmospheric response to the SST anomalies shown in Fig. 9.2 is Fig. 9.1 using the principles of mechanics and thermodynamics.

* The National Center for Atmospheric Research is sponsored by the National Science Foundation.

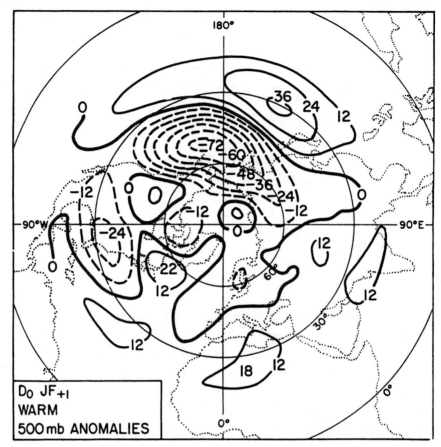

Fig. 9.1 Composite 500 mb height field of January and February following
oceanic El Niño (van Loon, 1989, personal communication).

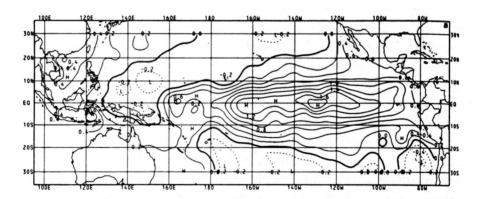

Fig. 9.2 Composite SST anomalies in equatorial Pacific. SST anomalies for
December–February following El Niño (from Rasmusson and Carpenter, 1982).

Southern oscillation and tropical circulations

Prior to discussing the anomalous atmospheric tropical response
to sea surface temperature anomalies induced by oceanic warm
and cold events, it is useful to analyze in simple terms the basic
east–west Walker circulation and north–south Hadley circulation.
Both such circulations are thermally direct in the sense that the
motion is driven by thermal heat input and thus heat energy is con-
verted into kinetic energy of atmospheric motion. This conversion
comes about because as a parcel of air is warmed it becomes buoyant with respect to its environment and rises due to Archimedean
forces. In its acceleration upward the buoyant parcel locally reduces the atmospheric pressure below which accelerates low level
air toward the center of low pressure. At the upper level the buoyant parcel eventually reaches equilibrium and spreads horizontally
outward, eventually cooling radiationally, sinking, and returning
to the surface to replace low level air which has accelerated towards the center of low pressure. Thus, a thermally direct cell like
the Walker circulation can be envisioned as in Fig. 9.3 with low
level convergence and ascent over the center of heating and low
pressure and upper level divergence and compensating downward
motion away from the center of maximum heating.

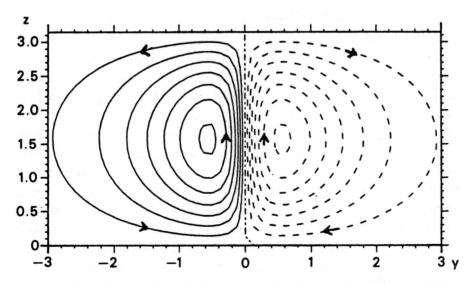

Fig. 9.3 Schematic direct circulation cell induced by latent atmospheric heating.

It is rather easy to see how such a direct circulation cell results from gradients in the surface temperature over the equatorial Pacific. Once upward motion over the warmest sea surface is initiated the convergence and ascent of moist air leads to convection and precipitation which releases the latent heat of vaporization, heating the atmosphere further and driving the circulation cell. As the low level air is accelerated toward lower pressure, it acquires moisture and heat from the warmer ocean surface which provides the fuel to drive the convection over the the warmest surface temperature, i.e., the center of low pressure and convergence.

In the normal state of the equatorial Pacific the sea surface temperatures are warmest in the western Pacific. Due to the presence of the maritime continental complex in this region, the atmosphere is heated near the surface initiating the above chain of events and driving the Walker circulation through the surface exchange of moisture and heat. Note that there is a positive feedback in this exchange since the wind stress implied by the low level easterly winds concentrates the warmest surface water along the western boundary of the Pacific, enhancing the surface temperature gradient (Bjerknes, 1969). During an oceanic warm event the warmest water no longer is in the western Pacific but broadly extends near the central equatorial Pacific, as discussed by Bjerknes and by Trenberth (Chapter 2) and Rassmusson (Chapter 10) in this volume. In this case the convection is less longitudinally localized and the climatological Walker circulation is weakened.

Both of the above atmospheric tropical circulation states (normal and warm event) have been successfully simulated using rather simple mathematical models of the atmosphere. These simple models replace the complicated boundary layer and precipitation processes described above with a specified heat source distributed throughout the interior of the atmosphere which drives a direct circulation. It should be noted that this is a gross simplification and somewhat inconsistent. As described above, the direct circulation is not only the result of atmospheric heating, but also cooperatively contributes to the flow which determines the heat source (cf. Webster (1981) for a self-consistent although more complicated formulation of the convective heating source). Notable computations of the mean zonal (Hadley) and time-averaged (combined Hadley and Walker) tropical circulation using this imposed internal heating simplification have been produced by Dickinson (1971) and Webster (1972), respectively. These studies also used the so-called linear approximation in which atmospheric variables are partitioned into a basic state and deviations from that basic state and the products of deviation are ignored. While such a procedure greatly simplifies the mathematical solution of these theoretical models, this simplification can only be strictly justified a posteriori, so that typically only qualitative information is usefully gleaned from simplified model analysis.

To illustrate the nature of such simplified models and the qualitative information available from such studies, a seminal work of Gill (1980) is partially reproduced in the appendix. The objective of this computation is to elucidate the dynamics of the anomalous tropical direct circulation induced by anomalous heating in the equatorial region. In this study Gill makes several simplifications *ab initio*, the validity of which depend upon the limited vertical and latitudinal extent of the induced circulations, as well as approximations made for mathematical simplicity and tractability. Gill analytically solves a simplified model of the atmosphere subject to localized heating at the equator. The solution obtained by Gill is shown in Fig. 9.4.

An extension of this work by Heckley and Gill (1984) using the same model equations and again with heating centered at the equator allows a more direct comparison with the observed low level flow due to anomalous heating. Despite the extreme simplicity of this model and its drastic approximations, the model reproduces many of the salient features of the observed anomalous atmospheric flow in the tropics during warm events as witnessed by a comparison of Figs. 9.4a and 9.5a with Fig. 9.5b, the surface wind analysis for 15 November 1982. Of particular note among the successes of this model is the prediction of a pair of low level cyclonic flow anomalies to the west of the localized heating source centered on the equator and eastward extension of the low pressure lobe, both in agreement with observed anomalies in the tropical belt. A side view (Fig. 9.4b,c) demonstrates the efficacy of driving the direct circulations (Hadley and Walker circulations) by the use of specified external heating. The simple model of Gill thus allows a mathematically tractable, reasonably realistic representation of the tropical tropospheric response to anomalous SSTs through the assumption that there exists a one-to-one correspondence between

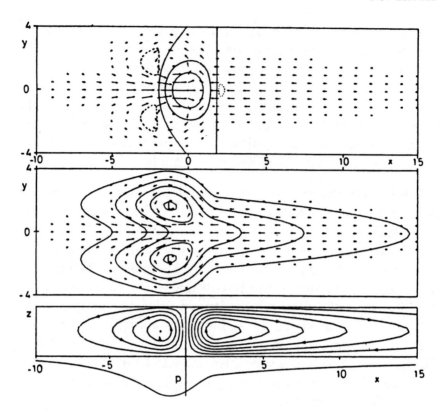

Fig. 9.4 (a) Top view of equatorial response to idealized heat source, vertical
velocity. Solid contours and vectors of low-level horizontal' velocity.
(b) As in Fig. 9.4a, for the low-level pressure and horizontal velocity.
(c) Side view of circulation of Fig. 9.4a in x–z plane after Gill (1980).

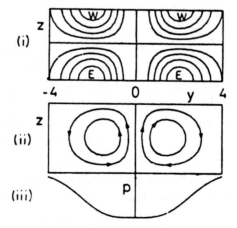

Fig. 9.4 (d) Side view of circulation of Fig. 9.4a in y–z plane after Gill (1980),
zonal wind, and meridional stream function.

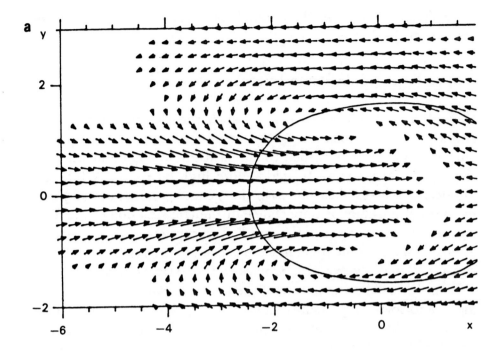

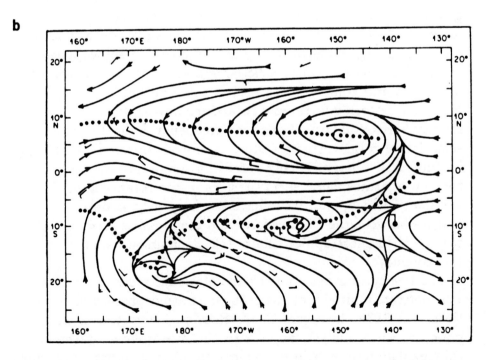

Fig. 9.5 (a) Low level flow associated with heating zone after Heckley and
Gill (1984). (b) Surface wind analysis for 15 November 1982 from
Sadler and Kilonsky (1983).

the SST anomalies and the anomalous internal heating which results from the induced precipitation anomaly.

Upper level midlatitude teleconnections

Having described the manner in which the thermal state of sea surface is communicated to the equatorial upper troposphere, our attention is next turned to the latitudinal communication of the equatorial response to middle and high latitudes. While the physical processes of turbulent exchange and condensational heating were of paramount importance in the transmission of the SST signal from the surface to the middle troposphere, only large-scale atmospheric dynamic processes are needed to propagate the ENSO signal horizontally from the tropics to the extra tropics.

From the oscillating nature of the anomaly pattern (alternating highs and lows of pressure) shown in Fig. 9.1, it might be reasoned that some form of wave phenomenon is responsible for the longitudinal and latitudinal extent of the upper level flow anomalies. It has been known since the late nineteenth century (Margules, 1893) that the atmosphere can sustain two distinct types of wave oscillations which maintain vertical hydrostatic balance indigenous to large-scale, nearly horizontal flow. These two types are gravitational waves, which have large horizontal divergence, high frequency, and compensating pressure fluctuation indicative of a buoyancy restoring force, and rotational waves, which have large relative horizontal vorticity and low frequency, indicative of a conservation of total vorticity restorative force. In middle and high latitudes it is observationally noted that an overwhelmingly large fraction of atmospheric energy resides in the rotational modes of variability, which, in order to have low frequency, maintain a near balance between the Coriolis force and the pressure gradient force, the so-called geostrophic balance (cf. Rasmusson, Chapter 10 in this volume).

The theory of gravitational and rotational modes of variation was motivated initially by the problem of atmospheric and oceanic tides. In 1939, however, C. G. Rossby demonstrated the full power of wave analysis applied to the atmosphere by theoretically explaining the semi-permanent circulation pattern in terms of rotational modes. In this work, Rossby not only explained the semi-permanent features but he also developed a simplified wave equation for atmospheric (rotational) waves which distilled their dynamics to their barest essence. Rossby reduced the dynamics of the most important modes of atmospheric variability to a single equation governing the evolution of atmospheric relative vorticity and isolated the restoring force for rotational modes as the variation of the vertical component of the earth's rotation with latitude. In honor of his elucidation of the dynamics of rotational waves and their general applicability in the study of atmospheric mid-latitude dynamics, these variations are called Rossby waves. The evolution equation which encapsulates their dynamics, using the Cartesian geometry approximation devised by Rossby, is

$$\xi'_t + \overline{U}\xi'_x + \beta\psi'_x = 0 \ , \tag{9.1}$$

in which x is eastward, y is northward, ξ' is the perturbation relative vorticity $(v_x - u_y)$, $\overline{U}$ is a basic state zonal wind field, $\beta = df/dy$, where f is the Coriolis parameter $2\Omega \sin\phi$ with ϕ latitude, $y = a\phi$ with a being the radius of the earth, and ψ' is the perturbation streamfunction for non-divergent flow satisfying $\psi'_{xx} + \psi'_{yy} = \xi'$. By assuming solutions independent of y of the form $\psi = A \sin[k(x - ct)]$, Rossby obtained a now famous relationship between the phase velocity c and the wavenumber $k = 2\pi/L$ where L is the wavelength:

$$c = c_R = \overline{U} - \frac{\beta}{k^2} \ . \tag{9.2}$$

Rossby and his collaborators applied the consequences of Rossby's theoretical development to dynamical meteorological problems for a decade after its introduction.

A particularly relevant (to the explanation of teleconnections) application of Rossby's theory to the problem of energy transmission in the atmosphere was produced by a student of Rossby, T. C. Yeh. Yeh (1949) utilized the concept of wave dispersion; i.e., the fact that Rossby waves of different wavelengths propagate with a phase velocity that is wavelength dependent, to describe the propagation of energy by Rossby waves at a speed which depends upon the differential phase velocity of waves with nearly identical wavelengths, the group velocity. Yeh's study demonstrated the manner in which an isolated initial vorticity disturbance in a westerly background flow develops a distance L toward the east at a time, $t = L/U_g$, where

$$U_g \equiv \frac{\partial}{\partial k}(kc_R) = \overline{U} + \frac{\beta}{k^2} \ . \tag{9.3}$$

In particular, for a stationary wave $c_R(k) = 0$, which is of special importance for a vorticity forcing that is steady in time, the downstream dispersion causes energy to propagate at a speed $U_g = 2\overline{U}$.

Yeh considered only propagation of energy in the x, or longitudinal, direction, using Rossby's planar approximation. A more complete discussion of energy propagation in the atmosphere requires the use of spherical geometry, which was made possible by the extension of Rossby's theory to the sphere along with its connection with tidal theory accomplished by Haurwitz (1940). It would take nearly four decades before the combined effects of dispersion and spherical geometry, which includes the spatial variation of $\beta = df/dy$, were studied by Hoskins et al. (1977), who examined how the latitudinal variation of β caused both the wavelength and direction of energy propagation to change with latitude.

Both Hoskins and collaborators and Haurwitz used the fact that Rossby's wave equation is a linearized version of the vorticity equation for a horizontally non-divergent fluid. The governing equation for such a fluid in spherical geometry, in the absence of forcing and dissipation, can be written as:

$$\xi_t + J(\psi, \xi + 2\Omega\mu) = 0 \ , \tag{9.4}$$

where $J(A, B) \equiv a^{-2}(A_\lambda B_\mu - A_\mu B_\lambda)$, $\mu \equiv \sin\phi$, and ψ is the streamfunction, and as in Eq. (9.1), $\xi = \nabla^2\psi$. If Eq. (9.4) is linearized about a simple mean westerly flow of solid body rotation (i.e., $\psi = \overline{\psi} + \psi', \overline{\psi} \equiv -\overline{\omega}\mu$), the equation for small deviations about this mean flow is:

$$\nabla^2\psi'_t + \overline{\omega}\nabla^2\psi'_\lambda + 2(1 + \overline{\omega})\psi'_\lambda = 0 \ , \tag{9.5}$$

where this equation has been non-dimensionalized using Ω^{-1} as the time scale and a as the length scale. Equation (9.5) can be solved using spherical harmonics which are then the equivalent of Rossby's plane waves in spherical geometry as was shown by Haurwitz. Hoskins et al., however, in addition to solving (9.5) as an initial value problem, examined the response to a steady vorticity source, $S(\lambda, \mu)$, added to the right-hand side. An example of the solution to the vorticity equation obtained by Hoskins et al. is shown in Fig. 9.6a. The mean flow used in this example is the solid-body rotation, as in Eq. (9.5).

In the analysis of the transient set-up of the response to steady forcing in a fully three-dimensional model of the atmosphere, Hoskins and Karoly (1981) used wave–ray theory similar to that developed by Lighthill (1965) and Whitham (1965) to demonstrate that packets of Rossby–Haurwitz waves propagate out of the source region in rays emitted primarily in the northeast and southeast sectors. As the wave groups progress poleward, the rays turn more and more easterly, eventually reaching turning latitudes where the rays are directed toward the east. The rays then turn equatorward, returning to the latitude of their emission (although in general at a different longitude) and then repeating the process for each emitted ray. In the presence of dissipation, the amplitude of secondary and tertiary rays at each progressive equatorial crossing are weaker and eventually ignorable. Hoskins and Karoly show further that ray paths in a basic state of solid body rotation are great circle paths on the surface of the spherical earth, which gives the characteristic arching path of northeasterly emission and southeasterly reflection near the pole, evident in Fig. 9.6a.

If one linearizes around a more realistic basic state mean flow with easterly winds in the tropics, the solution depicted in Fig. 9.6a is altered in such a way that it becomes hemispherically localized as in Fig. 9.6b. The reason for this is that the rays emitted by a stationary Rossby wave source cannot penetrate a region of easterly mean flow, nor are they reflected there. A region of easterly mean flow absorbs Rossby wave rays incident upon it analogous to the way in which a black surface absorbs incident light. This absorption of Rossby wave rays by easterly winds implies that a strong variation in the mid-latitudinal response to Rossby wave sources should be apparent as source regions move from region of mean easterly wind to region of mean westerly flow. The seasonality of the mid-latitude response noted by Rasmusson in Chapter 10 in this volume may be explained by the sensitivity to the location of an equatorial vorticity source relative to the seasonally varying mean easterly winds. In the Northern Hemisphere (NH) winter, the zonal mean easterlies extend only to 5°N at upper levels, while in the NH summer the easterlies extend nearly to 30°N.

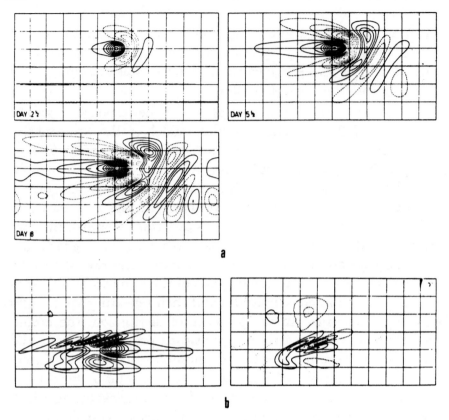

Fig. 9.6 (a) Solution to linearized non-divergent model of the atmosphere for
a steady vorticity source after Hoskins et al. (1977) for solid-body
rotation basic state. (b) As in Fig. 9.6a but for realistic basic state.

Thus, because of the absorptive property of easterly winds noted
above, little NH/SH mid-latitude teleconnection response might
be expected during the NH/SH summer while little absorption
and a larger response in the NH/SH winter.

In addition to the aforementioned ray studies, Hoskins and
Karoly examined the response to thermal forcing in a longitudinally invariant realistic basic state flow, using the hydrostatic
equations of motion in spherical geometry. The immediate application of the Hoskins and Karoly work was to the problem of atmospheric teleconnections in the Atlantic because of the strong struc-
tural similarity in the pattern depicted in the Hoskins and Karoly
and Hoskins et al. works and the western Atlantic teleconnection
patterns described by Wallace and Gutzler (1981; cf. Fig. 9.6c).
However, the Rossby wave propagation studies of Hoskins and
collaborators, together with the analysis of Gill described above,

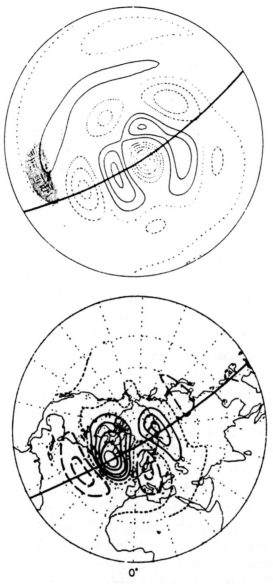

Fig. 9.6 (c) Comparison of results of linear model of Hoskins and Karoly
(1981) with observed western Atlantic teleconnection patterns of
Wallace and Gutzler (1981).

also provide a simple theoretical explanation of the ENSO teleconnection signal shown in Fig. 9.1: the anomalous sea-surface
temperature in the equatorial region (e.g., warm event) induces a
direct circulation anomaly with associated low level convergence
and upper level divergence. When the upper level divergence, δ, is
incorporated into the vorticity equation (derived in the appendix),

$$\xi_t + J(\psi, \xi + f) =$$

$$\underbrace{-(\xi + f)\delta}_{S_1} \underbrace{- \nabla \chi \cdot \nabla (\xi + f)}_{S_2} \underbrace{+ w_y v_z - w_x v_z - w\xi_z}_{S_3} = S ,$$

$$(9.6)$$

where χ is the velocity potential satisfying $\nabla^2 \chi = \delta$, and using the approximation $S \simeq S_1$, it acts as an effective source of vorticity at upper levels. To a first approximation, this source sets off a train of dispersing Rossby waves which become a stationary wave pattern like that shown in Fig. 9.6b.

More recent developments

Recently, work has centered on explaining more fully the dynamics of teleconnection patterns, particularly those aspects which are inconsistent with this simple Rossby wave propagation picture. The first inconsistency addressed was one noted by Geisler et al., (1985) in a follow-up of previous work (Blackmon et al., 1983) demonstrating the efficacy of imposed sea surface temperature anomalies in producing realistic teleconnection patterns which could be detected in 90-day averages of climate integrations using the climate model of the National Center for Atmospheric Research (NCAR). This earlier work used SST anomalies representative of the composite El Niño and reproduced a response in good agreement with the composite mid-latitude response and evidently producing wavetrains of dispersing stationary Rossby waves (cf. Figs. 9.7a,b). In the later work, the influence of the longitudinal location of the equatorial SST anomaly on the response pattern was investigated. Surprisingly, and at odds with the predictions of the Gill–Hoskins simple teleconnection theory, the response showed little sensitivity to the position of the anomalous SST. (Note that the Gill–Hoskins theory would predict that the response should move in conjunction with the vorticity source as determined the anomalous heating induced by the SST anomaly.)

This lack of sensitivity to the longitudinal location of the vorticity source was also noted by Simmons et al. (1983) in idealized experiments designed to elucidate the role of tropical vorticity sources in stimulating teleconnection patterns. They noted that, in contrast to experiments in which the background flow was

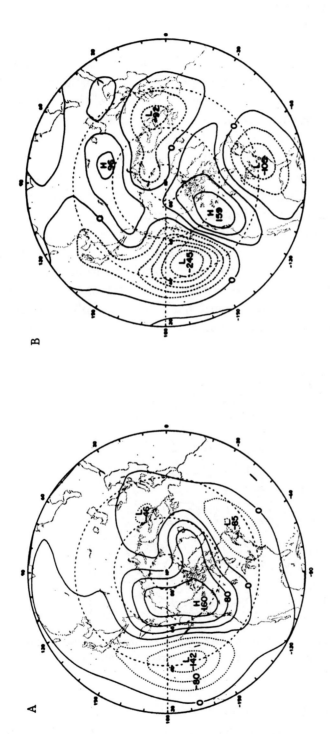

Fig. 9.7 (a) Mean response of three-dimensional nonlinear general circulation model to twice SST anomaly of Fig. 9.2 from Blackmon et al. (1983). (b) Composite observed 500 mb height anomaly associated with a high Pacific North America (PNA) index of Wallace and Gutzler (1981).

zonally symmetric, the response to a stationary vorticity source showed almost no sensitivity to the source position in the Pacific if the background flow was the zonally asymmetric climate mean flow for January. In the course of their analysis, Simmons et al. discovered the reason that the response to vorticity forcing in the Pacific was similar in each experiment, in spite of the dissimilarity in forcing, was that the mean January upper level flow used as a basic state was dynamically unstable. This implies that – within the context of linear models – disturbances superimposed on this basic state flow, such as those produced by vorticity sources, can amplify by extracting energy from the mean flow. The coherent structure that is most efficient in extracting energy from the basic state flow (the most unstable normal mode of the linearized model) is structurally nearly identical with the teleconnection patterns observed by Simmons et al. in their experiments and has been called the SWB mode in deference to the discoverers of this instability (cf. Fig. 9.8). This unstable mode has a period of phase repetition of about 50 days and an amplitude doubling time of about seven days.

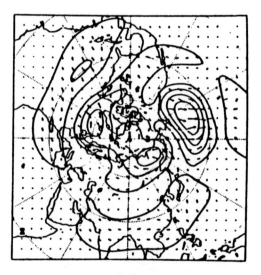

Fig. 9.8 Most unstable mode of barotropic flow linearized about climatological January mean flow. From Simmons et al. (1983), in terms of amplitude and phase.

Simmons et al. further showed that the SWB mode extracts energy only in specific local areas and only in certain fractions of its cyclic period when the amplitude is strongest on the east coasts of the NH continents. At other times and locales this mode behaves dynamically like a packet (or group) of stable free Rossby waves. This study then reconciles the Rossby wave aspects of teleconnections with the insensitivity to excitation locale, since almost any disturbance will eventually organize itself into this most unstable coherent pattern, the SWB mode.

Further analysis of the excitation of the SWB mode revealed some additional complications. Branstator (1985) examined the problem of determining the most efficient structure for exciting the SWB pattern and found, not surprisingly, that the most efficient forcing pattern (vorticity source–sink pattern) has largest amplitude off the east coasts of the continents where the SWB mode extracts its energy the mean flow. This is also a region of relatively strong wind velocity and shear. However, many SST anomalies associated with ENSO result in maxima and minima of SST which lie in the mid and eastern Pacific and the vorticity source computed from S_1 has almost no amplitude in the crucial energy extracting region for the SWB mode. This minor inconsistency was reconciled by Sardeshmukh and Hoskins (1985, 1988), who noted that while a comparison of the order of magnitude of the various terms in the full vorticity equation for anomalies from the climate mean, ξ_a, results in the leading order source term being $(\xi_c + f)\,\delta_a$ (see the appendix to this chapter), the convergence of the transport of mean vorticity, $(\xi_c + f)$, by the divergent wind induced by the SST anomaly, S_2, is non-negligible. This transport, because of the locally strong gradients of vorticity off the east coasts of the continents, can lead to large stimuli of the SWB pattern where the vortex stretching source term, $(\xi_c + f)\,\delta_a$, is negligibly small due to the remoteness of the SST from this region.

With the above emendations of the Gill–Hoskins picture of the mechanisms involved in producing teleconnection patterns, an updated theoretical explanation of these patterns would be:

(1) Anomalously warm (cold) ocean surface temperature leads to anomalous low level convergence (divergence) enhancing (reducing) precipitation.

(2) Enhanced (reduced) precipitation increases (decreases) mid level latent heat release inducing anomalous upper level divergent outflow (convergent inflow).

(3) Vorticity transport due to anomalous upper level outflow (in-flow) excites SWB instability in East Asian Jet region, ex-tracting energy from the climatological mean flow.

(4) Disturbance moves away from the region of energy exchange as a train of dispersing nearly stationary Rossby waves.

Postscript

The above outlines the current state of the theoretical understand-ing of the mechanisms underpinning atmospheric teleconnections excited by ENSO. As with most theoretical pictures there are re-maining problems which must be addressed further. Several such aspects which are under current study are noted in this postscript. First is the problem of the vertical mismatch between the divergent flow excited by the enhanced convection (2) and the extraction of energy from the mean flow (3) by the SWB instability. The di-vergent flow, and thus the vortex stretching source of vorticity, is essentially two-signed (+ −; see Fig. 9.9) in the vertical while the SWB energy conversion is one-signed, involving the only vertically averaged flow in its response. This problem has recently received attention (Kasahara and Silva Dias, 1986) and it is clear from that study that more subtle effects like the vertical shear of the mean wind must be included to complete the link implied in (2) and (3).

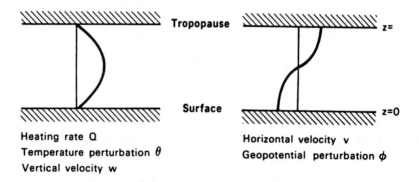

Heating rate Q
Temperature perturbation θ
Vertical velocity w

Horizontal velocity v
Geopotential perturbation ϕ

Fig. 9.9 (a) Vertical structure of heating rate Q, temperature θ, and vertical velocity w, in the model by Gill (1980). (b) The vertical structure of horizontal velocity V and geopotential ϕ in the model by Gill (1980).

A second question that arises in the above depiction of telecon-nection excitation relates to the importance of equatorial Pacific

SST anomalies in this process. Because an instability has been incorporated into the dynamics, divergent outflow due to precipitation anomalies should be considered one of a myriad of possible mechanisms stimulating a self-excited teleconnection pattern. Thus, while it is likely that a sufficiently large SST anomaly will excite a teleconnection pattern of a specific amplitude and phase, at any given time there will also be considerable variability in the SWB mode due to other mechanisms of excitation. Such considerations have practical ramifications for the importance of ENSO in extended range (month to a season in advance) forecasting. These issues are discussed in a recent study by Tribbia (1988).

Yet another complication has arisen in a study of Held and collaborators (1989) in which the SST anomaly associated with the composite difference between warm and cold events apparently stimulates a teleconnection pattern only indirectly. In the mechanism diagnosed in this study using linear and nonlinear versions of the Geophysical Fluid Dynamics Laboratory climate model, the SST anomaly driven heating rearranges the high frequency transient eddies (i.e., the daily weather producing high and low pressure disturbances) which in turn stimulate a planetary Rossby wave train teleconnection pattern. As noted by Held et al. (1989), this is not thought to be the typical circumstance in individual events, but the degree to which such mechanisms are instrumental in any particular ENSO event is as yet undetermined.

Lastly, it should be noted that while in general the mid-latitude teleconnection response in the NH during the NH summer is small, occasionally large responses are possible if the SST anomaly pattern is favorable. Such a favorable pattern, apparently, was instrumental in the stimulation of the anomaly pattern of June 1988 over North America, leading to a persistent drought in the middle of the continent as shown by Trenberth et al. (1988).

Appendix

This appendix serves to introduce the reader to the types of approximations made in simplified models of the atmospheric response to SST anomalies. The starting point is the Boussinesq hydrostatic system used by Gill (1980). A discussion of the dynamical approximations made in the use of this system can be found in Hoskins and Bretherton (1972).

The governing equations of this system are:

$$\frac{d\mathbf{V}}{dt} + f\mathbf{k} \times \mathbf{V} = -\nabla\Phi + \mathbf{F} \qquad \text{horizontal Newtonian force balance;}$$

$$\Phi_z = \frac{g}{\Theta_o}\theta \qquad \text{vertical (hydrostatic) force balance;}$$

(9.A1)

$$\frac{d\Phi_z}{dt} = Q \qquad \text{conservation of thermodynamic energy}$$

$$\nabla \cdot \mathbf{V} + w_z = 0 \qquad \text{fluid continuity}$$

In the above f, $\mathbf{V} = (u,v)$ is the vector of horizontal wind, Φ is the geopotential, θ is the potential temperature, w is the vertical velocity, Θ_0 is a reference potential temperature, $d/dt = \partial/\partial t + u\partial/\partial x + v\partial/\partial y + w\partial/\partial z$ is the total time derivative following a parcel of air, $\nabla = \mathbf{i}\partial/\partial x + \mathbf{j}\partial/\partial y$ is the horizontal gradient operator, $f = 2\Omega\sin\phi$ is the Coriolis parameter, ϕ is latitude, Ω is the rotation rate of the earth, $\mathbf{i}$, $\mathbf{j}$ and $\mathbf{k}$ are the unit vectors in the eastward, northward and vertical direction, respectively, and Q and $\mathbf{F}$ are the vector of frictional forces and the diabatic heating, respectively. Gill then makes the following approximations alluded to in the text.

(1) The system is specialized to the equatorial region by replacing $f = 2\Omega\sin\phi$ by $2\Omega\phi$ and defining $y = a\phi$ and $\beta = 2\Omega/a$ so that $f = \beta y$, a being the equatorial radius of the earth.

(2) A resting stratified basic state is assumed in which $(u,v,w) = (\bar{u},\bar{v},\bar{w}) = (0,0,0)$ and $\Phi = \bar{\Phi} = N^2 z^2/2$, with $N^2 > 0$ where N is the Brunt–Vaisäla frequency.

(3) The system is linearized about this basic state so that all dependent variables are written as the sum of the basic state variable and a deviation (e.g., $\Phi = \bar{\Phi} + \Phi'$, where $^-$ denotes a basic state quantity and $'$ denotes its deviation from the basic state, and $\Phi' = \Phi - \bar{\Phi}$). Linearization further implies that

products of deviation quantities are neglected in the equations of motion. Thus the governing system used by Gill becomes

$$\frac{\partial \mathbf{V'}}{\partial t} + \beta y \mathbf{k} \times \mathbf{V'} = -\nabla \Phi' + \mathbf{F'} ,$$

$$\Phi'_z = \frac{g}{\Theta_o} \Theta' ,$$

$$\frac{\partial \Phi'_z}{\partial t} + N^2 w' = Q' , \qquad (9.\mathrm{A}2)$$

$$\nabla \cdot \mathbf{V'} + w'_z = 0 .$$

(4) $Q' = h(x,y) \cdot \sin(z/H) - \epsilon \Phi'_z$, $\mathbf{F'} = -\epsilon \mathbf{V'}$ and only the steady state response is sought. The vertical profile of the heating and w' is shown in Figure 9.9a, while the form of u', v', ϕ' is shown in Figure 9.9b. The equations are then non-dimensionalized by Gill using H as the vertical length scale and $(HN/2\pi\beta)^{1/2}$ as the horizontal length scale. $h(x,y)$ in terms of the non-dimensional (x,y) coordinates is given by

$$h(x,y) = \begin{cases} e^{-1/4y^2} \cos kx & |x| < L \\ 0 & |x| > L \end{cases} \quad \text{where } k = 2\pi/L. \text{ Gill is}$$

then able to solve the system analytically in terms of parabolic cylinder functions in y and sines and cosines in x.

Rossby's wave equation can also be motivated by the original equation 9.A1. If one forms an equation governing relative vorticity $\xi = v_x - u_y$ by differentiating the y-component of the momentum equations by x and the x-component of the momentum equation by y and subtracting the result is

$$\frac{d\xi}{dt} + \beta v = -(\xi + f)\nabla \cdot \mathbf{V} + w_y u_z - w_x v_z ,$$

where $\beta = df/dy$. It can be shown that any horizontal vector field can be uniquely decomposed into a non-divergent component and an irrotational component. For the horizontal velocity vector this gives $\mathbf{V} = \mathbf{V}_\psi + \mathbf{V}_\chi$ where $\mathbf{V}_\psi$ is non-divergent and $\mathbf{V}_\chi$ is irrotational. It can also be shown that $\mathbf{V}_\psi = \mathbf{k} \times \nabla \psi$ and $\mathbf{V}_\chi = \nabla \chi$ where $\nabla^2 \psi = \xi$ and $\nabla^2 \chi = \nabla \cdot \mathbf{V} \equiv \delta$. Rossby noted that $\mathbf{V} \simeq \mathbf{V}_\psi$ and thus neglected all the terms in the vorticity equation which depended on δ, χ and w, and arrived at

$$\xi_t + J(\psi,\xi) + \beta \psi_x = -(\xi + f)\delta - \nabla \chi \cdot \nabla(\xi + f)$$
$$+ w_y u_z - w_x v_z - w\xi_z \simeq 0 , \qquad (9.\mathrm{A}3)$$

where $J(A, B) = A_x B_y - A_y B_x$. Linearizing this equation about $\bar{\psi} = -\bar{U}y$ and $\bar{\xi} = 0$ Rossby's equation is obtained:

$$\xi'_t + \bar{U}\xi'_x + \beta\psi'_x = 0 . \tag{9.A4}$$

Lastly, the linearized equation for anomalous relative vorticity, $\xi_a = \xi - \xi_c$ where ξ_c is the climate mean relative vorticity is given by:

$$\begin{aligned}
\xi_{at} &+ J(\psi_c, \xi_a) + J(\psi_a, \xi_c + f) = -(\xi_c + f)\delta_a \\
&-\nabla\chi_a \cdot \nabla(\xi_c + f) - (\xi_a\delta_c + \nabla\chi \cdot \nabla\xi_a) - w_a\xi_{cz} \\
&+w_{ay}u_{cz} - w_{ax}v_{cz} + w_{cy}u_{az} - w_{cx}v_{az} - w_c\xi_{az}
\end{aligned} \tag{9.A5}$$

in which the subscripts a and c refer to anomalous and climatological fields, respectively, and the terms on the right-hand side have been ordered approximately in decreasing order of magnitude and importance.

References

Bjerknes, J. (1969). Atmospheric teleconnections from the equatorial Pacific. *Monthly Weather Review*, **97**, 163–72.

Blackmon, M.L., Geisler, J.E. & Pitcher, E.J. (1983). A general circulation model study of January climate anomaly associated with interannual variation of equatorial Pacific sea surface temperatures. *Journal of the Atmospheric Sciences*, **40**, 1410–25.

Branstator, G. (1985). Analysis of general circulation model sea-surface temperature anomalies using a linear model II eigenanalysis. *Journal of the Atmospheric Sciences*, **42**, 2242–54.

Dickinson, R.E. (1971). Analytic model for zonal winds in the tropics. Part I. *Monthly Weather Review*, **99**, 501–10.

Geisler, J., Blackmon, M.L., Bates, G.T. & Muñoz, S. (1985). Sensitivity of January climate response to the magnitude and position of equatorial Pacific sea surface temperature anomalies. *Journal of the Atmospheric Sciences*, **42**, 1037–49.

Gill, A E. (1980). Some simple solutions for heat-induced tropical circulations. *Quarterly Journal of the Royal Meteorological Society*, **106**, 447–62.

Haurwitz, B. (1940). The motion of atmospheric disturbances on the spherical earth. *Journal of Marine Research*, **3**, 254–67.

Heckely, W.A. & Gill, A.E. (1984). Some simple analytic solutions to the problem of forced equatorial long waves. *Quarterly Journal of the Royal Meteorological Society*, **110**, 203–17.

Held, I.M., Lyons, S.W. & Nigam, S. (1989). Transients and the extratropical response to El Niño. *Journal of the Atmospheric Sciences*, **46**, 163–74.

Hoskins, B. & Bretherton, F.P. (1972). Atmospheric frontogenesis models: Mathematical formulation and solutions. *Journal of the Atmospheric Sciences*, **29**, 11–39.

Hoskins, B., Simmons, A.J. & Andrew, D.G. (1977). Energy dispersion in a barotropic spherical atmosphere. *Quarterly Journal of the Royal Meteorological Society*, **103**(438), 553–67.

Hoskins, B. & Karoly, D. (1981). The steady linear response of a spherical atmosphere to thermal and orographic forcing. *Journal of the Atmospheric Sciences*, **38**, 1179–96.

Kasahara, A. & Silva Dias, P. (1986). Response of planetary waves to stationary tropical heating in a global atmosphere with meridional and vertical shear. *Journal of the Atmospheric Sciences*, **43**, 1893–1911.

Lighthill, J. (1965). Group velocity. *Journal. Institute of Mathematics and its Application*, **1**, 1–28.

Margules, M. (1893). Luftbewegung in einer rotieranden Sphäroidschale. *Akademie der Wissenschaften in Wein der Mathematisch-Naturwissenschaftlichen Klasse*, **102**, Part IIA, 11–56.

Rasmusson, E.R. & Carpenter, T.M. (1982). Variations in tropical sea surface temperature and surface wind field associated with the Southern Oscillation/El Niño. *Monthly Weather Review*, **110**, 354–84.

Rossby, C.G. and collaborators (1939). Relation between variations in the intensity of the zonal circulation of the atmosphere and the displacements of the semi-permanent centers of action. *Journal of Marine Research*, **2**, 38–55.

Sardeshmukh, P.D. & Hoskins, B.J. (1988). The generation of global rotational flow by idealized tropical divergence. *Journal of the Atmospheric Sciences*, **45**, 1228–51.

Sardeshmukh, P. D. & Hoskins, B.J. (1985). Vorticity balances in the tropics during the 1982–83 El-Niño Southern Oscillation Event. *Quarterly Journal of the Royal Meteorological Society*, **111**(406), 261–78.

Sadler, J C. & Kilonsky, B.J. (1983). Meteorological events in the central Pacific associated with the 1982–1983 El Niño. *Tropical Ocean-Atmosphere Newsletter*, **21**, 3–5.

Shea, D.J. (1986). *Climatological atlas 1950–79 surface air temperature, precipitation sea level pressure and sea-surface temperature (45° S–90° N)*. NCAR Technical Note TN/–269+STR.

Simmons, A., Wallace, J.M. & Branstator, G.W. (1983). Barotropic wave propagations and instability and atmospheric teleconnective patterns. *Journal of the Atmospheric Sciences*, **40**, 1363–92.

Trenberth, K.E., Branstator, G.W. & Arkin, P.A. (1988). Origins of the 1988 North American Drought. *Science*, **242**, 1640–5.

Tribbia, J.J. (1988). The predictability of monthly mean teleconnection patterns. *Pontifica Academia Scientiarum Acta*, **69**, 567–92.

Wallace, J.M. & Gutzler, D. (1981). Teleconnections in the geopotential height field during the Northern Hemisphere winter. *Monthly Weather Review*, **109**, 784–812.

Webster, P.J. (1972). Response of the tropical atmosphere to local steady forcing. *Monthly Weather Review*, **100**, 518–41.

Wester, P.J. (1981). Mechanisms determining the atmospheric response to sea
    surface temperature anomalies. *Journal of the Atmospheric Sciences*, **38**,
    554–71.
Whitham, G.B. (1965). A general approach to linear and non-linear dispersion
    waves. *Journal of Fluid Mechanics*, **22**, 273–83.
Yeh, T. (1949). On energy dispersion in the atmosphere. *Journal of Meteorol-
    ogy*, **6**, 1–16.

10

Observational aspects of ENSO cycle teleconnections

EUGENE M. RASMUSSON
Cooperative Institute for Climate Studies
Department of Meteorology
University of Maryland
College Park, MD 20742

Introduction

A vast body of literature now exists on the observational and theoretical aspects of ENSO teleconnections. This chapter provides a mainly descriptive review and focuses on the general characteristics of teleconnection patterns, both tropical and extratropical, during opposite phases of the ENSO cycle. It is hoped that this general, "broad brush" treatment of ENSO cycle teleconnections will serve as a useful framework within which to view the more detailed regional descriptions found in other chapters of this volume. Fundamental questions dealing with ocean circulation and coupled ocean–atmosphere interactions have been addressed by Trenberth (Chapter 2, this volume).

The term "teleconnection," as used in this chapter, refers to a pattern of significant simultaneous correlations between seasonally averaged variations in meteorological parameters at widely separated "centers of action." Nearly two centuries before the term came into general use, a missionary from Greenland named Saabye noted in his diary the tendency for temperatures in northern Europe and Greenland to vary in opposite directions. We now know this temperature seesaw to be part of the regional North Atlantic Oscillation (van Loon and Rodgers, 1978).

The first evidence of global teleconnections emerged from analyses of surface pressure data available at the end of the nineteenth century (e.g., Hildebrandsson, 1897; Lockyer and Lockyer, 1904). Approximately two decades later, Sir Gilbert Walker (1924) identified three large atmospheric oscillations: two regional "Northern

Oscillations," the North Atlantic Oscillation and the North Pacific Oscillation, and a global-scale "Southern Oscillation" (SO), whose primary centers of action are in the tropical Southern Hemisphere. In a later paper (Walker and Bliss, 1932), Walker characterized the primary SO teleconnection in the following terms:

When pressure is high in the Pacific Ocean it tends to be low in the Indian Ocean from Africa to Australia; these conditions are associated with low temperatures in both these areas, and rainfall varies in the opposite direction to pressure. Conditions are related differently in winter and summer, and it is therefore necessary to examine separately the seasons of December through February and June through August.

The last sentence is often omitted, which is unfortunate, for it calls attention to a fundamental aspect of teleconnections, i.e., their seasonality.

Walker's conclusions were derived solely from correlations of surface pressure, temperature and precipitation between widely spaced observing stations. With no a priori hypotheses or theoretical support, his statistical results were understandably controversial. As recently as the 1960s, the SO was still largely dismissed as a climate curiosity. Then, Jacob Bjerknes (1969) put physical flesh on the bare statistical bones of Walker's SO. In a remarkable synthesis of diverse lines of research, he developed a conceptual framework, supported by plausible dynamic and thermodynamic reasoning, which linked basin-scale sea surface temperature (SST) variations in the tropical Pacific to Walker's SO and associated wintertime extratropical teleconnections in the Northern Hemisphere. In so doing, he clearly identified, for the first time, the coupled ocean–atmosphere rhythm that dominates interannual climate variability over much of the earth.

The primary SO teleconnection described by Walker is in the tropical Pacific–Indian Ocean sector. These changes in atmospheric heating give rise, in turn, to weaker but nevertheless important teleconnections that extend around the entire tropical belt and deep into the extratropics. The multi-year time scale on which this remarkably coherent global ensemble of teleconnections oscillates is now referred to as the El Niño/Southern Oscillation (ENSO) cycle. It has become common practice to use the Tahiti minus Darwin surface pressure anomaly difference as an index of the phase of this oscillation (Trenberth, Chapter 2, this volume, Fig. 2.1). Walker's previously quoted description reflects the ENSO "high index" (cold) state, but it is the opposite phase that has received most attention. These low index "warm episodes" are marked by above normal precipitation and sea surface temperatures (SST) in the central and eastern equatorial Pacific, and are usually associated with El Niño conditions in coastal regions of northern Peru and southern Ecuador.

The evolution of the ENSO cycle in the equatorial belt during the very active and relatively well observed decade of the 1980s is described in the following section. This multi-year "case study" serves as a valuable benchmark for relating equatorial belt anomalies to global teleconnection patterns. Salient steps in the development of our present conceptual framework for viewing ENSO cycle atmospheric teleconnections are also reviewed below.

Walker's early descriptive studies implied three pervasive features of ENSO teleconnections: global domain, seasonality (i.e., their links to the annual cycle) and linearity, (i.e., the tendency for anomaly patterns of reverse sign to appear during opposite phases of the ENSO Cycle). These general aspects of the ENSO cycle are re-examined in the context of the conceptual framework outlined in the previous section. Concluding remarks are contained in the last section.

Equatorial belt evolution: 1981–88

Because of significant improvements in data assimilation and analysis techniques, as well as some improvement in observations, e.g., satellite-observed outgoing longwave radiation (OLR) and SST, and a modest increase in in situ surface marine data, the description of interannual variability in the tropics improved dramatically during the 1980s. Fortunately, this decade has also been characterized by a series of well developed ENSO oscillations and teleconnection patterns, which serve as case studies which aid in the interpretation of teleconnection statistics derived from the historical database.

Equatorial time-longitude sections or "Hovmoeller diagrams" for SST, OLR (precipitation), lower troposphere (850 mb) and upper troposphere (200 mb) zonal wind anomalies are shown as

Figs. 10.1–10.5 for the period 1981–88. These sections are derived from the Climate Diagnostics Data Base (CDDB) and supporting climatological summaries of the NOAA Climate Analysis Center (CAC). The CDDB summaries were, in turn, derived from operational data and analyses acquired or prepared by the CAC, the U.S. National Meteorological Center (NMC), and the U.S. National Environmental Satellite, Data, and Information Service (NESDIS).

Sea surface temperature

The most striking feature of the SST section (Fig. 10.1) is the contrast between the relatively high SSTs (> 28°C) over the Indian Ocean–west Pacific sector and the lower values east of the dateline in the Pacific. SSTs in the western Pacific are among the highest found in the large ocean basins, while the lowest SSTs observed at low latitudes occur in the eastern equatorial Pacific upwelling region and along the Peru coast. This results in a pronounced east–west SST gradient across the eastern and central equatorial Pacific.

Major ENSO swings are associated with longitudinal extensions and contractions of the western Pacific warm water pool. The positive SST anomalies associated with the 1982–83 ENSO warming developed over the Indonesian–far western Pacific sector during 1981 and reached maximum amplitude over the central and eastern Pacific during 1982 and early 1983 (Fig. 10.2). The characteristics of this Pacific warming have been extensively documented, e.g., Rasmusson and Wallace (1983), Cane (1983), Rasmusson and Arkin (1985). Its basin-scale evolution was generally similar to the evolution of SST and zonal wind composites discussed by Rasmusson and Carpenter (1982), although the episode was unusually intense. The most significant differences were in the eastern Pacific, where the primary regional El Niño warming occurred during later rather than early stages of the episode. Consequently, SST anomalies appear to migrate eastward rather than westward as they did on the Rasmusson/Carpenter composite (Rasmusson and Carpenter, 1982). However, the migration of basin-scale wind and precipitation anomalies are more closely related to the movement of the SST isotherms, and these migrated eastward in both cases.

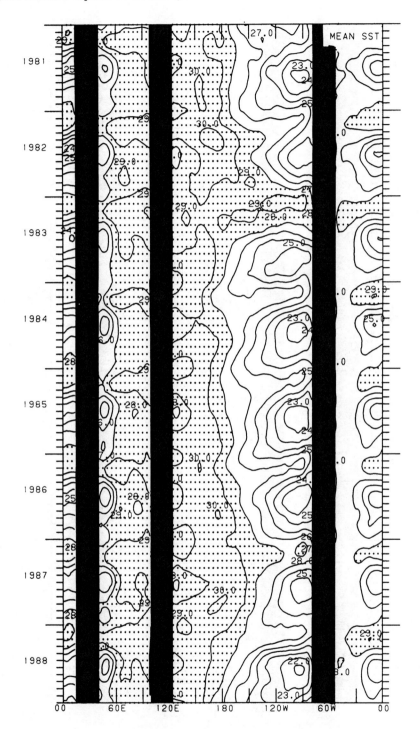

Fig. 10.1 Hovmoeller diagram (time–longitude section) of sea surface temperature (SST) averaged over the latitude belt 5°N–5°S. Contour interval: 1.0°C. Shading indicates SST greater than 28°C.

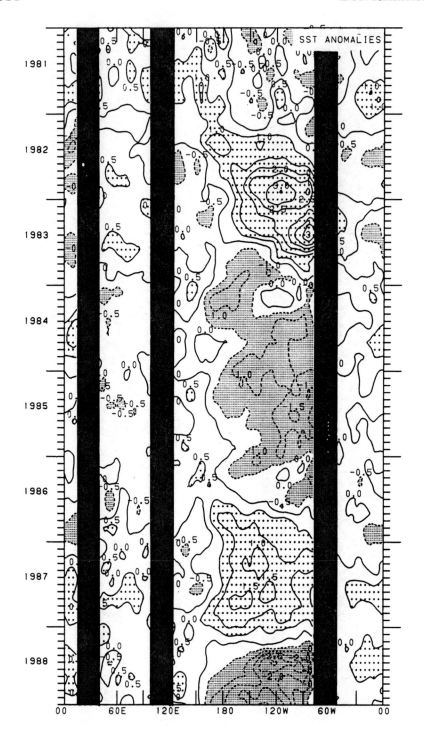

Fig. 10.2 Same as Fig. 10.1 but for SST anomalies. Contour interval: 0.5°C.
Coarse shading indicates anomalies greater than 0.5°C. Fine stippling
indicates values less than −0.5°C.

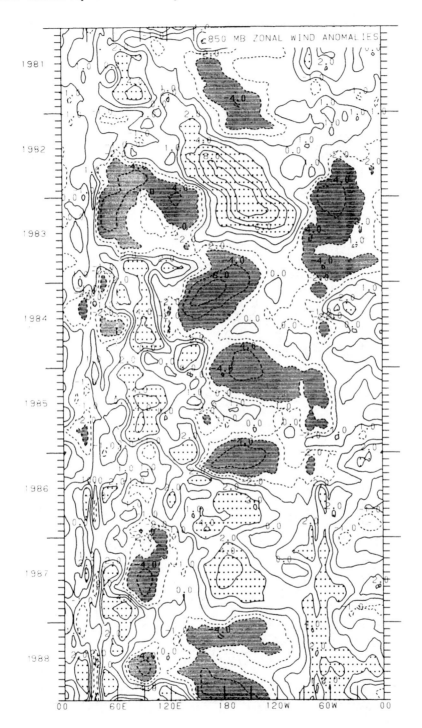

Fig. 10.3 Same as Fig. 10.1 except 850 mb zonal wind anomalies. Finely stippled
          areas indicate anomalies greater than 2 m/s (from the west); coarse
          shaded areas (dark) anomalies less than –2 m/s.

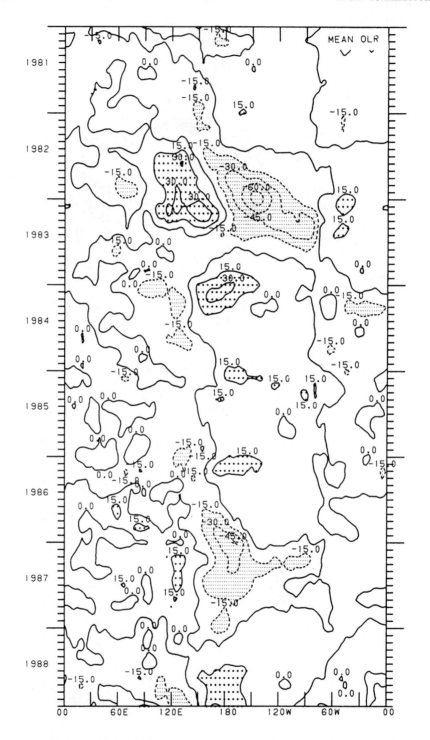

Fig. 10.4 Same as Fig. 10.1 except outgoing longwave radiation (OLR) anomalies. Finely stippled areas indicate anomalies greater than 10 W/m².
Coarse shaded areas indicated anomalies less than –10 W/m².

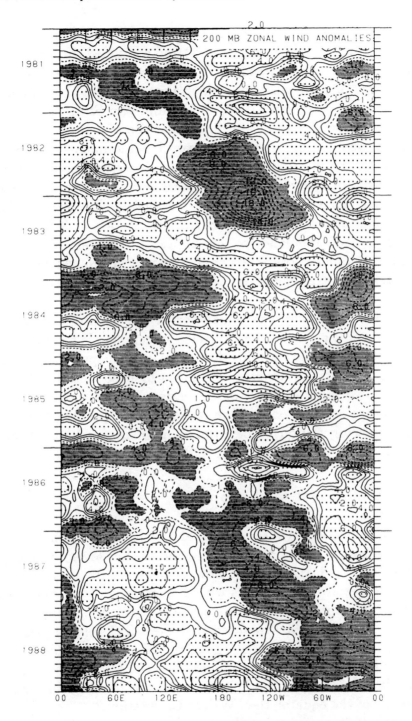

Fig. 10.5 Same as Fig. 10.1 except 200 mb zonal wind anomalies. Finely stippled areas indicate anomalies greater than 4 m/s; coarse-shaded areas (dark) anomalies less than –4 m/s. Note the tendency for the nodes of the global-scale standing oscillation to be near 150°E and 90°W.

Moderate negative SST anomalies developed over the central and eastern equatorial Pacific in 1983 and continued through 1985. A new positive trend appeared during 1986, marking the beginning of the 1986–87 warming, which peaked in 1987. Anomalies decreased rapidly in early 1988, and by mid-year SSTs in the mid-Pacific were lower than at any time since the mid-1970s.

There was a consistent pattern of interannual variability throughout the period, which can be characterized as an out-of-phase variation between SST anomalies over the central–eastern Pacific and the eastern Indonesian–western Pacific sector. This is the signature of the primary SO teleconnection in SST. While the larger SST anomalies by far appear in the eastern Pacific, the small anomalies over the climatologically warm waters of the western and central Pacific are of greater significance, as discussed later.

SST anomalies over the remainder of the equatorial belt are suggestive of a global pattern of teleconnections. For example, a modest west Atlantic warming during early 1983 appears to be associated with the mature stage of the warm episode in the Pacific. A second warming in the tropical eastern Atlantic during 1984 also appears to have roots in the earlier ENSO episode (discussed in the following section). The Atlantic anomalies of 1983–84 have been discussed by Horel et al. (1986).

The 850 mb zonal wind anomalies

When warming occurs in the central and eastern equatorial Pacific, the east–west SST gradient decreases and the low-level, thermally-driven, westward atmospheric flow decreases. But weaker surface easterlies are synonymous with a weakening of the east–west pressure gradient that drives the circulation in the vicinity of the equator. Bjerknes (1969) realized that these SST-related pressure changes correspond to oscillations of the SO pressure seesaw (see Trenberth, Chapter 2 this volume, Fig. 10.2). This linkage to Walker's SO prompted Bjerknes to name the zonal circulation in the plane of the equator the "Walker Circulation."

Figure 10.3 shows a striking series of eastward-migrating, "Walker type" zonal wind anomaly regimes in the Pacific sector associated with the 1982–3 warm episode. The nearly simultaneous appearance of below normal SST over the Pacific and positive

SST anomalies in the Indonesian–west Pacific sector during 1981 was associated with the development of low level easterly anomalies over the Pacific and westerly anomalies over the Indian Ocean. This zonal anomaly couplet migrated eastward, with the enhanced easterlies disappearing as positive SST anomalies spread across the Pacific during early 1982. By late 1983, the 850 mb zonal anomalies had reverted to the initial high index pattern: westerly anomalies over the Indian Ocean and enhanced easterlies over the Pacific.

A relatively stationary high index regime (easterly anomalies in the central and eastern equatorial Pacific, westerlies to the west) prevailed during the subsequent "cold episode" of 1984–85, but an eastward migratory component appeared again in 1986 with the onset of the next warm episode. With the cooling in early 1988, the anomalies appear to have again settled into essentially a high index standing mode.

The 850 mb zonal wind anomalies over the equatorial Atlantic sector during 1983–84 are noteworthy. The ENSO-related easterly anomalies, which prevailed during 1983, collapsed in early 1984. This is reminiscent of the well known build-up and relaxation sequence associated with El Niño development in the eastern Pacific (Cane, 1983), and it apparently led to the 1984 warming in the Atlantic (Philander, 1986). This appears to be an interesting example of a complex ocean–atmosphere–ocean teleconnection between the tropical Pacific and Atlantic. No similar situation has yet been reported in association with the 1986–87 warm episode.

OLR anomalies

Bjerknes' conceptual framework implies a direct relationship between SST gradient and low level circulation in the deep tropics, and this relationship has been demonstrated by Lindzen and Nigam (1987). However, the absolute value of SST also appears to be a significant factor in determining the seasonally averaged distribution of precipitation (Neelin and Held, 1987). Specifically, there is empirical evidence that a threshold value (around 27°C–28°C) may be a necessary though insufficient condition for recurrent, widespread precipitation (Gadgil et al., 1984; Graham and Barnett, 1987).

OLR data are used to infer changes in the distribution and intensity of tropical convective rainfall. This approach is based on the hypothesis, now amply demonstrated, that decreased OLR corresponds to increased coverage of cold (high) cloud tops and increased convective rainfall in the tropics. The largest OLR anomalies and the best relationship with SST are associated with shifts in the massive region of convection over Indonesia and the west Pacific, the region where climatological SST exceeds 28°C (Fig. 10.4). OLR and 850 mb zonal wind anomalies of opposite sign closely coincide over this sector. It is well to note that this coincidence of anomalies would not exist if the two-dimensional Walker Circulation were the sole mechanism for the low level moisture transport and convergence. In that case the OLR and low level zonal wind anomalies would be more or less in quadrature, i.e., about a quarter wavelength out-of-phase. This points up the role of anomalous north–south convergence, a feature clearly shown on the warm episode composite of Rasmusson and Carpenter (1982) (Trenberth, this volume, Fig. 2.3), and emphasizes the inadequacy of simple, two-dimensional "Walker Circulation" arguments when diagnosing atmospheric variations in the equatorial belt.

200 mb zonal wind anomalies

The large-scale zonal wind anomalies in the upper troposphere (Fig. 10.5) show a "noisier" pattern than that at 850 mb (Fig. 10.3), but careful comparison reveals a general tendency for the largest time–space scale anomalies at the two levels to be out-of-phase. This "baroclinic" vertical structure is a characteristic feature of the time-averaged circulation in the deep tropics.

There is again clear evidence of a global teleconnection pattern at 200 mb which includes both standing and eastward-migrating components. The migratory component is particularly evident over the Indian Ocean–Pacific sector during turnabouts in the ENSO cycle. The standing mode reflects a tendency for out-of-phase variations between the Pacific sector and the remainder of the equatorial belt, and is most easily identified during persistent high index regimes. The relationship of these features, first noted by Yasunari (1985), to the global pattern of 200 mb circulation anomalies is described in a later section.

In summary, the primary SO teleconnection is often viewed as a standing anomaly seesaw. A more complex picture emerged in the equatorial belt during the well documented 1980s. In addition to the primary standing oscillation, the turnabouts in the SST anomaly regime of the central equatorial Pacific seem to be associated with periods of eastward migrating wind and precipitation anomalies (Gutzler and Harrison, 1986). While it is not so apparent from these short timesections, there is also a tendency for local phasing of the anomalies with the annual cycle, a feature which will be considered in the next two sections.

Conceptual framework

The zonally symmetric part of the mean tropical circulation is obtained by averaging around a latitude circle. Except for the short transition seasons of October-November and April-May, one dominant direct thermal circulation cell appears in the zonally averaged vertical–meridional plane (Oort and Rasmusson, 1971). Its low level branch can be characterized as a seasonally varying, cross-equatorial flow from the cool winter hemisphere toward the belt of highest surface temperatures located in the low latitude summer hemisphere, i.e., toward the "thermal equator." This marks the latitude of the upward branch of the mean meridional circulation cell and associated precipitation maximum. The compensating return flow to the winter hemisphere takes place in the upper troposphere.

This mean meridional "Hadley" circulation is a fundamental component of the general circulation of the tropics. However, there are large regional departures from the zonal average which arise as a consequence of the seasonally varying surface thermal contrasts. These include the thermal forcing associated with large-scale SST gradients, particularly in the Pacific, as well as the land–ocean contrasts which give rise to the classical monsoon circulations.

Divergent circulations

The atmospheric wind field, like any vector field, can be separated into its rotational and divergent components. Atmospheric flow is predominantly rotational, so much so that above the lowest one to two kilometers, that is, in the free atmosphere where frictional effects are small, the seasonally averaged total wind field is hardly distinguishable from its rotational component. The divergent component is largest in the tropics, but even there it represents only a small "drift" or secondary flow superimposed on the dominant rotational circulation. Nevertheless, the divergent flow is crucial to our understanding of the tropical circulation, for it reflects the direct thermal circulations that are central to the ENSO cycle and teleconnection dynamics.

The divergent component can be represented in terms of a velocity potential field (Fig. 10.6). The divergent part of the flow is parallel and proportional to the potential gradient, i.e., directed normal to the potential contours, from low toward high values. Data and analyses are now adequate to describe semiquantitatively the seasonally averaged velocity potential at 200 mb, a level which reflects the upper branches of the direct thermal circulations of the tropics. Figure 10.6 shows the three-year average 200 mb velocity potential for December–February (DJF) and June–August (JJA), derived from the most recent NMC operational analyses (beginning June 1986). The meridional component of flow reflects the local contribution to the Hadley circulation. The zonal component reflects the upper branch of the Walker circulation, but as can be seen from the diagram, the circulation near the equator often departs substantially from that of the idealized Walker cell.

The features on these charts should be viewed in the context of the mean rainfall distribution in the tropics, as indexed by the OLR fields shown by Trenberth (this volume, Figures 2.5a and 2.10). OLR values less than 240 W/m have been found to delineate the regions of heaviest convective rainfall, which generally coincide with the regions of large-scale mean upward motion and associated upper level outflow shown on Fig. 10.6. The latent heat of condensation released in these areas is a primary forcing mechanism for the large-scale circulation of both the tropics and extratropics.

The major upper troposphere outflow regions show a north–south seasonal migration. Two of these are anchored over or near the summer hemisphere continental areas (Africa, Central/South America and adjacent waters). The third and most variable region of tropical convection is located over the warm waters of the western tropical Pacific–eastern Indian Ocean and adjacent land areas,

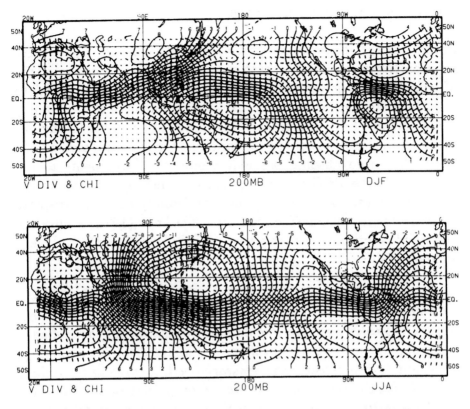

Fig. 10.6 Three-year average 200 mb velocity potential anomalies for Dec.–Feb. 1986–87, 87–88, 88–89 (upper) and Jun.–Aug. 1986, 87, 88 (lower). Arrows indicate direction and relative magnitude of divergent flow. Contour interval is 1×10^6 m^2/s.

with an eastward extension along the warm SST tongue marking the position of the North Pacific Intertropical Convergence Zone (ITCZ), and a southeastward extension along the South Pacific Convergence Zone (SPCZ) (see Trenberth, Chapter 2, this volume, Fig. 2.6). The pronounced upper troposphere divergent flow from the west Pacific–Indonesian convective region to the arid region of subsiding air over the cold waters of the eastern tropical Pacific is apparent, as is the outflow toward the subsiding airmass over the cold Asian continent in the northern winter, and the southwestward outflow toward the subsiding branch of the monsoon circulation over the south Indian Ocean during the southern winter.

Extratropical teleconnections

Because of the tendency for parcels of air to conserve their angular momentum about the earth's rotational axis, broad currents of equatorward/poleward flowing air in the tropics will tend to take on a stronger easterly/westerly component of flow. Hadley (1735) did not fully understand this principle but invoked angular momentum conservation to explain the tradewinds, and Bjerknes (1966) used absolute angular momentum arguments to explain the North Pacific extratropical response to tropical Pacific SST anomalies. Years before data and analyses were adequate to evaluate the divergent circulation, Bjerknes (1973) accurately described conditions over the western Pacific during the northern winter as follows:

> *All this air convergence into the Indonesian area must lead to upper tropospheric divergence not only by the two Walker cells, eastward over the Pacific and westward over the Indian Ocean, but also by an antimonsoon directed northward. As the antimonsoon leaves the equatorial belt, it turns to the right by Coriolis action and joins the westerly jetstream found on all wintertime upper-tropospheric streamline maps from the Himalayas to the western Pacific. It is in that sector of geographical longitude that the wintertime global Hadley circulation receives its strongest contribution of northward flux of angular momentum.*

Bjerknes reasoned that, as ENSO low index conditions develop, the west Pacific warm pool and associated heavy precipitation will extend eastward. The upper level outflow region will also elongate eastward, resulting in a weakening of the Walker circulation and an intensification of the upper troposphere Hadley outflow in the central Pacific. The enhanced poleward angular momentum transport leads in turn to an eastward elongation of the north Pacific subtropical jetstream (see Fig. 10.7), which sometimes extends into the eastern Pacific or even across the northern Gulf of Mexico.

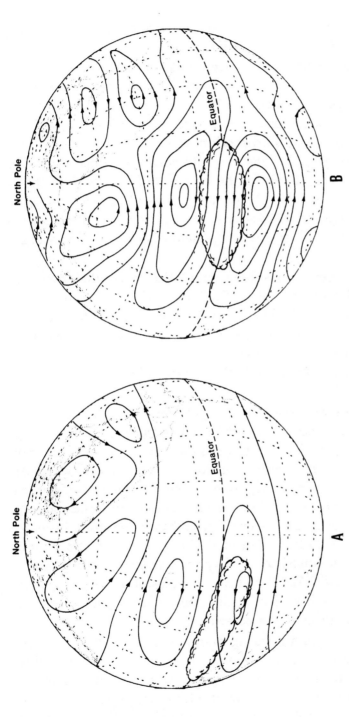

Fig. 10.7 During the mature phase (Dec.–Feb.) of a typical ENSO cycle warm episode (A), atmospheric anomalies in the upper troposphere (around 200 mb) include a pair of anticyclones to the north and south of the equator associated with the eastward extension of the cloud cover and rainfall (area enclosed by scalloped curve). A triplet of circulation anomalies occurs over the North Pacific, Canada, and the southern United States–western Atlantic. During the 1982–83 warm episode (B), the equatorial rainfall and anticyclonic anomalies extended much farther east, while the triplet of extratropical circulation anomalies remained much as it appears during a typical warm episode (from Rasmusson and Wallace, 1983).

Vorticity dynamics

Vorticity conservation, which provides a more complete and satisfying framework for viewing ENSO cycle teleconnections, largely
supplanted angular-momentum arguments in the 1980s. It was
previously noted that the atmospheric circulation can be separated into a large rotational and a small divergent component.
The vertical component of relative vorticity is a measure of the
rotational component of fluid motion around a vertical axis relative to the rotating earth. The absolute vorticity, which is a more
conservative quantity, includes the rotation of the earth-oriented
coordinate system as well. The primary tropical sources and sinks
of vorticity in the upper troposphere arise from the nonlinear relationship between the divergent component of flow and the absolute
vorticity.

The typical upper troposphere circulation anomalies over the
Pacific during the mature phase (DJF) of an ENSO warming cycle
are shown in Fig. 10.7a. The dominant low latitude feature is a
huge pair of anticyclonic circulation anomalies: a clockwise gyre
to the north and a counterclockwise circulation to the south of the
equator. These features straddle the equatorial region of enhanced
rainfall.

The anomalous westerlies on the poleward flank of the northern anticyclone are associated with the eastward extension of the
east Asian subtropical jetstream first noted by Bjerknes. Poleward
and eastward of the jetstream extension is a teleconnection triplet:
a cyclonic circulation anomaly over the North Pacific, an anticyclonic anomaly over western Canada and another cyclonic anomaly
normally located over the eastern United States–western Atlantic.
Horel and Wallace (1981) identified this pattern as the ubiqui-
tous Pacific–North American (PNA) teleconnection described by
Wallace and Gutzler (1981). These "centers of action," together
with the northern subtropical anticyclonic anomaly, trace out an
apparent wave train with the polarity shown in Fig. 10.7a.

A vorticity source in a rotating fluid system such as the earth's
atmosphere generates Rossby waves. These are the "meteorological waves" which appear, for example, as the meanders in the
upper tropospheric westerly flow at midlatitudes that are associated with surface cyclones and anticyclones. The dynamics of
Rossby wave trains emanating from the tropics has been inten-

sively investigated during the past decade, using a variety of simple models. Under idealized conditions, i.e., zonally uniform background flow, the ray-path connecting anomaly centers traces out a great circle trajectory (Hoskins and Karoly, 1981). Considering the geometry of their composite teleconnection, Horel and Wallace (1981) concluded that the pattern reflects a quasi-stationary Rossby wave train propagating from an upper troposphere vorticity source in the region of enhanced rainfall. However, further examination revealed that the teleconnection pattern for individual events sometimes departed significantly from the composite. The intense 1982–83 episode is a good example (Fig. 10.7b). In that case, the enhanced convection and associated upper level anticyclonic couplet was located far to the east of its usual position, but the seasonally averaged PNA triplet did not show a comparable eastward shift.

Simmons et al. (1983) provide an alternate hypothesis which better fits this situation than does the Hoskins–Karoly model. Their model experiments indicate that during the Northern Hemisphere winter, when the climatological stationary long waves in the upper tropospheric circulation are of large amplitude, the response to tropical forcing may take the form of a geographically fixed "normal mode," which resembles the PNA teleconnection. Under these circumstances, an eastward or westward movement of the region of enhanced equatorial rainfall may well produce the same extratropical PNA response, although the amplitude, and perhaps the polarity, of the anomalies might change (Rasmusson and Wallace, 1983). This selective and rather superficial summary of some of the early theoretical and modeling studies of teleconnection dynamics is meant to serve simply as an introduction. Tribbia (Chapter 9, this volume) presents a thorough review of the subject.

It must be kept in mind that while ENSO cycle teleconnection patterns exhibit broad similarities, each warm or cold episode also departs to a greater or lesser extent from a "canonical" description. Improvements in tropical observations and analyses now make it possible to diagnose some of the more important case-to-case differences, and this provides further insight into the nature and causes of some of the puzzling deviations from an idealized response.

Sardeshmukh and Hoskins (1985, 1987) examined the vorticity balance at 150 mb, which is representative of the level of cumulus outflow in the upper troposphere. They evaluated the magnitude of the various terms in the vorticity equation from analyses produced as part of the operational data assimilation–analysis–forecast cycle of the European Center for Medium-Range Prediction (ECMWF). Their analyses reveal that the vorticity source region encompasses a much larger area than simply the region of precipitation and latent heat release. In fact, the most intense source regions are often associated with the descending rather than the ascending branches of the regional Hadley circulations (Rasmusson, 1988). Consequently, the regional Hadley circulation, which was the key to Bjerknes' explanation of the North Pacific teleconnection, remains a crucial element of teleconnection dynamics in the vorticity balance framework.

Annual cycle control

Figure 10.6 shows that the outflow/inflow regions, i.e., the upward and downward branches of the climatological regional Hadley circulations, change hemispheres with the seasons. This annual cycle in the divergent circulation, coupled with the seasonal changes in the rotational component of flow, results in pronounced seasonal variations in the position and intensity of the climatological vorticity sources, and thus in the planetary circulation patterns.

Because of the non-linear nature of the vorticity source terms, and the fact that the divergent circulation anomalies are usually small perturbations superimposed on these stronger climatological features, the position of a vorticity source anomaly can be relatively insensitive to the longitudinal position of the divergence anomaly from which it arises (Sardeshmukh and Hoskins, 1988). Consequently, the mean annual cycle imposes a seasonality on ENSO cycle teleconnections, which is further enhanced by a tendency for phasing of the low latitude precipitation and divergence anomalies with the mean annual cycle.

The seasonally related eastward migration of ENSO cycle anomalies in the equatorial belt was described earlier. The meridional migration of the basin-scale anomaly pattern is another very important, although not widely recognized aspect of ENSO evolution. This feature is illustrated by a meridional OLR section

averaged across the longitude sector 100°W–150°E (Fig. 10.8).
Isolines are total OLR, rather than anomalies, in order to more
clearly reveal the relationship between the mean annual cycle and
precipitation anomalies.

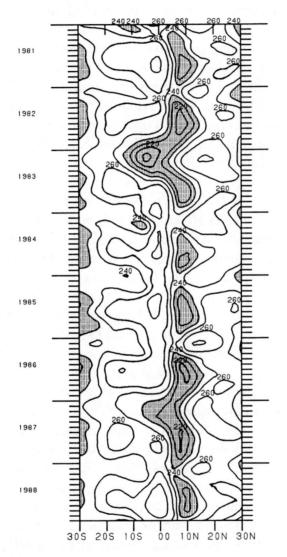

ZONALLY AVERAGED OLR   (150E-100W)

Fig. 10.8 Time-latitude section for OLR, averaged over the longitude sector
100°W–150°E. Contour interval 10 W/m^2. Shading indicates values
less than 240 W/m^2.

The primary climatological convergence zone (the ITCZ) in the
central and eastern Pacific remains north of the equator through-

out the year. This is reflected as a persistent OLR minimum (precipitation maximum) north of the equator on Fig. 10.8. However, the convergent meridional component of the low latitude Pacific tradewind regimes has a large annual cycle, much larger than that observed in the dominant easterly component of flow (Horel, 1982). This is reflected as a *relative* north–south seasonal migration of the low-level convergence and precipitation belt that is superimposed on the annual mean pattern. This relative Hadley cycle is clearly reflected in Fig. 10.8 as an enhancement/diminution of rainfall in each hemisphere during its local warm/cold season. During high index periods, the presence of cold upwelling water suppresses precipitation near the equator, and to the south the Southern Hemisphere summer rainfall maximum is relatively short and often indistinct.

The modification of the annual cycle during the 1982–83 and 1986–87 warm episodes is striking. There was a general increase in precipitation, particularly near and south of the equator during the Southern Hemisphere summer, i.e., during the mature phase of a typical warm episode. The longitudinally averaged precipitation anomalies migrated with the seasonal migration of the low level convergence zone, i.e., with the seasonally varying Hadley component, and the higher equatorial SSTs allowed a continuous interhemispheric seasonal progression of the rainfall maximum.

The pattern implies a significant strengthening of the Pacific Hadley circulation during the warm episodes, particularly during the northern autumn and winter. These changes are reminiscent of Bjerknes' statement that "a warmer than normal equatorial ocean over a wide span of longitude will make the Hadley circulation run faster than normal" (Bjerknes, 1966). Since the precipitation enhancement is most pronounced in the summer hemisphere, it seems likely that it will most strongly affect the distribution and strength of vorticity sources and extratropical teleconnections in the winter hemisphere descending branch of the Hadley circulation.

Global teleconnections

Because of the tendency for annual cycle phase-locking, originally noted by Walker (1924), a useful picture of the typical ENSO anomaly patterns can be obtained by compositing or stratifying

the data according to season and phase of the ENSO Cycle. This
has been done by Rasmusson and Carpenter (1982) and many
other investigators. Broadscale aspects of the global pattern of
ENSO teleconnections will be illustrated in this section using sea
level pressure and 200 mb circulation composites.

Sea level pressure

Van Loon (1986) constructed seasonal composites of global sea
level pressure anomalies for various phases of a warm episode.
The composites were constructed from three different data sets
which differ in both length and quality. The number of events
making up the composites varies from 19 north of 20°N to as few as
four south of 10°S. Anomaly fields for the two solstice seasons are
shown in Figs. 10.9 and 10.10 to illustrate some broadscale aspects
of teleconnection seasonality and linearity. The composites for
JJA(0) and D(0)JF(1) are for the warm (low index) phase of the
oscillation. Composites for JJA(-1) and D(-1)JF(0) are for a year
earlier, but because of the strong biennial component of ENSO
variability (Kiladis and Diaz, 1989; Rasmusson et al., 1990), these
composites can be broadly interpreted as characteristic of the cold
(high index) phase of the cycle.

Seasonality may appear as systematic seasonal differences in the
intensity, position or character of the anomaly patterns. One clear
reflection of seasonality is the overall tendency for larger anomalies
in each hemisphere during its cold season. This seasonal difference is far more pronounced in the Northern Hemisphere, which
also exhibits a more pronounced mean annual cycle. For example,
the Northern Hemisphere PNA teleconnection is essentially a cold
season phenomenon (compare Figures 10.9 and 10.10). Seasonal
changes in the position of the primary Indian Ocean–Pacific teleconnection are also apparent, suggesting caution in interpreting
annual averaged representations.

Broadly speaking, the large-scale anomaly pattern for a particular season tends to switch sign between low index and high
index states, i.e., it exhibits at least a crude degree of linear variation with the SO index. This is true over much of the tropics
and Southern Hemisphere during both solstice seasons, and over
the Northern Hemisphere extratropics in winter, where the PNA
triplet switches polarity between high and low index.

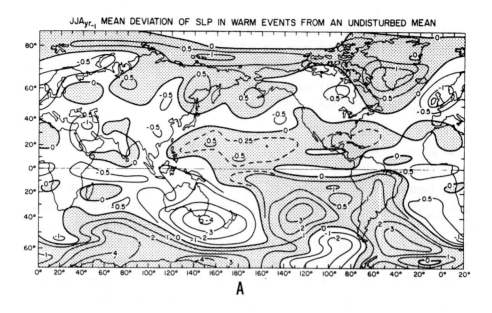

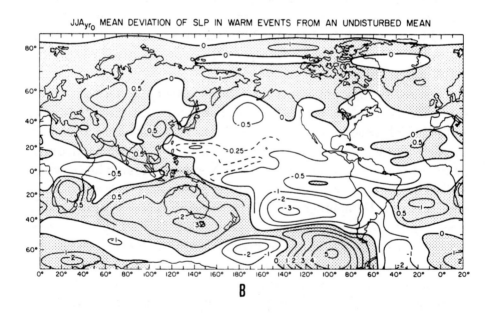

Fig. 10.9 Mean sea level pressure anomalies (mb) for Jun.–Aug. of the year
before (a) and the year of (b) a warm episode (from van Loon, 1986).

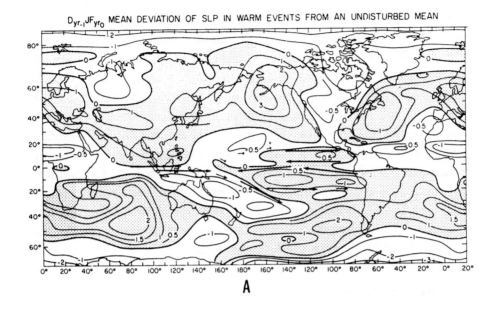

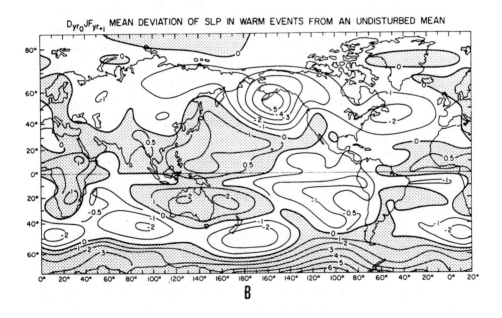

Fig. 10.10 Mean sea level pressure anomalies for Dec.–Feb before (a) and during
(b) a warm episode (from van Loon, 1986).

Ropelewski and Halpert (1989) and Kiladis and Diaz (1989) have demonstrated that many, perhaps most, of the significant regional ENSO temperature and precipitation anomalies appear as departures of opposite sign during opposite phases of the cycle. There are, of course many variations on this simple theme. For example, in the coastal region of northern Peru, precipitation is near zero most of the time, with occasional pronounced, event-like El Niño wet periods. There are large areas where the composite ENSO signal is essentially non-existent, or too weak to be unambiguously identified. For several areas, particularly in the Southern Hemisphere, it is hard to say whether the apparent lack of linearity and/or seasonality in the sea level pressure composites is real, or simply due to inadequate data.

200 mb circulation anomalies

The reliability of the NMC 200 mb divergent wind analyses has improved steadily during the 1980s, but several changes in analysis procedures during the period make it difficult to construct consistent composites. However, analysis procedures have not changed substantially since the significant improvements in 1986, and this allows comparison of the contrasting divergent flow patterns that prevailed during the warm episode of 1986–87 and the subsequent high index conditions of 1988–89. These changes are illustrated in Figs. 10.11 and 10.12 as departures from a three-year mean. One must clearly keep in mind that these are simply departures from a very short base period mean, rather than anomalies from a long term mean.

Figure 10.11 contrasts the seasonally averaged DJF departures for 1986–87 (low index) and 1988–89 (high index). There is general reversal of the large-scale pattern over the Pacific that reflects an intensification (weakening) of the upper tropospheric outflow from the equatorial Pacific during the low (high) index regime. This is particularly evident as a strengthening (weakening) of the Northern Hemisphere Hadley circulation in the central Pacific. A modulation of the Walker components, centered south of the equator during this season, also occurred, but in this particular case the changes were small east of about 140°W.

Many features on the departure charts can be related to well known ENSO precipitation teleconnections. For example, the

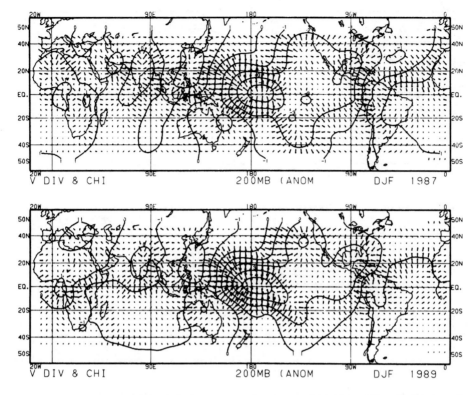

Fig. 10.11 Velocity potential departures for Dec. 1986–Feb. 1987 (upper) and
Dec. 1988–Feb. 1989 (lower) from the three-year mean (Dec. 1986–
Feb. 1989). Arrows indicate direction and relative magnitude of di-
vergent flow. Contour interval is 1×10^6 m^2/s.

equatorial inflow–outflow departures correspond well to the pattern of negative (positive) OLR anomalies over the Pacific sector,
i.e., enhanced (diminished) central equatorial Pacific precipitation
in 1986–87 (1988–89), with reverse anomalies over the Indonesian sector (see Fig. 10.4). The divergent pattern over the South
American sector suggests relatively wet conditions over the eastern equatorial Pacific and dry conditions over northeastern South
America during 1986–87, and the reverse during 1988–89. The
divergence anomalies over the southwest Pacific clearly reveal the
westward shift of the SPCZ from 1986–87 to 1988–89.

The JJA charts (Fig. 10.12) reflect the contrasting departures
during the well-developed warm episode of 1987 and the pronounced high index/cold equatorial pattern of 1988. Comparison
with Fig. 10.6 shows an eastward (westward) extension (contrac-
tion) of the west Pacific outflow region, resulting in an enhanced
(diminished) Hadley component of outflow concentrated in the

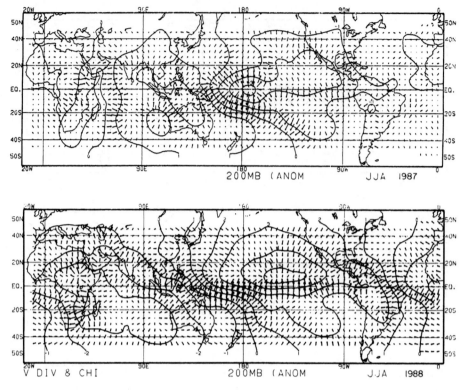

Fig. 10.12 Velocity potential departures from the 1986–88 Jun.–Aug. mean for
1987 (upper) and 1988 (lower). Arrows indicate direction and relative
magnitude of divergent flow. Contour interval is 1×10^6 m^2/s.

central Pacific during the warm episode, but broadly distributed
during the cold conditions.

The JJA departure pattern can again be related to many well-
known teleconnection features of the ENSO cycle. The equatorial
outflow (inflow) corresponds to the characteristic negative (positive) OLR anomaly couplet over the central Pacific–eastern Indian
Ocean sector discussed above (Fig. 10.4). The pattern over the
Pacific is also characteristic of the enhanced (diminished) intensity of the ITCZ and its equatorward (poleward) shift during low
(high) index conditions. There is enhanced (diminished) inflow
and subsidence over the Bay of Bengal (India) region, consistent
with the characteristic drought (flood) variation of the monsoon
observed in 1987 and 1988. The changes in the outflow pattern
east of Australia also reflect the characteristic east–west shift of
the SPCZ.

Figures 10.13 and 10.14 are 200-mb composites of the total circulation anomalies for five recent high index/cold and low index/warm seasons (P. Arkin, personal communication, 1989). These composites can be related to features appearing on the equatorial belt 200 mb zonal wind anomaly section (Fig. 10.5), thus placing the equatorial anomalies in a global context.

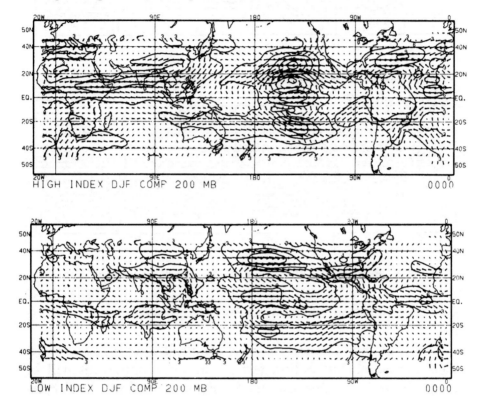

Fig. 10.13 Dec.–Feb. 200 mb composite wind anomalies for five high index/cold SST (a) and five low index/high SST seasons. Contour interval: 2 m/s. Arrows indicate direction and relative magnitude of the anomalous flow.

A striking feature of the 200-mb equatorial time section (Fig. 10.5) is the tendency for out-of-phase zonal wind variations over the Pacific and Atlantic. Figs. 10.13 and 10.14 show these anomalies to be the common equatorial branches of huge subtropical circulation couplets of opposite sense over the Atlantic and central Pacific. The poleward branches of these circulations are associated with variations in the subtropical westerlies of both hemispheres, and with higher latitude teleconnections, e.g., the

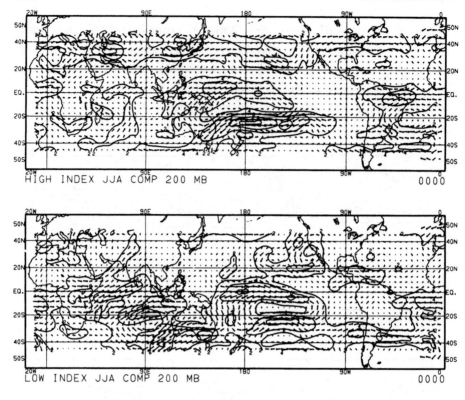

Fig. 10.14 Same as Fig. 10.13 for Jun.–Aug. Contour interval: 3 m/s.

PNA teleconnection, which are incompletely displayed on the figures.

There are many features of these composites which deserve detailed discussion, but only a few of the more important will be
noted. If we focus first on the Pacific sector of the DJF composite (Fig. 10.13), the dominant feature is the high index (low
index) cyclonic (anticyclonic) couplet associated with diminished
(enhanced) central equatorial Pacific precipitation. A low index
DJF composite for a different ensemble of warm episodes was illustrated in Fig. 10.7. The anticyclonic couplet is quite variable
from one warming to the next in both longitude and intensity.
The result is a weaker and less distinct composite pattern than
that typically found on individual charts. The high index cyclonic couplet is more consistent from event to event, which accounts for the stronger and more distinct composite. Note that
circulation anomalies in the western Pacific are relatively weak,

suggesting that the central Pacific anomalies arise primarily from east–west extensions/contractions of the west Pacific circulation regime rather than from eastward displacements.

The anomaly features over the Pacific during JJA appear somewhat west of the DJF features. The individual cells of both the Pacific and the somewhat weaker Atlantic couplet are strongest during the local winter. In fact, the composite summer circulation anomalies in the Northern Hemisphere are, for the most part, rather insignificant.

The anomaly pattern in the Afro-Indian Ocean sector is less distinct and more difficult to interpret. Features which appeared in one form or another during at least four of the seasons making up the composite are (1) the broad belt of easterly anomalies centered north of the equator during low index DJF, (2) the tendency for easterly anomalies over southern Asia during high index JJA, and (3) the cyclonic anomaly west of Australia, and easterly anomalies across southern Australia during low index JJA.

When comparing the 200 mb and sea level pressure composites, it should be kept in mind that upper and lower troposphere circulation anomalies tend to be out-of-phase near the equator (compare again Figs. 10.3 and 10.5). One would therefore expect the 200 mb anomalous flow to be directed toward regions of increasing sea level pressure anomalies in the equatorial belt. This seems to be true over the Pacific and to a lesser extent over the Atlantic.

Upper and lower troposphere circulation anomalies in the extratropics are typically in-phase, i.e., equivalent barotropic. Using this as a criterion, wintertime sea level pressure and 200 mb circulation anomaly patterns over the extratropical Northern Hemisphere are in general qualitative agreement over the North Pacific where anomalies are large, but less so over the North Atlantic, and conclusions are hard to draw for other areas of the Northern Hemisphere. Comparison seems poor over the extratropical South Pacific, where both analyses are more questionable. Karoly (1989) has constructed 200 mb composites for three ENSO warm episodes between 1972 and 1983. Major features of his anomaly charts and van Loon's sea level composites are broadly similar poleward of 40°S.

Concluding remarks

Walker's early work (Walker, 1924) implied three basic features of SO teleconnections: global domain, seasonality, and quasi-linearity. Time-sections and composites have been used to illustrate these features, as well as other broadscale aspects of ENSO cycle global teleconnection patterns.

The deluge of empirical studies and model simulations during the past two decades has greatly increased our knowledge and understanding of the global linkages and lag relationships associated with the ENSO cycle, and significantly broadened our understanding of this coupled ocean–atmosphere rhythm. There is far more complexity in ENSO interactions than can be accurately represented by a simple composite, statistical relationship, or model. The response of the atmosphere during any particular swing of the ENSO cycle will vary with the amplitude and position of the tropical Pacific SST anomalies, and may be strongly influenced by the superposition of non-ENSO variability, particularly that generated in the extratropics. Thus, any "canonical" description of ENSO cycle teleconnection patterns should be viewed in the abstract, much as the mean annual cycle, which is never strictly observed.

In order to determine whether a particular ENSO teleconnection pattern is consistent enough to be useful as a forecast tool, we must continue to resort to simple statistical correlations between anomaly patterns and some ENSO index. In contrast to Walker, however, we now have a well developed conceptual framework within which to interpret the statistical relationships. We may be able to apply this insight to stratify the total ensemble of individual episodes in a way that reveals subsets or classes of occurrences. However, the ultimate hope lies in the use of coupled ocean–atmosphere models to skillfully predict the complex case-to-case differences.

Acknowledgments

This work was supported by NSF grant ATM-8806447. In addition, it is a pleasure to acknowledge the generous technical support from the NOAA Climate Analysis Center, and from individual staff members, in particular the constructive discussions and other help provided by K. Mo and P. Arkin, who provided Figs. 10.6, 10.11, and 10.12, and the efforts of M. Halpert in preparing Figs. 10.1 through 10.5. I am indebted to H. van Loon for permission to use Figs. 10.9 and 10.10.

References

Bjerknes, J. (1966). A possible response of the atmospheric Hadley circulation to equatorial anomalies of ocean temperature. *Tellus*, **18**, 820–9.

Bjerknes, J. (1969). Atmospheric teleconnections from the equatorial Pacific. *Monthly Weather Review*, **97**, 163–72.

Bjerknes, J. (1973). Atmospheric teleconnections from the equatorial Pacific during 1963–67. Final Report, NSF Grant GA 27754, UCLA Department of Meteorology.

Cane, M.A. (1983). Oceanographic events during El Niño. *Science*, **222**, 1189–94.

Gadgil, S., Joseph, P.V. & Joshi, N.V. (1984). Ocean–atmosphere coupling over monsoon regions. *Nature*, **312**, 141–3.

Graham, N.E. & Barnett, T.P. (1987). Sea surface temperature, surface wind divergence, and convection over the tropical oceans. *Science*, **238**, 657–9.

Gutzler, D.S. & Harrison, D.E. (1986). The structure and evolution of seasonal wind anomalies over the near-equatorial eastern Indian and western Pacific Oceans. *Monthly Weather Review*, **114**, 285–94.

Hadley, G. (1735). On the cause of the general trade-winds. *Philosophical Transactions*, **39**, (437), 58–62.

Hildebrandsson, H.H. (1897). Quelques recherches sur les centres d'action de l'atmosphère. *Kungliga Svenska Vetenskapakadiems*. Avhandlinger i Naturskyddarenden, **29** (entire issue).

Horel, J.D. (1982). On the annual cycle of the tropical Pacific atmosphere and ocean. *Monthly Weather Review*, **110**, 1863–78.

Horel, J.D., Kousky, V.E. & Kagano, M.T. (1986). Atmospheric conditions in the Atlantic sector during 1983–84. *Nature*, **322**, 248–51.

Horel, J.D. & Wallace, J.M. (1981). Planetary scale atmospheric phenomena associated with the Southern Oscillation. *Monthly Weather Review*, **109**, 813–29.

Hoskins, B.J. & Karoly, D.J. (1981). The steady linear response of a spherical atmosphere to thermal and orographic forcing. *Journal of the Atmospheric Sciences*, **38**, 1179–96.

Karoly, D.J. (1989). Southern Hemisphere circulation features associated with El Niño-Southern Oscillation events. *Journal of Climate*, **2**, 1239–52.

Kiladis, G.N. & Diaz, H.F. (1989). Global climatic anomalies associated with extremes in the Southern Oscillation. *Journal of Climate*, **2**, 1069–90.

Neelin, J.D.& Held, I.M. (1987). Modeling tropical convergence based on the moist static energy budget. *Monthly Weather Review*, **115**, 3–12.

Lindzen, R.S. & Nigam, S. (1987). On the role of sea surface temperature gradients in forcing low level winds and convergence in the tropics. *Journal of the Atmospheric Sciences*, **45**, 2440–58.

Lockyer, N. & Lockyer, W.J.S (1904). The behavior of the short–period atmospheric pressure variation over the earth's surface. *Proceedings of the Royal Society of London*, **73**, 457–70.

Oort, A.H. & Rasmusson, E.M. (1971). On the annual variation of the monthly mean meridional circulation. *Monthly Weather Review*, **98**, 423–42.

Philander, S.G.F. (1986). Unusual conditions in the tropical Atlantic Ocean in 1984. *Nature*, **322**, 236–38.

Rasmusson, E.M. (1988). Atmospheric teleconnection dynamics during the 1986–89 ENSO cycle. *Proceedings, Third International Conference on Southern Hemisphere Meteorology and Oceanography*. Boston: American Meteorological Society, 361–5.

Rasmusson, E.M., Wong, X. & Ropelewski, C.E. (1990). The biennial component of ENSO variability. *Journal of Marine Systems*, **1**, 71–96.

Rasmusson, E.M. & Arkin, P.A. (1985). Interannual climate variability associated with the El Niño/Southern Oscillation. Coupled Ocean–Atmosphere Models, ed. J.C.J. Nihoul. *Elsevier Oceanographic Series*, **40**, 697–725.

Rasmusson, E.M. & Wallace, J.M. (1983). Meteorological aspects of the El Niño/Southern Oscillation. *Science*, **222**, 1195–1202.

Rasmusson, E.M. & Carpenter, T.H. (1982). Variations in tropical sea surface temperature and surface wind fields associated with the Southern Oscillation/El Niño. *Monthly Weather Review*, **110**, 354–84.

Ropelewski, C.F. & Halpert, M.S. (1989). Precipitation patterns associated with the high index phase of the Southern Oscillation. *Journal of Climate*, **2**, 268–84.

Sardeshmukh, P.D. & Hoskins, B.J. (1985). Vorticity balances in the tropics during 1982–83 El Niño–Southern Oscillation event. *Quarterly Journal of the Royal Meteorological Society*, **468**, 261–78.

Sardeshmukh, P.D. & Hoskins, B.J. (1987). On the derivation of the divergent flow from the rotational flow: the "chi" problem. *Quarterly Journal of the Royal Meteorological Society*, **113**, 339–60.

Sardeshmukh, P.D. & Hoskins, B.J. (1988). The generation of global rotational flow by steady idealized tropical divergence. *Journal of the Atmospheric Sciences*, **45**, 1228–51.

Simmons, A.J., Wallace, J.M. & Branstator, G.W. (1983). Barotropic wave propagation and instability, and atmospheric teleconnection patterns. *Journal of the Atmospheric Sciences*, **40**, 1363–92.

van Loon, H. (1986). The characteristics of sea level pressure and sea surface temperature during the development of a warm event in the Southern Oscillation. *Namias Symposium*, ed. J.O.Roads. *Scripps Institution of Oceanography Reference Series*. 86–117, 160–73.

van Loon, H. & Rogers, J.C. (1978). The seesaw in winter temperatures between Greenland and Northern Europe. *Monthly Weather Review*, **106**, 296–310.

Walker, G.T. (1924). Correlation in seasonal variations of weather IX: A further study of world weather. *Memoirs of the Royal Meteorological Society*, **24(9)**, 275–332.

Walker, G.T. & Bliss, E.W. (1932). World Weather V., *Memoirs of the Royal Meteorological Society*, **4**, 53–84.

Wallace, J.M. & Gutzler, D.S. (1981). Teleconnections in the geopotential height field during the Northern Hemisphere winter. *Monthly Weather Review*, **109**, 784–811.

Yasunari, T. (1985). Zonally propagating modes of the global east–west circulation associated with the Southern Oscillation. *Journal of the Meteorolological Society of Japan*, **63**, 1013–29.

11

Forecasting El Niño with a geophysical model

MARK A. CANE

Lamont-Doherty Geological Observatory
Palisades, New York 10964

Introduction

The weather this season is not the same as it was a year ago, and common experience leads us to expect that it will be different still a year hence. None of us, including the experts in long-range forecasting, have a reliable idea of how it will differ.

Some of the year-to-year variations in climate are the result of random sequences of events, just as a series of coin flips will occasionally produce a long run of heads. A region may experience a dry spell because no storms happen to pass that way for a time. Prediction of such stochastic events is not possible. However, many climatic variations are part of patterns that are coherent on a large scale. Skillful prediction is a possibility in such cases, especially when the patterns are forced by observable changes. It appears that the El Niño/Southern Oscillation (ENSO) phenomenon is such a case.

This chapter gives an account of the state of the art of prediction of El Niño events using computer models which simulate the evolution of the climate system. This numerical technique has become the dominant approach to weather prediction. For El Niño the coupled ocean–atmosphere system in the tropical Pacific region must be simulated. Though El Niño may be defined in purely oceanographic terms, it is not possible to predict the ocean behavior alone. The ocean circulation is forced by the atmosphere, and its future state will depend on the future state of the atmosphere. To predict El Niño, one must predict both atmosphere and ocean.

The model considered here (Cane et al., 1986; Zebiak and Cane, 1987) is, for the time being, the only one which predicts El Niño by numerical integration of equations describing the dynamics and thermodynamics of the ocean and atmosphere. It would be hard to assert that its predictions are clearly superior to those of a purely statistical scheme (Barnett et al., 1988). However, the example of numerical weather prediction encourages the expectation that predictions based on a physical model will ultimately prove superior. *Inter alia*, this approach creates an experimental apparatus for studying the physics of the ENSO cycle.

Regardless of how well it seems to work, it is difficult to feel confident about a prediction scheme whose basis is as mysterious as a fortune teller's crystal ball. Therefore, we begin with an account of a theory for the genesis and maintenance of the ENSO cycle. The existence of such a theory may help to accommodate the reader to the notion that the forecasting results are neither mysterious nor a fluke. In addition, the forecasting scheme was developed out of the theory and this accounts for some of its apparently arbitrary aspects.

In some circumstances the power of statistical methods makes it possible to predict a phenomenon without understanding it at all. With ENSO even quite sophisticated statistical schemes cannot be carried out without some a priori insights to provide a physical basis (see Graham et al., 1987). For a procedure based on a physical model, it is an unavoidable requirement. By the same token, any success in forecasting may be taken as supporting the theory underlying it.

A theory for ENSO

The foundation for our present understanding of the ENSO cycle was laid down in a series of papers by Bjerknes (1966, 1972, and especially 1969). Other chapters in this book testify to the importance of his work for all subsequent studies; Rasmussen's chapter, for example, points out the seminal role Bjerknes' observational work played in establishing the link between the oceanic El Niño and the atmospheric Southern Oscillation.

As noted by Rasmusson (Chapter 10), Bjerknes went beyond the purely descriptive to propose a hypothesis for the origins of ENSO which depends on a two-way coupling between the equatorial ocean and atmosphere in the Pacific. (He also pointed out the atmospheric teleconnections between the tropical Pacific and midlatitudes). Bjerknes began by noting a striking fact about the "normal" state of the equatorial Pacific: the sea surface temperatures (SST) at the eastern side are unusually cold for low latitudes. The contrast with the very warm western Pacific drives a direct thermal circulation in the atmosphere along the equator; the relatively cold, dry air above the cold waters of the eastern equatorial Pacific flows westward along the surface toward the warm western Pacific. Having been heated and supplied with moisture in its travels, the air can then join in the moist ascent characteristic of the Australasian low. Some of this air joins the poleward flow at upper levels which is part of the Hadley Circulation, and some returns to the east to sink over the eastern equatorial Pacific, thereby completing the circuit of the Walker Circulation (Trenberth, Chapter 2). Bjerknes named this the "Walker Circulation" because he felt that fluctuations in this circulation initiated pulses in Walker's Southern Oscillation (Trenberth, Chapter 2). He argued that it could have such global consequences because its operation engenders a large tapping of the tremendous energy source of the large-scale moist ascent, evident in the copious rainfall over the "maritime continent".

Bjerknes also pointed out that even while the surface winds are being driven westward along the equator by the zonal SST gradient, they are acting to create the cold ocean temperatures in the east responsible for that gradient. In a search for the mechanisms responsible, differences in surface heating can quickly be ruled out; while heat flux estimates are quite uncertain, there is no doubt that more heat is going into the equatorial Pacific at the east than at the west (e.g., Weare et al., 1980; see also Bjerknes, 1966). The causes of the unusually cold SSTs are to be found in wind-driven ocean dynamics:

(1) The easterly winds drive westward currents along the equator, advecting the cold waters from the South American coast.

(2) The Coriolis force turns ocean currents to the right in the Northern Hemisphere and to the left in the Southern Hemisphere. Consequently, the surface flow at the equator is deflected poleward, and the poleward flow must be fed by waters which upwell along the equator, waters which are colder than the surface.

(3) The tropical ocean can usefully be viewed as a two-layer fluid, consisting of a warm upper ocean layer and the layer of the cold abyssal waters. In the real ocean the two are separated

by the thermocline, a narrow (50–100 m) region of strong temperature change (10°C or more). The easterlies along the equator push the waters of the warm upper layer to the west, pulling the thermocline to the surface in the east. As a result, the water upwelled there is colder than it would be if the upper layer waters were more evenly distributed with longitude.

The limitations of ocean theory and observation in his time made it impossible for Bjerknes to decide which of these three factors is most important (cf. Bjerknes, 1969). Even today there is considerable uncertainty as to their relative roles, although it seems to be the case that none are negligible. (cf. Seager et al., 1988).

Thus, the oceanic and atmospheric circulations over the tropical Pacific are mutually maintained by what Bjerknes (1969) referred to as a "chain reaction"; "an intensifying Walker Circulation also provides for an increase of the east–west temperature contrast that is the cause of the Walker Circulation in the first place". He also noted that the interaction could operate in the opposite sense: a decrease of the equatorial easterlies diminishes the supply of cold waters to the eastern equatorial Pacific (by any of the three mechanisms); the lessened east–west temperature contrast causes the Walker Circulation to slow down.

Bjerknes thus provided us with an explanation for the association of the low phase of the Southern Oscillation with El Niño as well as the association of the high phase with the normal cold state of the eastern Pacific. In each phase a positive feedback operates – in other words, an instability of the coupled-atmosphere system. However, the central question for prediction is the causes of the onset of El Niño, and Bjerknes stopped short of offering an explanation for the turnabout from one state to the other.

Nonetheless, Bjerknes' elegant scenario, in which both the tropical ocean and the atmosphere are active participants, is the foundation which subsequent researchers have built on to create a more complete theoretical structure, one which includes an explanation of the oscillation. Here we only touch on the subset of this work important in the sequel (for a more complete account see Cane, 1986).

Wyrtki (1975, 1979) seized on the point that during El Niño the ocean response is dynamical rather than thermodynamic (i.e., due to variations in surface heat flux). He shifted attention from

SST to sea level. SST variations are readily apparent only in the eastern part of the ocean, and, as noted above, even after one recognizes that SST changes are dynamically caused, it is a far more complex response than sea level and therefore more difficult to decipher. By collecting and charting sea level data, Wyrtki was able to show that the oceanic changes during El Niño are basin-wide. He also showed that the initial changes in the wind were in the central and western Pacific, far from the locale of the SST changes. Finally, he suggested that the signal could propagate eastward from the area of the wind change to the South American coast through the equatorial wave guide in the form of equatorial Kelvin waves. These ideas were amplified by a number of investigators and verified in a set of numerical experiments. (Busalacchi and O'Brien, 1981; Busalacchi et al., 1983). In these experiments a linear shallow water model was driven by nearly two decades of monthly surface wind stress fields, and the model thermocline anomalies showed a significant correlation with sea level observations.

Having mentioned Kelvin waves, it may be helpful to pause and summarize the relevant theory. Our interest here is in the variations in the upper tropical ocean over timescales of a few years or so. In this context, it is adequate to regard the ocean as a two-layer system: an active upper layer separated by the thermocline from a deep abyssal layer of greater density. Motions in the lower layer are quite slow, and may be disregarded relative to the active layer above. Making the useful assumption that linear physics govern the evolution of vertically integrated characteristics of this upper layer such as thermocline displacements and upper-layer transports, the governing equations reduce to the shallow water equations of tidal theory.

The total response may be analyzed into a sum of free and forced waves. For the time and space scales relevant to El Niño, only two types of wave motions matter: long Rossby waves and equatorial Kelvin waves. The latter are strongly trapped to the equator and owe their existence to the vanishing of the Coriolis force there. Except within a few degrees of the equator, the geostrophic balance between Coriolis and pressure gradient forces dominates the dynamics of the ocean. This balance is characteristic of Rossby waves and strongly constrains their propagation speeds and their amplitude in response to wind driving. The Kelvin wave is the fastest of the low-frequency ocean motions. It can cross the Pacific in less than three months, whereas the fastest Rossby wave is three times slower.

Long Rossby waves propagate energy westward, while Kelvin waves travel to the east. When a Kelvin wave hits the eastern boundary, its reflection is made up of an infinite sum of Rossby waves, which collectively act to extend the equatorial wave guide up and down the eastern boundary to high latitudes. However, the faster Rossby waves at low latitudes carry much of the mass and energy brought east by the Kelvin waves back toward the west. The reflection process at the west is more efficient; all of the mass flux which the Rossby waves carry into the boundary is collected by boundary currents and brought equatorward, where it is returned eastward in the form of Kelvin waves.

The special properties of equatorial waves mentioned above are essential to the El Niño phenomenon. Only at low latitudes can low-frequency waves cross the ocean in times matched to the seasonal variations in the winds. A given wind change generates a stronger response at the equator than at higher latitudes, and equatorial waves are less susceptible to the destructive influences of friction and mean currents. Finally, we shall see in a moment that the asymmetries in the waves and their reflections are essential to the ENSO cycle.

In the 1980s a number of models were built which incorporated enough physics to be able to simulate the mechanisms singled out by Bjerknes and Wyrtki (Cane and Zebiak, 1985; Zebiak and Cane, 1987; Battisti, 1988; Schopf and Suarez, 1988).* These models have shown themselves able to simulate many of the features characteristic of the observed ENSO cycle. An example is given in Figs. 11.1 and 11.2. The model SST anomaly in the eastern Pacific shows peaks of varying amplitude occurring at irregular intervals, but typically three to four years apart. They tend to be phase-locked to the annual cycle, with major events reaching their maximum amplitude at the end of the calendar year and decaying rapidly thereafter. The pattern of the model SST anomaly bears a

---

* The author is in a position to assure the reader that the first of these, the Zebiak–Cane model, was a *conscious* attempt to make a numerical model of the Bjerknes hypothesis with due attention to the ocean dynamics needed to incorporate Wyrtki's amendments.

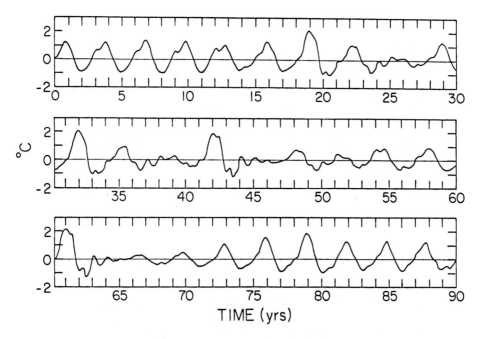

Fig. 11.1 SST anomalies averaged over the eastern equatorial Pacific region
NINO3 (5N-5S, 90W-150W) for 90 years of coupled model integration
(from Cane & Zebiak, 1985).

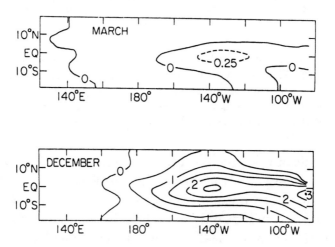

Fig. 11.2 Coupled model SST anomalies for March and December during the
model El Niño event in year 31 (note that the contour interval for
March is 0.25°C and that for December is 0.5) (from Cane & Zebiak,
1985).

strong resemblance to the observed one (cf. Fig. 11.2 with Fig. 2.3
of Trenberth, this volume).

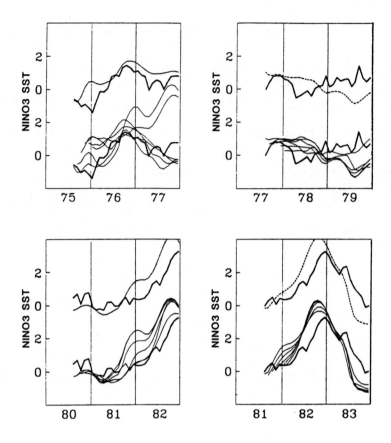

Fig. 11.3 SST anomalies (°C) for selected periods averaged over the NINO3
region. The light curves in each panel are from forecasts initiated in
the six successive months from August through the following January.
The dashed curve is the average of the six (the consensus forecast)
and the heavy curves are the observed values.

On the basis of their model results, the investigators mentioned
above have converged on a theory for the ENSO cycle. Convergence is considerable, but not complete; the account below is mine,
but is intended to include only ideas held in common. Distillations
of these ideas to simple idealized models are given by Schopf and
Suarez (1988, 1990), Battisti and Hirst (1989), and Cane et al.
(1990). The last of these touches on some of the subtleties and
possible shortcomings of the theory.

We begin with a quick reprise of the amended Bjerknes hypothesis. If the eastern equatorial SST warms, the thermal contrast
between this region and the west Pacific is reduced, thereby reducing the strength of the trades along the equator. The change

in ocean dynamics induced by the reduction in wind stress enhances the original warming; a positive feedback, resulting in the large anomaly referred to as El Niño. The opposite states, with cold SSTs and strong trades, are maintained by the same feedback loop operating in the opposite sense.

In both the models and nature the major wind changes are in the central and western Pacific, while the major SST change are in the eastern Pacific. Since the SST change is east of the winds, the warming signal must be transmitted through the ocean as Kelvin waves, which will act to both: (1) reduce westward advection; and (2) to deepen the thermocline. Observational data cannot decide which factor is more important; the models clearly show that while both are important at some point in the ENSO cycle, it is the change in thermocline depth which initiates the temperature rise at the start of a warm event (Zebiak, 1984; Battisti, 1988). (The significance of this result is emphasized in Cane et al., 1990.)

An eastward wind stress perturbation in the central or western Pacific produces a depression of the thermocline in the east, increasing the temperature of the upwelled water and thus warming the SST. The atmosphere responds by enhancing the original westerly anomaly. By itself, this positive feedback, which is at the core of Bjerknes' scenario, would give only the pure growth Bjerknes envisioned and there would be no oscillation. But it is not the whole story; mass conservation requires that the increase of warm water in the eastern equatorial Pacific be compensated by a decrease somewhere else. The nature of equatorial ocean dynamics dictates that the negative-thermocline depth signal propagates westward from the region of the wind change in the form of Rossby waves. When this signal reaches the western boundary, it is reflected back along the equator as Kelvin waves. At the eastern end this negative Kelvin signal begins to compete with the directly forced, positive Kelvin signal.

If this indirect, delayed signal dominates, then the thermocline will rise at the east, the amplification of the wind stress will be arrested, and an oscillation is possible. Model results and analysis show that if the coupling between ocean and atmosphere is too strong, then the direct Kelvin signal, having benefited from the added growth during the delay period, will be too strong. If the coupling is somewhat weaker, then the indirect signal can eventually overcome the direct Kelvin signal, reducing the amplitude at the east, and eventually bringing it to zero. The ocean–atmosphere system is then about to begin the cold phase of the ENSO cycle.

Our account thus far has ignored the Rossby waves generated as the reflection of Kelvin waves at the eastern boundary. Cane et al. (1990) show that, were the reflection process at this boundary as efficient as the one at the west, then these waves would interfere with the ability of the forced Rossby waves to turn things around. The asymmetry in the reflections characteristic of equatorial dynamics appears to be a crucial ingredient for the oscillation.

It is important for the prediction problem to appreciate the decisive role of this indirect Rossby wave signal for the onset of a warm event. Consider the moment when the height anomaly at the east is changing from negative to positive, presaging the next warm event. Since, in this scenario at least, the wind depends on this height, the wind stress anomaly will also be changing, from easterly to westerly. Only then will the forced Kelvin wave begin to carry a *positive* height signal to the east. Thus a positive Rossby signal could only be generated at some earlier time when the wind anomalies were easterly. One signature which may indicate that the Rossby waves have prepared the way for the next event is that the mean thermocline depth anomaly in the vicinity of the equator becomes positive. Zebiak (1989a) describes how the required transfers of mass in and out of the equatorial zone take place through the course of a typical ENSO cycle.

Forecasting considerations

A dynamical model forecasts by simulating the evolution of the coupled ocean–atmosphere system into the future. The scenario presented above suggests that it may be sufficient to treat only the tropical Pacific sector of the coupled system. While there is ample evidence (for instance, presented throughout this book) that ENSO affects climate globally, the suggestion here is that the mechanisms responsible for the genesis of ENSO events all act within the tropical Pacific.

A clear implication of the foregoing is that when the dynamical model is initialized with a description of the present state of the system it will be crucial to include the distribution of the Rossby and Kelvin waves. It is a consequence of equatorial ocean dynamics (especially the geostrophic constraint) that doing so is equivalent to specifying the field of heat content variations in the ocean. In terms of the very useful approximation which allows the tropical ocean to be viewed as a two-layer fluid, this is equivalent to specifying the thickness of the layer of warm water above the thermocline.

Ideally, this specification would be done from an adequate network of observing stations. While the situation is rapidly improving under the auspices of the TOGA (Tropical Ocean–Global Atmosphere) program, such a network does not yet exist and certainly was not available in the past. Since a forecasting system must be exercised over past events in order to establish its credibility, we must face up to the need to proceed without such data. The strategy adopted is to create the oceanic initial conditions needed for a forecast by using a numerical ocean model driven by observed surface wind fields.

Unfortunately, the wind data available for the tropical Pacific is far from adequate, even today (Reynolds et al., 1989). Most of the data are obtained from merchant ship reports, and while the accuracy of an individual report is questionable, a greater difficulty is the sparse coverage of the vast Pacific by a limited number of ship tracks. Since the model requires a complete field of observations, the gaps must be filled by some kind of analysis procedure (Goldenberg and O'Brien, 1981). This analysis cannot substitute for hard information; it has been estimated that the errors attributable to wind errors in simulations of monthly variations of heat content are one-half to two-thirds of the total variations in these fields (Miller and Cane, 1989). The reader is entitled to wonder how one could hope to forecast successfully with such data; we postpone an answer to this concern until after the forecast results are presented.

The numerical forecasting model simulates the coupled evolution of the ocean and atmosphere in the tropical Pacific region. Model variables evolve deterministically, according to the physical laws governing the ocean and atmosphere. However, the model was constructed for abstract studies of large-scale ocean–atmosphere interactions in the tropics (of the kind discussed above), and greatly simplifies the suite of physics found in nature. In that context the simplifications served a didactic purpose: insofar as the model simulations of ENSO are judged to be correct, one may conclude that the omitted processes are non-essential. When prediction is the goal, then faithful, detailed simulation is desirable and the simplifications are a drawback. While some simplification is forced on us by ignorance or limited computing resources, this model is far less complex than the state-of-the-art numerical models used, for example, in operational weather forecasting.

Such shortcomings in the data and the model limit the skill that ENSO forecasts can attain in practice. In addition, forecasts of ENSO at long lead time may be impossible in principle. The interval between El Niño events is irregular – were it otherwise, predicting the next event would be trivial. The simplest versions of the theory outlined above allow a regular cycle of recurring El Niño events. One possibility is that nature modifies this regular cycle by adding noise (cf., e.g., Schopf and Suarez, 1988); in this context, "noise" would be all the phenomena at shorter time and space scales which might alter the basic ENSO cycle. Another possibility is that nonlinearities which are intrinsic to the ENSO cycle are responsible; that is, ENSO is an instance of a deterministic dynamical system exhibiting chaotic behavior. (Münnich et al., 1989, show examples of such irregular behavior in a highly idealized ENSO model in which the only nonlinearity is the rate at which ocean temperature varies with depth, reflecting the slower temperature changes above and below the narrow thermocline region.) It is likely that both "noise" and "deterministic chaos" contribute to the irregularity observed in nature, but in any case it follows that the ENSO cycle cannot be predicted infinitely far ahead. Still, predictions at useful lead times – one or two years ahead, for example – are not precluded.

Forecasting results

Detailed accounts of the forecasting model and procedure have been given elsewhere (see especially Zebiak and Cane, 1987 and Cane et al., 1986) and will not be repeated here. The model is in terms of anomalies and makes use of some empirical information about mean conditions. Otherwise, the only data input into the forecasting procedure is the surface wind field. From this information, a full set of initial conditions is created by driving the ocean model up to the time when the forecast is to begin. From that time onward, the coupled ocean–atmosphere model evolves on its own without further data inputs.

A new forecast is made every month. As evident in the examples shown in Fig. 11.3, there are often considerable month-to-month differences in these forecasts. These differences originate in the month-to-month variations in the initial conditions, which are partly a consequence of errors in the wind field and partly real, attributable to monthly variability such as the 30–60 day waves observed in the western Pacific. It is clear from the figure that the initial differences can amplify, an instance of the sensitivity to initial conditions characteristic of chaotic dynamical systems. With the aim of increasing the reliability of our predictions, we adopted the practice of averaging the forecasts initiated in six consecutive months; thus forecasts which nominally begin in, say, December, are actually the average of the six individual forecasts from December, November, October, September, August, and July.

The model forecasts were first made in 1985, and first published in *Nature* in the spring of 1986 (Cane et al., 1986). Fig. 11.4, taken from that article, compares a map of observed SST anomalies in January 1983, the peak of the 1982–83 event, with that from the forecast initiated two years earlier in January 1981. It is typical of the more successful forecasts; the gross character of the eastern Pacific El Niño anomaly is reproduced, although its meridional and westward extent is underestimated.

That article also reported the results of a large number of "retrospective forecasts" of the period 1970–85. (Wind data from earlier years were thought to be too poor to allow meaningful forecasts.) The forecasting procedure was generally successful in predicting warm events when they occurred and predicting nonevents when they did not (Fig. 11.3). A number of tests were described which assess the statistical significance of the model's ability to predict whether or not an El Niño event will occur. Based on an event/non-event (yes/no) criterion applied to that limited sample, it is extremely unlikely that the model's apparent forecasting skill was just a chance occurrence – something like one chance in 10 to one chance in 20, depending on the test used.

Figure 11.5 summarizes the retrospective forecasts in a different way by presenting the correlations of the forecast and observed SST anomaly averaged over the eastern equatorial Pacific region called NINO3 (5S to 5N; 90W to 150W). The NINO3 index was devised by the Climate Analysis Center of NOAA (National Oceanic and Atmospheric Administration), because a warming in this re-

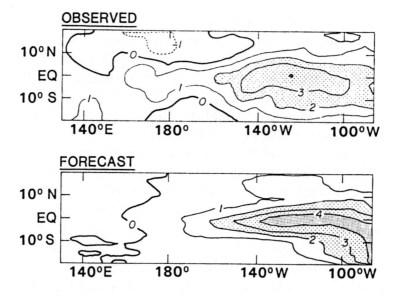

Fig. 11.4 SST anomalies (C) in January 1983. Top: observed, based on the analysis of the Climate Analysis Center (CAC) for NOAA. Bottom: predicted by the model forecast initiated in January 1981, two years earlier.

gion is thought to influence the global atmosphere strongly. It is probably the best single indicator of an ENSO episode likely to affect global climate.

The correlation coefficients in Fig. 11.5 offer another way of evaluating whether the skill of the forecasting procedure is significantly better than chance. For example, on samples as small as these a correlation of approximately 0.52 is significant at the 95 percent level, according to the usual statistical tests. (Though it is far from obvious that the assumptions underlying such tests apply to the ENSO process.)

It is more difficult for the forecasts to pass this test than the event criteria tests, because it effectively considers the ability to predict month-to-month and seasonal variations as well as the longer-term variations associated with ENSO events. The model appears to have no skill with these higher-frequency fluctuations. In fact, there is evidence that even an ocean model forced by the available wind data and capable of simulating the principal ENSO signatures has no skill at simulating higher frequencies (Seager, 1989). It would be useful to be able to simulate and forecast variations on all these time scales, but it seems that the sub-ENSO periods are out of the reach of present models and data.

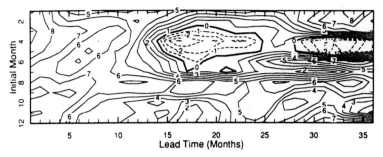

Correlation of Individual Forecasts with Observed NINO3
1970 - 1985

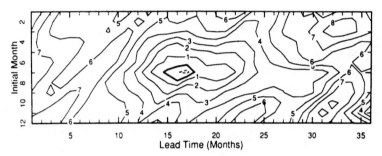

Correlation of Latest 6 Forecasts Averaged with Observed NINO3
1970 - 1985

Fig. 11.5 Correlation coefficients of the model forecast NINO3 index with observed values, 1970–85. The ordinate gives the calendar month of forecast initiation, with January at the top and December at the bottom. The abscissa is the lead time in months.

Another null hypothesis to test against can be created out of the common notion that an El Niño happens every three or four years, a "fact" which may be used to guess the next year's outcome with some skill. We elaborated that idea by using the historical record for the last hundred years to define the probability of a strong, moderate or weak El Niño event (following Quinn et al., 1987) given the number of years since the last event. Knowing when the last El Niño was, this table of probabilities can be used to generate predictions by using a random number generator. Doing so a large number of times for the 1970–85 period builds a distribution of the performance of this forecasting scheme. It turns out that there is less than a one-in-10 chance that this sophisticated guessing scheme would have performed as well as the model-based prediction system.

In summary, the retrospective forecasts show a statistically significant ability to predict such ENSO signatures as SST anomalies in the eastern equatorial Pacific, but their skill is not high. The forecasting scheme shows some ability to predict the event-like (low-frequency) features, but none for higher-frequency fluctuations. Fig. 11.6, a record of forecasts from 1970 leads, provides a view of overall forecasting performance.

Although the retrospective forecasts were carried out in a manner intended to mimic the true forecast protocol, there is some possibility that the results were influenced by the forecasters' knowledge of the events of the period. In mid-1985 we began a series of true forecasts; that is, genuine predictions of the future. An example is shown in Fig. 11.7, which compares the observed tropical Pacific SST in January 1987 with the forecast based on data through January 1986, one year earlier. This is the same forecast as that published by Cane, Zebiak and Dolan some months in advance of the verification. The forecast has clearly succeeded in predicting the moderate El Niño warming at this time. In general, the model does best in the (boreal) winter period. This performance is discussed below: (see also Figs. 11.3 and 11.4).

Figure 11.8 gives further forecast results for the 1986–87 event and its aftermath. It shows the forecasts of SST anomalies for the NINO3 region in the eastern Pacific at various lead times. The nine-month lead forecast starts the warming too soon, is correct for the winter of 1986–87, and then cools slightly in the spring while the ocean actually remains warm. Its good performance continues until summer 1988; then it underestimates the cooling event. The six-month forecast is generally similar, but the three-month one is markedly worse; it misses the 1987 warming, indicating instead a spring–summer cooling. Significantly, this behavior is shared by the zero-month lead, which is essentially the initial conditions (recall that no SST data is used in the initialization). Thus, the ocean model in hindcast mode is trying to cool while the world warms. This behavior appears to be shared by more elaborate ocean models (i.e., GCMs), and a case study of summer 1987 might help to identify serious flaws in the models or wind forcing data.

It may seem puzzling that the model forecasts can actually improve at longer lead time. One clue to this behavior is suggested

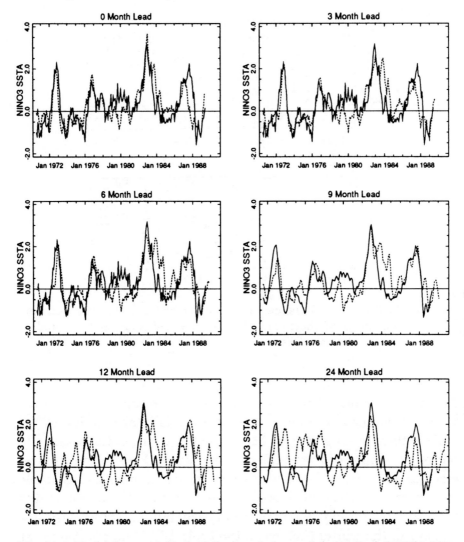

Fig. 11.6 Model forecasts (dashed line) and observed (solid line) NINO3 SST
anomalies (°C) from 1970 to 1989). The forecasts are at the various
lead times indicated; a zero-month lead forecast is actually the initial
condition generated by the ocean model forced by observed winds.

in the preceding paragraph: the initialization may be so poor that
the model benefits from having a longer lead time to recover (this
is similar to the "initialization shock" phenomenon of numerical
weather prediction). Implicit in this explanation is the idea that
the essential ENSO signal is low-frequency (cf. Barnett et al.,
1988).

A related clue is provided by Fig. 11.5, which shows that there
is a strong seasonal dependence to the forecasting skill. Forecasts

Observed (CAC/NOAA)

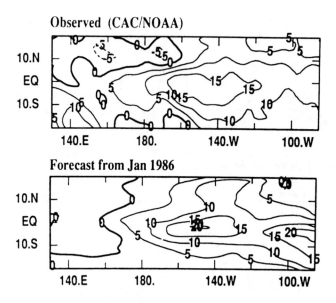

Forecast from Jan 1986

Fig. 11.7 SST anomalies in January 1987 as observed (top) and as forecast from
January 1986 (bottom). The observed field is actually the product
produced by the Climate Analysis Center (CAC)/NOAA.

for the winter period tend to be more skillful, and forecasts initi-
tiated in winter also tend to be best, especially at long leads. We
believe these results may be understood in terms of the seasonally
varying stability of the coupled ocean–atmosphere in the tropical
Pacific. The system is most unstable in the summer and is stable
in winter, a pattern evidenced in the fact that ENSO anomalies
tend to grow rapidly in summer and early fall, and are nearly stationary in the winter (i.e., when the typical El Niño anomaly is at
a maximum). All our initial conditions have considerable noise.
For forecasts initiated in summer, this noise is amplified by the
unstable conditions, often leading to a poor or erroneous forecast.
With a start in winter when the atmosphere–ocean coupling is
weakest, noise is readily dispersed by wave motions in the ocean,
thus allowing the lower-frequency ENSO signal to assume control
during the upcoming summer growth season.

   Figure 11.9, which resembles Fig. 11.8, shows the latest forecasts at the time of writing. Forecasts in this format will now be
included as an experimental product in the Climate Diagnostics
Bulletin issued monthly by the CAC. The interested reader may
make use of the other chapters of this book to judge whether such

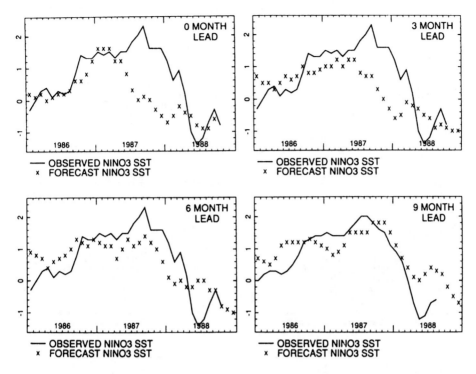

Fig. 11.8 SST anomalies in the NINO3 region (90W-150W, 5N-5S) from the analysis product (solid lines) compared with the forecasts at lead times of zero, three, six, and nine-months (crosses). For the nine-month lead each point represents the three-month average. Each forecast value is first normalized by the model mean and standard deviation for the period 1970–85 (as in Barnett et al., 1988) and then multiplied by the observed standard deviation for the same period.

forecasts have as yet attained a level of skill which makes them of some practical value.

Prospects for prediction

The degree of forecasting skill obtained despite the crudeness of the model is telling. It suggests that the mechanism responsible for the generation of El Niño events and, by extension, the entire ENSO cycle, is large-scale, robust and simple; if it were complex, delicate or dependent on small-scale details, this model could not succeed. Neither the ocean nor the atmosphere is the prime mover in the ENSO cycle. It is essential in modeling ENSO that the two-way coupling between ocean and atmosphere is active in both the El Niño and the non-El Niño phase of the cycle. The only

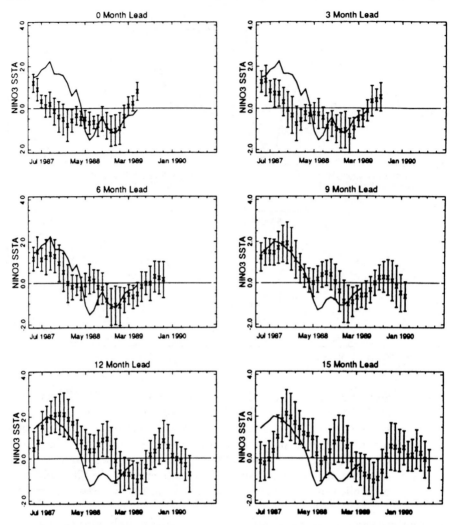

Fig. 11.9 Forecasts of NINO3 SST Anomalies. Forecasts are made using the model and procedures outlined in Cane et al. (1986) and Barnett et al. (1988). Results are presented for six different lead times ranging from zero months to 15 months. Each forecast is actually the mean of forecasts from six consecutive months, adjusted to have the same mean and standard deviation as the observed. Note that zero-month lead can differ from the observed because SST data is not used in initialization. For each lead time, forecast values are indicated by crosses and observed values by a solid line (for nine, 12 and 15-month lead times, observed values are three-month averages; otherwise, they are monthly averages). Error bars represent one root-mean-square error, based on the years 1972–87. The latest forecast included in these results (from May initial conditions) is provisional. The results generally indicate a warming trend throughout most of 1989, but do not show a major warming. Of late, individual monthly forecasts have been unusually inconsistent.

region where the model does a creditable job of simulating the requisite interactions is near the equator. Its success with ENSO is a confirmation of Bjerknes' emphasis on the interactions which take place in the equatorial plane.

The dynamics of the upper layers of the tropical ocean had to be added to the basic Bjerknes hypothesis. The tropical Pacific region remains the locus for all of the physics responsible for the existence of the ENSO cycle. This is not to deny that events in the Indian Ocean region, in midlatitudes or elsewhere, may influence the evolution of an El Niño event. While we do mean to assert that the essence of the oscillation takes place in the tropical Pacific, we expect that forecasts would be improved by simulating the rest of the globe, especially the Indian Ocean sector.

It seems clear that the predictability limit for ENSO is years rather than weeks or months. We cannot yet state it more precisely, and it cannot be calculated from forecasts alone, because the intrinsic lack of predictability of the coupled ocean–atmosphere system is not the only reason for inaccurate forecasts; poor data and model errors also contribute. We presume these failings account for the cases where the forecasts from many consecutive months are both consistent and wrong (for example, the forecasts for 1973 from 1971). It is also likely that the predictability is highly variable: at some times, the future evolution of the coupled system is insensitive to small changes; at others, the system passes so close to a bifurcation point that even small-amplitude disturbances may alter its future behavior substantially, perhaps making the difference between El Niño and non-El Niño states. We conjecture that the latter times are those for which forecasts from consecutive months are inconsistent (for example, the forecasts from 1983). Here we are implicitly assuming that month-to-month variations in initial conditions are sufficient to reveal all of these uncertain times. Further studies may reveal a better indicator of periods of unpredictability.

Some features are doubtless more predictable than others, and it must be kept in mind that our forecasts were scored catagorically, El Niño vs. non-El Niño. The predictability of more subtle features would likely be considerably shorter. The presence of the 30–60 day waves in the tropical Pacific suggests that it may be impossible to predict the early stages of an El Niño event on a month-to-month basis. It seems likely that these waves could advance or retard the discernable onset of an event by a month. If so, accurate prediction of monthly SST anomalies would require that the pattern of 30–60 day activity be predicted as well.

Little attention has been given to the model's ability to predict details of the evolution and spatial pattern of particular El Niño events. Many straightforward steps could be taken to improve the model, and such comparisons would be more rewarding with a model designed for forecasting. More important, there have been only four El Niño events since 1970. It would be extremely valuable to have more events to practice on; more are certainly needed to compute meaningful statistics. Weather forecasting would doubtless have developed more slowly had there been only four storms to work with. An optimistic estimate is that suitable data sets could be extended back to the mid-1950s; earlier times are problematical (cf. Cardone et al., 1989).

Apart from climatologies, the only data presently used in the model forecasts are derived from ship reports of surface winds. Coverage of the tropical Pacific remains poor (e.g., Reynolds et al., 1989), and studies have shown that the differences among model hindcasts of SST and heat content when driven by different wind analyses are often comparable to the signal, with only the strong signal of an ENSO event standing clearly above the noise. In view of this, the success of the forecasts is somewhat surprising. It suggests that it is the large-scale, low-frequency features of the surface wind field – the only part one would expect to be correct – which are important for ENSO. This is consistent with our emphasis on the integrated heat content, a part of the oceanic response which integrates the surface wind stress field, allowing errors to cancel. (In contrast, the ridges and troughs in the dynamic topography, which depend on the spatial variations of the wind stress, demand more accuracy at smaller scales.) It is to be expected that better knowledge of this most crucial variable would improve the forecasts, but a significant enhancement of the surface wind observations probably will have to wait for the advent of wind sensing satellites.

Though coverage is quite sparse, observations of SST, currents and ocean thermal structure are also available. As yet no attempts have been made to factor these into the initial fields. It could be done in a straightforward manner, but ultimately will require more elaborate procedures for data assimilation and model initialization, comparable in complexity to the methods now used in operational weather forecasting. The model has a different climatology when run in an unforced forecasting mode than when forced with observed winds in order to generate initial conditions. The discrepancy gives rise to an initialization shock when it is switched over from initialization mode to forecasting mode. This doubtless leads to a reduction in forecasting skill.

The problem could be addressed by constructing a more sophisticated initialization procedure. For example, the wind fields derived from observations have more structure than the model winds, a difference which could be eliminated with an additional analysis scheme. However, a deeper cause is the shortcomings of the model, its limited ability to achieve realistic simulations of the climate system.

The model was originally developed for theoretical studies, not forecasting; it is at the level of complexity of the weather forecasting models of 30 years ago. The model's forecasting performance must suffer from the lack of midlatitude variability and the absence of the 30–60 day waves. The model is unable to reproduce episodes of strong cold anomalies (e.g., 1975) and overstates easterly wind anomalies. The success of the forecasts in the face of these flaws – and many others – must be taken as an indication of the robust, large scale character of the ENSO phenomenon.

A particular concern is the model's understatement of the variability in the western equatorial Pacific. This prevents it from reproducing the observed eastward propagation of westerly anomalies during the early stages of El Niño. The absence in the model of warm anomalies west of the dateline in early 1986 may have kept the atmosphere component from generating the stronger than normal easterlies observed during that period. Strong easterlies would retard the development of an El Niño; note that the model developed its event too early.

The strongest conclusion to be drawn from the results reviewed here bears on the predictability of the coupled ocean–atmosphere system. It indicates that El Niño is generally predictable at least a year in advance. Without going beyond the present state of the art, there is considerable room for improvement in the observing system, the data assimilation procedure, and the model itself. It seems clear that with better observing networks and more refined models, operational climate forecasts of definite utility can be expected in the near future. The potential social and economic benefits of such a forecasting scheme is considered in several other chapters in the book.

Acknowledgments

I am grateful for the collaboration with Steve Zebiak and Sean Dolan on the development of the forecasting model. The writing of this article was supported financially by grant no. NA-87-AADAC081 from the U.S. TOGA Project Office of NOAA and logistically by Virginia DiBlasi.

References

Barnett, T., Graham, N. Cane, M.A., Zebiak, S.E., Dolan, S., O'Brien, J.J. & Legler, D. (1988). On the prediction of the El Niño of 1986-1987. *Science*, **241**, 192–6.

Battisti, D.S. (1988). The dynamics and thermodynamics of a warming event in a coupled tropical atmosphere–ocean model. *Journal of the Atmospheric Sciences*, **45**, 2889–2919.

Battisti, D.S. & Hirst, A.C. (1989). Interannual variability in the tropical atmosphere/ocean system: Influence of the basic state and ocean geometry. *Journal of Atmospheric Sciences*, **46**, 1687–1712.

Bjerknes, J.H. (1966). A possible response of the atmospheric Hadley circulation to equatorial anomalies of ocean temperature. *Tellus*, **18**, 820–9.

Bjerknes, J.H. (1969). Atmospheric teleconnections from the equatorial Pacific. *Monthly Weather Review.* **97**, 163–72.

Bjerknes, J.H. (1972). Large-scale atmospheric response to the 1964–65 Pacific equatorial warming. *Journal of Physical Oceanography*, **15**, 1255–73.

Busalacchi, A.J. & O'Brien, J.J. (1981). Interannual variability of the equatorial Pacific in the 1960s. *Journal of Geophysical Research*, **86**, 10901–7.

Busalacchi, A.J., Takeuchi, K. & O'Brien, J.J. (1983). Interannual variability of the equatorial Pacific – revisited. *Journal of Geophysical Research*, **88**, 7551–62.

Cane, M.A. (1986). El Niño. *Annual Review of Earth Planetary Science*, **14**, 43–70.

Cane, M.A., Münnich, M. & Zebiak, S.E. (1990). A study of self-excited oscillations of a tropical ocean–atmosphere system; Part I: linear analysis. *Journal of the Atmospheric Sciences*, **47**, 1562–77.

Cane, M.A. & Zebiak, S.E. (1985). A theory for El Niño and the Southern Oscillation. *Science*, **228**, 1085–7.

Cane, M.A., Zebiak, S.E. & Dolan, S.C. (1986). Experimental forecasts of El Niño. *Nature*, **321**, 827–32.

Cardone, V.J., Greenwood, J.G. & Cane, M.A. (1989). On trends in historical marine wind data. *Journal of Climate*, **3**, 113–27.

Goldenberg, S.B. & O'Brien, J.J. (1981). Time and space variability of tropical Pacific wind stress. *Monthly Weather Review*, **109**, 1190–2007.

Graham, N.E., Michaelsen, J. & Barnett, T.P. (1987). An investigation of the El Niño-Southern Oscillation cycle with statistical models; 1. Predictor field characteristics. *Journal of Geophysical Research*, **92**, 14251–70.

Miller, R.N. & Cane, M.A. (1989). A Kalman filter analysis of sea level height in the tropical Pacific. *Journal of Physical Oceanography*, **19**, 773–90.

Münnich, M., Cane, M.A. & Zebiak, S.E. (1990). A study of self-excited oscillations of a tropical ocean-atmosphere system; Part II: the nonlinear case. *Journal of the Atmospheric Sciences*, submitted.

Quinn, W.H. & Neal, V.T. (1987). El Niño occurrences over the past four and a half centuries. *Journal of Geophysical Research*, **92**, 14449–61.

Reynolds, R.W., Arpe, K., Gordon, C., Hayes, S.P., Leetmaa, A. & McPhaden, M.J. (1989). A comparison of tropical Pacific surface wind analyses. *Journal of Climate*, **2**, 105–11.

Schopf, P.S. & Suarez, M.J. (1988). Vacillation in a coupled ocean–atmosphere model. *Journal of the Atmospheric Sciences*, **45**, 549–66.

Schopf, P.S. & Suarez, M.J. (1990). Ocean wave dynamics and the timescale of ENSO. *Journal of Physical Oceanography*, in press.

Seager, R. (1989). Modeling tropical Pacific sea surface temperature: 1970–1987. *Journal of Physical Oceanography*, **19**, 419–34.

Seager, R., Zebiak, S.E. & Cane, M.A. (1988). A model of the tropical pacific sea surface temperature climatology. *Journal of Geophysical Research* **93**, 1265–80.

Weare, B.C., Strub, P.T. & Samuel, M.D. (1980). *Marine Climate Atlas of the Tropical Pacific Ocean*. Davis, CA: University of California, Department of Land, Air and Water Resources.

Wyrtki, K. (1975). El Niño–the dynamic response of the equatorial Pacific Ocean to atmospheric forcing. *Journal of Physical Oceanography*, **5**, 572–84.

Wyrtki, K. (1979). The response of sea surface topography to the 1976 El Niño. *Journal of Physical Oceanography*, **9**, 1223–31.

Zebiak, S.E. (1984). *Tropical Atmosphere-Ocean Interaction and the El Niño/Southern Oscillation Phenomenon*. Ph.D. Thesis.

Zebiak, S.E. (1989a). Ocean heat content variability and El Niño cycles. *Journal of Physical Oceanography*, **19**, 475–86.

Zebiak, S.E. (1989b). On the 30–60 day oscillation and the prediction of El Niño. *Journal of Climate*, **2**, 1381–7.

Zebiak, S.E. & Cane, M.A. (1987). A model El Niño-Southern Oscillation. *Monthly Weather Review*, **115**, 2262–78.

12

Use of statistical methods in the search for teleconnections: past, present, and future

BARBARA G. BROWN and RICHARD W. KATZ
Environmental and Societal Impacts Group
National Center for Atmospheric Research*
Boulder, CO 80307

The number of satisfactorily established relationships between weather in different parts of the world is steadily growing ... and I cannot help believing that we shall gradually find out the physical mechanism by which these are maintained, as well as learn to make long-range forecasts to an increasing extent.

—Sir Gilbert T. Walker
(Walker, 1918, p. 223)

Introduction

Teleconnections research has been a scientific endeavor among atmospheric scientists and oceanographers since the early part of this century. Sir Gilbert T. Walker, a mathematician by training and an early proponent of studies of the connections between weather conditions around the globe (i.e., what Walker called "World Weather"), pioneered in the application of statistical methods in such investigations (e.g., Walker, 1923). Lacking any well-developed physical theory, early searches for teleconnections naturally relied upon an empirical approach.

Interest in teleconnections research has peaked and ebbed and peaked during the past eighty years, but the statistical methods that have been used, and that are commonly used today, remain

* The National Center for Atmospheric Research is sponsored by the National Science Foundation.

basically unchanged from the approach taken by Walker and his colleagues in the early twentieth century. In fact, some of the methods that Walker used were fairly sophisticated, in some cases more sophisticated than methods that are applied today. Thus it may be instructive to reexamine the early history of teleconnections research in an attempt to determine why research in this area has not progressed as much as might have been expected over the past eight decades. In addition, an investigation of the experiences of the early empiricists who pioneered in research related to global teleconnections may uncover certain lessons for future research in this area.

As in the past, many recent studies devoted to the search for atmospheric/oceanic teleconnections are based on empirical analyses of climatic and oceanic data that rely extensively on statistical techniques. However, in some cases inappropriate statistical methods have been applied or statistical assumptions have been ignored. Consequently, the strength of some alleged linkages may have been exaggerated. Moreover, it has even been argued (e.g., by Ramage, 1983; Pittock, 1984) that some teleconnection relationships have either weakened, reversed sign, or vanished.

Based on lessons drawn from this historical review, the statistical approaches that are currently being taken in teleconnections studies are evaluated. In particular, the problems of "multiplicity" and autocorrelation (i.e., temporal dependence) are considered. Multiplicity arises, for example, when many correlation coefficients are computed but only the largest are selected as relevant. With only a few exceptions (e.g., Livezey and Chen, 1983; Preisendorfer and Barnett, 1983), the implications of multiplicity have not been fully appreciated within the meteorological community. The issue of autocorrelation, considered by Katz (1988) in the specific context of teleconnections research, concerns how temporal correlation may affect the identification of leading, lagging, or feedback relationships, as well as the reliability of estimated cross correlations. To demonstrate the magnitudes of the problems resulting from autocorrelation and multiplicity in the search for teleconnections, the results of simulation studies are described.

Historical review of teleconnections research

Goals

The primary goal of early teleconnections research was to establish relationships between the weather at distant points on the globe that could be used for seasonal forecasting, or seasonal "foreshadowing" as it was sometimes called (Walker, 1930). Of particular interest was the development of relationships that could be used to forecast Indian monsoon rainfall (e.g., Walker, 1910; Mossman, 1924; Normand, 1932; Walker, 1924b). In fact, it appears that the earliest official interest in seasonal forecasting arose in the 1880s as a result of a directive by an Indian Famine Commission (Normand, 1932). Among the other phenomena for which seasonal forecasting equations were sought are rainfall in Northeast Brazil (Walker, 1928a; Ferraz, 1929), rainfall in Australia (Walker, 1910; Hunt, 1929; Walker and Bliss, 1930), and rainfall in Southern Rhodesia (Bliss, 1928). Each of these regions has been plagued by drought and related agricultural and other societal impacts.

A secondary goal of early teleconnections research was to use empirical methods to obtain an understanding of physical relationships and processes in the atmosphere. In fact, Walker used some of the relationships that he discovered to describe the Southern Oscillation, as well as the North Atlantic and the North Pacific Oscillations (Walker, 1924a). From his understanding of the circulation in the Southern Hemisphere, obtained through examination of many correlations of worldwide weather, Walker formulated the first Southern Oscillation Index (SOI) (Walker and Bliss, 1932).

The goals of modern teleconnections research differ only slightly from the goals of the early studies. For example, the first objective of the Tropical Ocean and Global Atmosphere Program (TOGA), sponsored by the World Meteorological Organization and many national governments, is "to gain a description of the tropical oceans and the global atmosphere as a time-dependent system, in order to determine the extent to which this system is predictable on time scales of months to years, and to understand the mechanisms and processes underlying its predictability" (World Climate Research Programme, 1985, p. 3). Similarly, the U.S. National Academy of Sciences Climate Research Committee has stated a practical objective for the U.S. TOGA–related research program as "the development of improved schemes for prediction of short-term climate variability" (Climate Research Committee, 1983, p. 23).

As in earlier teleconnection studies, the development of seasonal forecasts of Indian monsoon rainfall and other regional rainfall phenomena is the goal of many modern investigations (e.g., Khandekar, 1979; Nicholls, 1981; McBride and Nicholls, 1983; Rasmusson and Carpenter, 1983; Shukla and Paolino, 1983; Bhalme and Jadhav, 1984; Hastenrath et al., 1984; Parthasarathy and Pant, 1984, 1985; Hastenrath, 1987a,b). However, in contrast to the earlier studies, a stronger linkage between empirical and physical investigations exists in many studies today. For example, many recent teleconnection studies devote considerable effort to investigating the physical processes that may lead to the observed relationships (e.g., Rasmusson and Carpenter, 1983). This difference exists primarily because of the availability of much more data, including satellite observations, that can be used to verify and understand relationships discovered through an empirical approach. Both of the TOGA research programs described earlier emphasize the need for physical studies (including computer simulations) in combination with empirical investigations.

Early methods (before 1930)

Prior to 1900, efforts at seasonal forecasting using relationships between the weather at two distant points were generally based on qualitative methods, consisting simply of observing the coincidence of two phenomena (Normand, 1932). Apparently Walker was the first researcher to apply statistical methods to the problem of finding and describing teleconnections (though he did not actually use the term "teleconnection"), and using these relationships to obtain quantitative predictions of the phenomena of interest. In particular, Walker introduced the use of correlation coefficients to the study of teleconnections, and the use of multiple regression to the problem of seasonal forecasting (Normand, 1932).

Walker used these methods to evaluate a very large number of relationships, including those involving time lag, between the weather at many points all over the globe (e.g., Walker, 1910, 1914, 1923, 1924a, 1928b). These relationships were used to develop prediction formulas for Indian monsoon rainfall (e.g., Walker, 1914, 1924b) and other seasonal weather events around the world (e.g.,

Walker, 1910, 1928a, 1930; Walker and Bliss, 1930). Generally these prediction formulas consisted of multiple regression equations containing six or seven terms. The variables included in the equations were selected on the basis of the magnitude of their simple correlations with the predictand (e.g., Walker, 1928a).

Following Walker's lead, many other researchers applied methods of correlation and regression in their search for relationships. For example, Mossman correlated many pairs of weather phenomena, such as Nile flood and rainfall at Santiago, Chile; temperatures at Cordoba, Argentina and Alice Springs, Australia; and so on (Mossman, 1913). Bliss investigated relationships between winter temperatures in Great Britain and world weather (Bliss, 1926a), and the Nile flood and world weather (Bliss, 1926b). In addition, he developed a formula for predicting the height of the Parana River in Argentina (Bliss, 1928). Groissmayr (1929) considered relationships between summer weather in India and the following winter weather in Canada. Others also searched for such relationships.

However, no one contributed as much as Walker to the application of statistics in the search for teleconnections. One of Walker's earliest accomplishments was the introduction of multiple regression and the multiple correlation coefficient into common usage among teleconnections researchers (Walker, 1910). These techniques encouraged investigators to consider relationships among groups of variables, rather than evaluating each pair of variables separately.

Walker also was cognizant of at least some of the pitfalls of his approach. Specifically, he recognized the problem of "multiplicity"; that is, simply by computing a large number of correlation coefficients, he was likely to find a certain number that would appear to be large, or "significant," even if no true relationships actually existed (Walker, 1914). The bias associated with selecting the largest among a group of correlation coefficients was of particular concern to Walker, who described and applied a method to compensate for this effect. This method, described in more detail later in this chapter, basically involved computing the "probable value" of the largest single correlation that could be expected among a group of many correlations. Only those correlations that were larger than the probable value were accepted as representing significant relationships (Walker, 1914).

Other researchers also made contributions to the use of statistics in studies of teleconnections. For example, Dines (1916) advocated the use of partial correlation coefficients and later initiated a debate regarding the significance of simple and multiple correlation coefficients (Dines, 1917). Wishart (1928) contributed a paper to the meteorological literature which considered significance levels of multiple correlation coefficients.

Reactions of the meteorological community (before 1960)

The response to Walker's work between 1910 and 1935 appears to have generally been positive, in that many other scientists adopted his approaches to the investigation of empirical relationships in the atmosphere. Walker also seems to have been a highly respected member of the meteorological community, and his work was well praised in discussions at the Royal Meteorological Society (e.g., Walker, 1930).

However, some skepticism was voiced by Dines (1916) in a published review of Walker's papers on "Correlation in Seasonal Variations of Weather" (Walker, 1910, 1914). Dines particularly criticized the use of correlations for forecasting, saying that the resulting reduction in standard error would be too small to be of much use. It was Dines' opinion that "Forecasts based on correlation coefficients under 0.70 or 0.80 can hardly be much more than pure guesses." Dines believed that correlation coefficients are useful only for investigating physical relationships. Walker responded by stating his belief that "qualitative" forecasts (i.e., forecasts of particular categories, such as below normal, near normal, above normal, rather than specific numerical forecasts) could be very successful (Walker, 1917). He also demonstrated that small reductions in the standard error of the predictand could be quite useful for forecasting. In further correspondence, Dines stated that, "purely from the algebraic side, the prospect of forecasting by means of correlation does not seem to me to be hopeful." Dines' concerns related to the accuracy with which correlations could be known and predictions could be made (Dines, 1917).

Later criticisms of Walker's work primarily involved reevaluations of the forecasting formulas that were developed during the course of Walker's studies of world weather, and these critiques may have led, in part, to the decline in the use of the statistical approach in the middle part of this century. Sellick (1932) was concerned that the correlation coefficients between the weather at various points on the globe had shown little improvement since Walker initiated his correlation studies, despite the fact that more data were available from more locations. Sellick also presented the results of a study in which correlations of various factors with Indian monsoon rainfall, originally computed by Walker, were updated using ten additional years of data. In general, the higher-valued correlation coefficients decreased in magnitude, whereas the smaller coefficients maintained their values or increased somewhat. Sellick concluded that the true values of the correlations described by Walker were probably somewhat lower than Walker had estimated. Today we know that such behavior could have been anticipated and, at least in part, simply reflects the bias introduced by Walker in his original selection of the largest correlation coefficients (i.e., because of the multiplicity problem).

Montgomery (1940a) described and critiqued the physical aspects of Walker's work in some detail. He also presented a reevaluation of Walker's formulas for Indian monsoon rainfall (Montgomery, 1940b). In particular, he updated the regression coefficients in these equations using data from various more recent periods. Montgomery found that the correlation coefficients between observed and predicted rainfall gradually deteriorated for later periods, and that the correlation changed sign for the latest period (1922–36) for two of the equations. Of the eight factors included in the equations, four factors had negligible correlations with Indian rainfall, one had reversed sign, and three factors had maintained the strength of their correlation in recent years.

Another reevaluation of Walker's monsoon rainfall forecasting equations was undertaken by Normand (1953). As a result of his analysis, Normand concluded that the formulas themselves were still fairly well related to monsoon rainfall. However, he criticized the fact that the equations seemed to be unable to adequately predict periods of drought. It was Normand's opinion that the formulas should not be used until this deficiency could be overcome. Normand also discussed some of the difficulties of the correlation approach to forecasting, and of the empirical approach taken by Walker to investigating relationships between weather around the world. He commended Walker for his care in using the correlation approach and his awareness of its limitations.

Grant (1956) examined the frequency distribution of the correlation coefficients presented in one of Walker's studies of monsoon
rainfall (Walker, 1924b). By comparing this empirical distribution
to the theoretical distributions associated with true correlations of
0.0, 0.1, and 0.2, she concluded that the true correlations were on
the order of 0.1, and that the relationships could not be of any
use for forecasting. Grant also discussed the bias associated with
selecting the largest coefficients from a group of correlations, an
issue of which Walker was well aware, as discussed earlier. In
addition, she considered the impact of trends and periodicities
on the computed values of correlation coefficients, demonstrating
that such systematic variations may seriously inflate the computed
coefficients above their true values.

Even Walker's obituary, which appeared in 1959 in the *Quarterly
Journal of the Royal Meteorological Society*, was somewhat critical
of Walker's empirical approach to the study of global weather relationships. In describing Walker's correlation studies, the author
of the obituary stated:

> *Walker's hope was presumably not only to unearth relations
> useful for forecasting but to discover sufficient and sufficiently
> important relations to provide a productive starting point for
> a theory of world weather. It hardly seems to be working out
> like that.*

Like so many forecasts of the future, this statement turned out to
be completely wrong.

Renaissance of teleconnections research (after 1960)

Interest in teleconnections has apparently followed a cyclical
course, ranging from intense hope and belief in the possibility
of formulating useful forecasting equations and explaining physical processes, to complete disregard for the possibility that useful
information is obtainable from the empirical approach. The possible explanations for these variations in interest are many, and the
actual reasons cannot be identified with certainty. The waning of
interest in the 1940s and 1950s may have been a result of the studies by Montgomery (1940a,1940b), Normand (1953), Grant (1956),
and others, which cast doubt on the long-term possibility of using
correlation-derived relationships as forecasting tools, due to their

apparent degradation in strength over time. Other possible factors include an increased reliance on physical/numerical methods in meteorological research, and a greater effort toward short-range (rather than long-range) weather prediction in this period. Studies by Troup (1965) and Bjerknes (1969) of the Southern Oscillation and teleconnections may have been the catalyst for the recent resurgent interest in teleconnections. These studies generally took a more physical (i.e., less empirical) approach than was taken in most of Walker's studies.

In spite of the criticisms of Walker's empirical approach, and in spite of the advances made in the theoretical understanding of the global climate system, empiricism is still strongly relied on in modern studies of teleconnections. Furthermore, the simple correlation coefficient is the tool most frequently used in the search for relationships. For example, this approach has recently been used by, among others, Elliott and Angell (1988), Rogers (1988), Van Heerden et al. (1988), and Wright et al. (1988).

Another approach that is commonly applied involves dividing the time series of the variable of interest (e.g., rainfall) into groups based on the concomitant value of another variable (e.g., occurrence/non-occurrence of an El Niño event). Then the mean values of the groups are compared using a t-test, or the median values are compared using a nonparametic test such as the Mann–Whitney U. For example, Yarnal and Diaz (1986) compared rainfall and temperature averages in the northwestern U.S. and western Canada between "warm event" and "cold event" years, using the t-test to test for significant differences. This approach would be more appropriate than ordinary correlation if the relationship is nonlinear or if extremes are of particular importance (see the following section).

The problems of autocorrelation and multiplicity are of concern in any event. However, in many cases these issues are still not adequately taken into account. For example, persistence is completely ignored in some studies (e.g., Andrade and Sellers, 1988; Rogers, 1988; Van Heerden et al., 1988). When autocorrelation is taken into account in determining the significance of the computed cross correlations (e.g., Gordon, 1986; Elliott and Angell, 1988; Wright et al., 1988), this adjustment generally is accomplished by estimating the number of independent samples using methods described by Quenouille (1952) and Bartlett (1955).

Modern investigators (including those who adjust significance tests for autocorrelation) are apparently unappreciative of the influence that autocorrelation can exert on the structure of the cross correlation function (Katz, 1988). As explained in the following section, the existence of autocorrelation can lead to apparent leading or lagging relationships when none is actually present. In modern studies, cross correlation functions are frequently presented for variables without first removing or correcting for the effects of autocorrelation (e.g., Angell et al., 1984; Wright et al., 1988). Although Walker was definitely aware of the persistence of the Southern Oscillation (e.g., Walker and Bliss, 1937), he and his colleagues were also unaware of its effects on cross correlation.

In contrast, Walker clearly recognized the problem of multiplicity – in particular, the bias associated with selecting only the factors with the largest correlations – and he attempted to take this problem into account in his teleconnection studies (e.g., Walker, 1914). This issue is generally disregarded or not appropriately handled in most recent investigations. For example, many studies compute large numbers of correlation coefficients and report their significance according to the tabled critical values listed for "5 percent" or "1 percent" significance levels (e.g., Andrade and Sellers, 1988; Rogers, 1988; Van Heerden et al., 1988). These values are appropriate for testing the significance of a single correlation, but are not appropriate when the significance of a large number of correlations is considered (see the following section).

Furthermore, because multiplicity has so regularly been ignored in these studies, an adequate methodological approach has not yet been developed to counteract its effects in teleconnection studies. For example, Wright et al. (1988) recognized the problem of multiplicity, noting that the true "a posteriori significance levels" of their correlation coefficients were actually much lower than the tabled values. However, they were unable to determine "how to estimate them accurately." Their solution was to select a critical correlation value that was larger than the value indicated by standard statistical tables; the computed correlations had to exceed this selected value to be considered significant. The intention of this approach was to use a critical value that would provide a conservative measure of the significance of the computed correlations.

With regard to the desire to develop methods for long-range forecasting, Walker also applied methods that are more sophisticated than those used in most modern studies. In particular, Walker used multiple regression and the multiple correlation coefficient to evaluate relationships with more than one variable (e.g., Walker, 1910). Current teleconnection studies, with some exceptions (e.g., Hastenrath et al., 1984; Hastenrath, 1987a,b), have not taken advantage of multiple correlation or regression. Of course, such methods must be used with even greater caution than the simple correlation coefficient, due to the increased difficulty of interpreting physically the results of regressions involving more than one variable and the increased potential for problems related to multiplicity.

Why haven't the statistical techniques used in modern teleconnections research improved much, if any, over the methods that were used by Walker and his contemporaries? One possible explanation is that the lessons learned by Walker and others have simply been forgotten with the passage of time. Moreover, the current situation may reflect a lack of appreciation by physical scientists of the complexities that arise in any empirical approach. Taking into account all of the factors that should be of concern in studying teleconnections, such as multiplicity and time dependence, may not be entirely straightforward. This consideration may have inhibited progress and improvement in the statistical methodologies used in teleconnections research, which in turn may have contributed to a less than optimal use of available information.

Current statistical issues

Before discussing current statistical issues in teleconnections research, it is perhaps helpful to review the role correlation coefficients play in assessing relationships between variables. It is not always appreciated that the correlation coefficient measures only the degree of *linear* relationship between two variables (in fact, the square of the correlation coefficient represents the proportion of the variation in one variable that can be "explained" by predictions based on the assumed linear relationship with the other variable). Consequently, the correlation coefficient is not an appropriate statistic when relationships between variables are highly nonlinear (e.g., when only extreme events such as excursions above or below a threshold matter). In such cases, other approaches such as contingency tables (e.g., categorizing years according to whether

or not an intense ENSO event occurred and as to whether or not
an extreme climatic anomaly occurred in the region of concern)
would be more appropriate (e.g., Quinn et al., 1978). Here we
restrict attention to only those situations in which the use of the
correlation coefficient is justified.

Autocorrelation

Statistical inferences about correlation coefficients are commonly
based on the assumption that the individual time series being
related are not correlated over time (i.e., no autocorrelation is
present). In fact, climatic/oceanic time series do possess substantial autocorrelation; generally, a positive dependence, reflecting a
tendency of climatic/oceanic anomalies to persist over time. The
effects of such temporal correlation on the distribution of the sample cross correlation coefficient are well established (e.g., Haugh,
1976), and techniques for adjusting for these effects are commonly
employed in fields such as economics (e.g., Haugh and Box, 1977).
Besides affecting the reliability of the estimates of the cross correlation between variables, autocorrelations complicate the search for
predictive relationships between time series (i.e., cases in which
one variable "leads" the other). Katz (1988) has demonstrated
how these problems arise in the specific context of teleconnections
research dealing with climatic/oceanic variables.

Consider a pair of climatic/oceanic variables $\{(X_t, Y_t) : t = \ldots, -1, 0, 1, \ldots\}$, where X_t and Y_t denote the observations of
the variables X and Y, respectively, at time t. Let $\rho_X(k) = \text{corr}(X_t, X_{t+k})$ and $\rho_Y(k) = \text{corr}(Y_t, Y_{t+k})$, $k = 1, 2, \ldots$, denote
the autocorrelation functions for the two time series, representing
the ability of an individual variable to predict its future behavior using only its own past and current states. In teleconnections
research, attention is commonly directed toward the cross corre-
lation function $\rho_{X,Y}(k) = \text{corr}(X_t, Y_{t+k})$, $k = \ldots, -1, 0, 1, \ldots$.
This function measures the extent of linear relationship between
the two variables, when one time series leads/lags the other by k
time steps.

A formal bivariate time series model is considered by Katz
(1988), in which $\{X_t\}$ and $\{Y_t\}$ are assumed to be first-order au-
toregressive processes and the temporal dependencies of the two
series are characterized by the parameters $\rho_X(1)$ and $\rho_Y(1)$ (i.e.,

the first-order autocorrelation coefficients), respectively. As mentioned earlier, the fact that climatic/oceanic time series are autocorrelated complicates the identification of leading/lagging relationships.

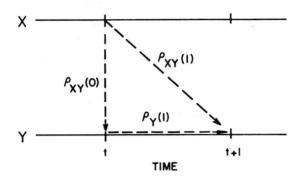

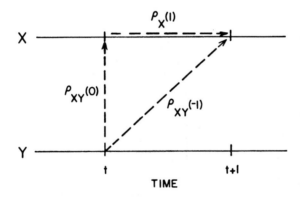

Fig. 12.1 Schematic diagram showing how cross correlation coefficients at
nonzero lags, $\rho_{X,Y}(1)$ and $\rho_{X,Y}(-1)$, depend on contemporaneous
cross correlation $\rho_{X,Y}(0)$, as well as on first-order autocorrelation
coefficients $\rho_Y(1)$ and $\rho_X(1)$, respectively.

   Figure 12.1 illustrates how the presence of autocorrelation
will automatically result in apparent leading/lagging relationships, even when only a contemporaneous relationship is actually
present. More specifically, Fig. 12.2 shows the theoretical cross
correlation function for a special case of the bivariate time series
model considered by Katz (1988). The parameters of this particular model are such that only a contemporaneous relationship is

present between X and Y with no leading, lagging, or feedback (i.e., the past and present states of X (Y) are of no help in predicting the future behavior of Y (X)). However, as is evident in Fig. 12.2, the effect of autocorrelation is to "smear out" the cross correlation function, producing nonzero cross correlations at nonzero lags (i.e., apparent leading, lagging, or feedback when none is actually present). One technique for dealing with this problem involves "prewhitening"; that is, transforming the original autocorrelated time series into an underlying uncorrelated time series. For the cross correlation function between the prewhitened time series, the identification of leading, lagging, or feedback relationships is much more straightforward (Katz, 1988).

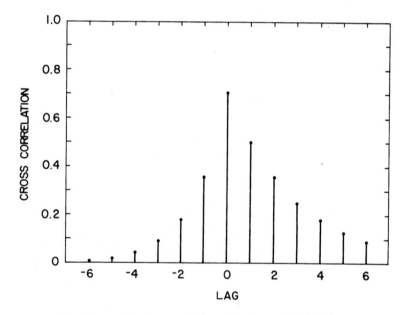

Fig. 12.2 Theoretical cross correlation function when $\rho_X(1) = 0.5$, $\rho_Y(1) = (0.5)^{1/2}$, and $\rho_{X,Y}(0) = (0.5)^{1/2}$ (see Katz, 1988).

Autocorrelation also has an effect on the sampling distribution of cross correlation estimates. In particular, for the bivariate time series model considered by Katz (1988), the standard deviation of the sample contemporaneous cross correlation coefficient is inflated over the case of no autocorrelation by the approximate factor

$$f = \left[\frac{1 + \rho_X(1)\rho_Y(1)}{1 - \rho_X(1)\rho_Y(1)}\right]^{1/2} \tag{12.1}$$

(Bartlett, 1955, p. 289). This effect is analogous to the adjustment for time averages termed the "effective number of independent samples" (e.g., Madden, 1979). Since $\rho_X(1)$ and $\rho_Y(1)$ are positive for most climatic/oceanic time series, f is greater than one. In heuristic terms, the positive autocorrelations produce higher frequencies of relatively long runs of much above or much below normal values for the two time series, resulting in more variable behavior for cross correlations obtained from such data. Consequently, failure to make the adjustment (Eq. 12.1) for autocorrelation would result in too frequently erroneously concluding that teleconnections exist when, in fact, they do not. Alternatively, if the time series are first prewhitened as mentioned earlier, then the effects of autocorrelation would be removed and no adjustment to the standard deviation of the sample contemporaneous cross correlation coefficient would be necessary (Katz, 1988).

It is popular to smooth climatic/oceanic time series (e.g., using moving averages) before calculating cross correlation functions (e.g., Niebauer, 1984; Gordon, 1986). However, smoothing has an effect on cross correlation similar to that of autocorrelation, because the process of smoothing serves to introduce or increase persistence (see Katz, 1988). Another practice that is common in teleconnections research is the search for changes in relationships between climatic/oceanic variables over time (Pittock, 1984). Sometimes running cross correlations are employed to examine the stability of such relationships (Ramage, 1983; Elliott and Angell, 1988). Like moving averages, the use of running cross correlations acts as a smoother and can introduce apparent patterns when none really exist. Katz (1988) provides compelling evidence that plots of running cross correlations should not be relied on as more than an exploratory tool of data analysis.

Multiplicity

Multiplicity is a well-known statistical problem that arises when a large number of statistical hypotheses are tested (e.g., when the results of an experiment are stratified in many ways). In the case of teleconnections research, the null hypothesis for a single significance test generally is that the true correlation between two variables is zero (i.e., no teleconnection). A Type I error consists of concluding that the correlation is nonzero (i.e., a teleconnection exists) when, in fact, the true correlation is zero (i.e., no teleconnection exists). The effect of multiplicity is such that the probability of rejecting at least one null hypothesis in error increases geometrically as more hypotheses are tested (Tukey, 1977). For example, the probability that at least one of ten independent tests of hypotheses will be judged in error to be significant, given that each is tested at the 0.05 level, is about 0.40, i.e. much greater than 0.05.

Multiplicity is an important issue in teleconnection studies in which a large number of correlation coefficients are computed. The typical approach taken in these studies has been to report those correlations that are significant according to the criteria used to evaluate the significance of individual correlation coefficients. Often only the maximum correlation (in absolute value) in a set is reported.

Some techniques, termed "multiple comparisons" (Miller, 1981), are available to counteract the effect of multiplicity. In these approaches, the important distinction needs to be made between the level (i.e., probability of a Type I error) of the individual tests, denoted by α, and the overall level, denoted by α_o, of the combination of tests. Here α_o represents the probability that it is concluded that one or more teleconnections exist, when, in fact, no teleconnections are present.

To make clear the distinction between the two probabilities α and α_o, Table 12.1 provides a hypothetical example of two independent tests. Suppose that we are interested in correlating the SOI with the weather at two locations, A and B, for example. Further, suppose that there is actually no correlation between the SOI and the weather at either location A or location B. For this situation, Table 12.1 outlines the calculation of the probability α_o of erroneously concluding that either the SOI and the weather at location A, or the SOI and the weather at location B, or both, are correlated (i.e., the probability of rejecting one or both of the null hypotheses). For instance, if $\alpha = 0.05$, then $\alpha_o \simeq 0.10$; if $\alpha = 0.01$, then $\alpha_o \simeq 0.02$.

It is desirable to select values of α for the individual tests that will produce an acceptable α_o (e.g., $\alpha_o = 0.05$) for the set of tests. For example, a method that is frequently applied in practice (e.g., Andrade and Sellers, 1988; Rogers, 1988; Van Heerden et al., 1988) is to arbitrarily show results obtained using values of

Table 12.1 Hypothetical example of multiplicity with two independent tests in which the correct decision is to not reject the null hypothesis: determination of overall level α_o.

Outcomes of:		Decision	Probability
Test No. 1	Test No. 2		
Do not reject	Do not reject	Correct	$(1-\alpha)^2$
Reject	Do not reject	Incorrect	$\alpha(1-\alpha)$
Do not reject	Reject	Incorrect	$(1-\alpha)\alpha$
Reject	Reject	Incorrect	α^2

$$
\begin{aligned}
\alpha_o &= \text{Probability of incorrect decision} \\
&= \alpha(1-\alpha) + (1-\alpha)\alpha + \alpha^2 \\
&= 1 - (1-\alpha)^2.
\end{aligned}
$$

α (e.g., $\alpha = 0.01$) that are less than the commonly used level of 0.05. The objective of using smaller values of α is to reduce the value of α_o and the number of correlations selected as being significant. However, with this technique the actual value of α_o is not explicitly considered.

The Bonferroni inequality (a statistical inequality related to the combination of probabilities) leads to another method of counteracting multiplicity in which the desired overall significance level is predetermined and the possibility of dependence among the tests is taken into consideration (e.g., Neter et al., 1983). Tests may be dependent, for instance, when the same variable is correlated with two or more other variables. With this approach,

$$
\alpha = \frac{\alpha_o}{k} \, ,
\tag{12.2}
$$

where k is the number of correlations tested, would be used to test the significance of the individual correlations. In other words, if $k = 10$ correlations between pairs of variables are calculated, then an individual test level of $\alpha = 0.005$ should be employed to achieve an overall level of $\alpha_o = 0.05$. This procedure is conservative, guaranteeing that the overall level is actually less than or equal to the desired value α_o.

Walker (1914) also suggested and applied a method for taking into account the problem of multiplicity. Walker's method consisted of determining a critical correlation value c satisfying, under the null hypothesis that all the correlations are actually zero,

$$\Pr\{ \max_{1 \le i \le k} |r_i| > c \} = \alpha_o \ , \qquad (12.3)$$

where r_i denotes the ith sample correlation coefficient. Applying the well-known formula for the exact distribution of the maximum value of a set of independent random variates (e.g., Mood et al., 1974, pp. 182–184) or using elementary probability theory, c is the value of the correlation coefficient that satisfies

$$\Pr\{|r_i| > c\} = 1 - (1 - \alpha_o)^{1/k}, \quad 1 \le i \le k \ . \qquad (12.4)$$

Hence the appropriate value of α is simply given by the right-hand side of (12.4). In contrast to the Bonferroni approach, this method assumes independence among the tests.

Because Walker's goal was to estimate the "probable value" of the largest in a set of correlation coefficients (i.e., what we would now call the *median* value of the maximum), he used a value of 0.50 for α_o. Walker computed estimates of the probable value using assumptions regarding the distribution of the correlation coefficient and an approach that would be considered rather crude today, although they were quite reasonable and perhaps even sophisticated at the time. In essence, Walker applied the large-sample normal approximation for the distribution of the sample correlation coefficient. Today we know that the correlation coefficient for small samples follows a t distribution (under the assumptions that the true correlation is zero and that the observations have a joint bivariate normal distribution). Thus, in a modernized version of Walker's approach to counteracting problems of multiplicity, we would set α_o equal to some relatively small value (e.g., $\alpha_o = 0.05$) and we would use the t distribution to determine the appropriate value of c.

Simulations to investigate the implications of multiplicity were based on a multivariate generalization of the bivariate time series models described by Katz (1988) and briefly mentioned earlier. In these simulations, sets of between two and 50 time series were created under specific assumptions regarding the autocorrelation and cross correlation structure of the series. Characteristics of the distributions of the sample cross correlations between the simulated series were studied: (i) to determine the effects of multiplicity on the distribution of the maximum (absolute value) sample cross correlation in the set; and (ii) to evaluate the ability of various approaches (e.g., the Bonferroni method) to counteract the effects of multiplicity.

In particular, two situations were considered: (i) the case in which autocorrelation is present in all of the series (i.e., $\phi = (0.5)^{1/2}$, where ϕ denotes the first-order autocorrelation coefficient common to all the time series); and (ii) the case in which none of the series are autocorrelated (i.e., $\phi = 0$). Nonzero contemporaneous cross correlation, γ say, was allowed to exist only between the first two time series (i.e., $\gamma = 0.5$); for all other pairs of time series, the cross correlations were always assumed to be zero. The case of no contemporaneous cross correlation between the first two series (i.e., $\gamma = 0$) also was considered. The simulated series each had a length of 50, a value that is similar to the number of years typically included in modern teleconnection studies, and each simulation experiment (i.e., for each combination of values of ϕ and γ) was repeated 500 times for each set of s time series. The sample contemporaneous cross correlations between series 1 and all the other series were computed for each simulation (that is, not all possible pairs of correlations were considered), implying that the number of correlations computed is $k = s - 1$. This approach mimics that taken in many teleconnection studies in which a particular time series (e.g., SOI) is correlated with a large number of other time series.

Distributions of the maximum sample contemporaneous cross correlations (in absolute value) obtained in each simulation, for the case of no actual contemporaneous cross correlation (i.e., $\gamma = 0$), are displayed in the form of box plots in Fig. 12.3. These plots indicate the strong effect of multiplicity in leading to spuriously large maximum values as the number of time series increases. For example, the median value of the maximum sample cross correlation with 50 time series and $\phi = 0$ is 0.35 (i.e., at least one correlation larger than 0.35 in absolute value would be expected to occur roughly half the time simply by chance). The impact of autocorrelation is to increase the magnitude of this multiplicity effect to a large degree.

For the case of an actual contemporaneous cross correlation of $\gamma = 0.5$, the distributions of sample maximum contemporaneous cross correlations (not shown) are relatively insensitive to the ef-

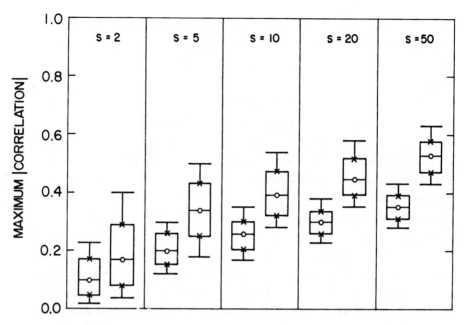

Fig. 12.3 Box plots depicting distributions of values of maximum sample contemporaneous cross correlations (in absolute value) for set of s simulated series with no contemporaneous cross correlation (i.e., $\gamma = 0$). First plot in each pair is for $\phi = 0$ (i.e., no autocorrelation); second plot is for $\phi = (0.5)^{1/2}$. The plots show (from top to bottom) the 0.90th, 0.75th, 0.50th, 0.25th, and 0.10th quantiles of the distributions.

fects of multiplicity. However, the probability that the sample cross correlation that is maximum in absolute value will have the wrong sign (i.e., will be negative) increases as the number of time series increases, up to 0.21 for $s = 50$ and $\phi = (0.5)^{1/2}$. Similarly, the probability that the maximum sample cross correlation will not be the correlation between time series 1 and 2 is as high as 0.55 for $s = 50$ and $\phi = (0.5)^{1/2}$. Both of these probabilities are larger for autocorrelated time series than for series with no persistence.

A Type I error occurs if at least one of the computed correlations in a simulation is deemed significant, or equivalently, if the maximum correlation is significant (since no cross correlation is actually present). Seven approaches for testing the significance of a cross correlation were applied to the maximum correlation values in each simulation experiment, with the objective of determining which methods are most successful at attaining the desired α_o (i.e., probability of Type I error) in practice. In this case the desired overall significance level is 0.05 (except for approach (iv)). The seven approaches are summarized in Table 12.2. The term "variance inflation" refers to an adjustment for the effect of auto-correlation on the variability of the sample cross correlation coefficient discussed earlier, with the adjustment for this effect being based on Eq. 12.1.

Table 12.2 Approaches for testing the significance of a set of $k = s - 1$ cross correlations, in which the desired overall significance level is $\alpha_o = 0.05$ (except in the case of approach (iv), where $\alpha_o = 0.50$).

Approach	Taken into account: Multiplicity?	Autocorrelation?	Nominal α
(i) Conventional	no	no	0.05*
(ii) Conventional with reduced α	somewhat	somewhat	0.01*
(iii) Bonferroni	yes	no	0.05/k*
(iv) Original Walker	yes	no	$1 - (0.5)^{1/k}$*
(v) Modernized Walker	yes	no	$1 - (0.95)^{1/k}$*
(vi) Conventional with variance inflation factor	no	yes	0.05
(vii) Modernized Walker with variance inflation factor	yes	yes	$1 - (0.95)^{1/k}$

* Only attained if no autocorrelation is present.

Similarities exist among some of these approaches. In particular, the α-values associated with the Bonferroni and modernized Walker approaches ((iii) and (v)) are quite close in spite of being based on entirely different assumptions concerning the dependence among tests. Nevertheless, it can be shown that both techniques produce approximately the same values of α when α_o is relatively small, and Miller (1977) has commented that the Bonferroni inequality performs impressively in this case.

Estimated probabilities of a Type I error (i.e., α_o) for the case when no cross correlation is actually present (i.e., $\gamma = 0$) are displayed in Figs. 12.4 and 12.5. It is evident from these diagrams that many of the approaches to testing significance do not adequately handle multiplicity. For example, the conventional approach (i) of applying the $\alpha = 0.05$ significance level in individual tests rapidly leads to a very large α_o as the number of series is

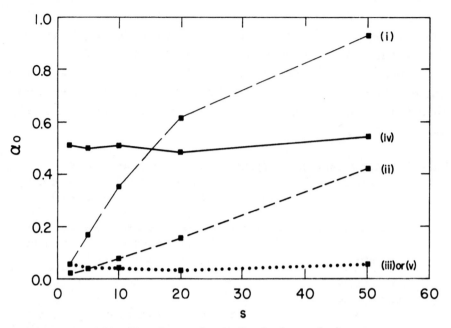

Fig. 12.4 Probability that the maximum (in absolute value) contemporaneous
sample cross correlation is significant for simulations with no cross
correlation or autocorrelation (i.e., $\gamma = 0$, $\phi = 0$), according to various approaches (defined in Table 12.2) as a function of the number
of time series s.

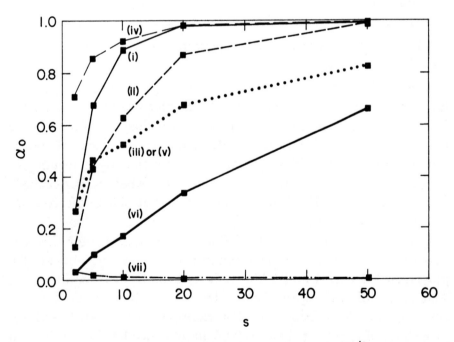

Fig. 12.5   As in Fig. 12.4, with $\gamma = 0$, $\phi = (0.5)^{1/2}$.

increased (e.g., for $s = 10$, the probability is about 0.35 for the case of no autocorrelation). Moreover, the effect of multiplicity is considerably worsened when the time series are autocorrelated (e.g., the probability associated with the $\alpha = 0.05$ approach (i) for $s = 10$ is about 0.89). With regard to counteracting the problem of multiplicity, the Bonferroni approach (iii) (approximately equivalent to the modernized Walker approach (iv)) appears to lead to adequate results (i.e., α_o somewhat less than 0.05) in the case of no autocorrelation. However, when the series are autocorrelated, the overall significance values are increased to unacceptable levels. This discrepancy between the observed and desired values of α_o arises because the nominal levels for the individual tests are no longer correct in the presence of autocorrelation. In that case, the combination of the variance-inflation approach with the modernized Walker (or Bonferroni) method (i.e., approach (vii)) leads to overall significance levels that actually decrease somewhat with increases in the number of time series, being substantially smaller than the desired value $\alpha_o = 0.05$ for large s. This departure can be attributed to the fact that (12.1) provides only an approximate expression for the effects of autocorrelation.

Prescription for future teleconnections research

Important statistical problems, such as time dependence and multiplicity, still need to be addressed in modern teleconnections research, much as some of them were of concern to Walker and his contemporaries. The results of our simulation studies indicate that the effects of autocorrelation and multiplicity in teleconnection studies can be extensive. Among other things, they can lead relatively frequently to observed maximum correlations that are spuriously large when in fact no cross correlation actually exists. In particular, the results indicate that use of a small significance level (e.g., $\alpha = 0.01$) may not always provide adequate compensation for the effects of multiplicity. A positive result of our study is the evidence suggesting that multiplicity may be successfully counteracted in the presence of autocorrelation by incorporating a variance-inflation technique along with the standard methods for dealing with multiple comparisons (e.g., the Bonferroni or modernized Walker approaches).

Although the adjustments to the individual level of significance prescribed by the Bonferroni or Walker approaches may appear to be drastic, they simply constitute the penalty for the lack of a well-formulated hypothesis. Multiplicity can be combatted by first recognizing the selection bias that results from computing large numbers of coefficients and selecting only the largest, and secondly by attempting a "confirmatory" study in which only the selected relationships are reevaluated. Another approach to avoiding multiplicity is to evaluate only those relationships for which an accepted physical relationship is known. Of course, this approach would severely limit the number of relationships that could be considered, and might prevent the use of correlations as an aid in discovering and/or describing important physical relationships (e.g., in Walker's case, the Southern Oscillation).

Recent developments in statistics can be expected to contribute to empirical teleconnections research in two basic ways. The obvious way is that the application of statistical techniques that are more appropriate for climatic/oceanic time series (and that are, consequently, probably more sophisticated) would naturally be expected to lead to better models for relationships among these variables (in particular, ones that produce more accurate forecasts). The less obvious way concerns techniques for assessing how well a given model will actually perform when employed to produce genuine forecasts. One currently perplexing problem relates to the observation that empirically based models invariably forecast more poorly than the statistics of the fitted model (e.g., correlation coefficient) would apparently indicate.

Modern statistical theory reveals two primary reasons to explain why the forecasting performance of empirically based models should be less than anticipated. One source of performance degradation relates to the fitting of a model in which parameters are estimated from the available data. It is well established in the statistics literature (e.g., Efron and Gong, 1983) that such an approach has a "tuning" effect, resulting in overestimates of forecasting skill. This phenomenon, on occasion, has also been recognized in the climatic/oceanic literature and is sometimes referred to as "artificial skill" (Davis, 1978). Simple approximate expressions are available to correct for this effect, which increases as the number of parameters estimated increases (e.g., Akaike, 1969).

A more serious overestimation of forecasting performance is related to the multiplicity problem discussed earlier. When several potential models are considered (e.g., several correlation coefficients are calculated) and the apparent best one is selected, a bias again results. Although this selection bias can be removed in a hypothesis-testing context in the manner previously outlined, it is still present in the conventional statistics used to anticipate forecasting skill. To correct for this bias, the use of more sophisticated techniques such as the "bootstrap" or "cross validation" would be necessary (Efron, 1982).

New, more advanced statistical methods may be the best practical solution to problems in empirical teleconnections research. For example, formal multiple time series analysis (Granger and Newbold, 1986) may be the only really satisfactory way of modeling the time-dependent characteristics of sets of meteorological time series. Moreover, this approach would allow a relatively straightforward interpretation of the cross correlation functions between the time series. Although the multiple time series approach is relatively complex, particularly in comparison with the methods that are currently in wide use, it would allow for simultaneous consideration of more than two variables, and for modeling of more complicated time dependencies on various time scales.

The role of empiricism in atmospheric research has oscillated greatly during the twentieth century. Dependence on statistical approaches peaked with Walker's extensive attempts to describe world weather using large sets of correlation coefficients, and it waned dramatically during the middle part of the century when there was a relative dearth of, and disregard for, that type of empirical work. It has apparently peaked once again in the past two decades, with the current widespread fascination with the computation of correlation coefficients relating all possible combinations of variables. Such great shifts in approaches to the same problems cannot lead to the most efficient progress in solving the problems. For example, at least partly as a result of the shifts in emphasis either toward or away from empiricism, progress in improving statistical techniques for teleconnections research has been slow at best, in spite of the great progress in increasing computer power and in developments in theoretical statistics. Such oscillations in approach allow us to "forget" lessons that have been learned in the past.

On the other hand, overreliance on statistical methods in recent years has allowed us to often neglect the importance of evaluating the physical basis for statistically determined relationships, as was emphasized by Walker (1936) and more recently by Ramage (1983). Statistical relationships, however, also have a role to play in guiding physical studies. For example, teleconnections derived by empirical methods have already been useful for designing experiments using global circulation models (e.g., Blackmon et al., 1983; Palmer and Mansfield, 1984; von Storch et al., 1988). Such experiments may even help to confirm, in a physical sense, the derived relationships. It thus appears that statistical and physical methods have complementary roles to play in the study of teleconnections, and that great oscillations in emphasis from one method to another are not productive. What is really needed is a research environment conducive to the development of a genuine "feedback" relationship between these two approaches.

References

Akaike, H. (1969). Fitting autoregressions for predictions. *Annals of the Institute of Statistical Mathematics*, **21**, 243–7.

Andrade, E.R., Jr. & Sellers, W.D. (1988). El Niño and its effect on precipitation in Arizona and western New Mexico. *Journal of Climatology*, **8**, 403–10.

Angell, J.K., Korshover, J. & Cotton, G.F. (1984). Variation in United States cloudiness and sunshine, 1950–1982. *Journal of Climate and Applied Meteorology*, **23**, 752–61.

Bartlett, M.S. (1955). *Stochastic Processes*. Cambridge: Cambridge University Press.

Bhalme, H.N. & Jadhav, S.K. (1984). The Southern Oscillation and its relation to the monsoon rainfall. *Journal of Climatology*, **4**, 509–20.

Bjerknes, J. (1969). Atmospheric teleconnections from the equatorial Pacific. *Monthly Weather Review*, **97**, 163–72.

Blackmon, M.L., Geisler, J.E. & Pitcher, E.J. (1983). A general circulation model study of January climate anomaly patterns associated with interannual variation of equatorial Pacific sea surface temperatures. *Journal of the Atmospheric Sciences*, **40**, 1410–25.

Bliss, E.W. (1926a). British winters in relation to world weather. *Memoirs of the Royal Meteorological Society*, **I**, No. 6, 87–92.

Bliss, E.W. (1926b). The Nile flood and world weather. *Memoirs of the Royal Meteorological Society*, **I**, No. 5, 79–85.

Bliss, E.W. (1928). Correlations of world weather and a formula for forecasting the height of the Parana River. *Memoirs of the Royal Meteorological Society*, **II**, No. 14, 39–45.

Climate Research Committee (1983). El Niño and the Southern Oscillation, A Scientific Plan. Washington, DC: National Academy Press.

Davis, R.E. (1978). Predictability of sea level pressure anomalies over the North Pacific Ocean. *Journal of Physical Oceanography*, **8**, 233–46.

Dines, W.H. (1916). Review of "Correlation in Seasonal Variations of Weather." *Quarterly Journal of the Royal Meteorological Society*, **42**, 129–32.

Dines, W.H. (1917). Correlation in seasonal variations of weather. *Quarterly Journal of the Royal Meteorological Society*, **43**, 333–5.

Efron, B. (1982). *The Jackknife, the Bootstrap, and Other Resampling Plans*. Philadelphia, PA: Society for Industrial and Applied Mathematics.

Efron, B. & Gong, G. (1983). A leisurely look at the bootstrap, the jackknife and cross-validation. *American Statistician*, **37**, 36–48.

Elliott, W.P. & Angell, J.K. (1988). Evidence for changes in Southern Oscillation relationships during the last 100 years. *Journal of Climate*, **1**, 729–37.

Ferraz, S. (1929). Sir Gilbert Walker's formula for Ceará's droughts. Suggestions for its physical explanation. *The Meteorological Magazine*, **64**, 81–4.

Gordon, N.D. (1986). The Southern Oscillation and New Zealand weather. *Monthly Weather Review*, **114**, 371–87.

Granger, C.W.J. & Newbold, P. (1986). *Forecasting Economic Time Series* (second edition). San Diego, CA: Academic Press.

Grant, A.M. (1956). The Application of Correlation and Regression to Forecasting. Meteorological Study No. 7. Melbourne, Australia: Bureau of Meteorology.

Groissmayr, F. (1929). Relations between summers in India and winters in Canada. *Monthly Weather Review*, **57**, 453–5.

Hastenrath, S. (1987a). On the prediction of India monsoon rainfall anomalies. *Journal of Climate and Applied Meteorology*, **26**, 847–57.

Hastenrath, S. (1987b). Predictability of Java monsoon rainfall anomalies: a case study. *Journal of Climate and Applied Meteorology*, **26**, 133–41.

Hastenrath, S., Wu, M.-C. & Chu, P.-S. (1984). Towards the monitoring and prediction of north-east Brazil droughts. *Quarterly Journal of the Royal Meteorological Society*, **110**, 411–25.

Haugh, L.D. (1976). Checking the independence of two covariance-stationary time series: A univariate residual cross-correlation approach. *Journal of the American Statistical Association*, **71**, 378–85.

Haugh, L.D. & Box, G.E.P (1977). Identification of dynamic regression (distributed lag) models connecting two time series. *Journal of the American Statistical Association*, **72**, 121–30.

Hunt, H.A. (1929). A basis for seasonal forecasting in Australia. *Quarterly Journal of the Royal Meteorological Society*, **55**, 323–34.

Katz, R.W. (1988). Use of cross correlations in the search for teleconnections. *Journal of Climatology*, **8**, 241–53.

Khandekar, M.L. (1979). Climatic teleconnections from the equatorial Pacific to the Indian monsoon – analysis and implications. *Archiv für Meteorologie, Geophysik und Bioklimatologie*, **A28**, 159–68.

Livezey, R.E. & Chen, W.Y. (1983). Statistical field significance and its determination by Monte Carlo techniques. *Monthly Weather Review*, **111**, 46–59.

Madden, R.A. (1979). A simple approximation for the variance of meteorological time averages. *Journal of Applied Meteorology*, **18**, 703–6.

McBride, J.L. & Nicholls, N. (1983). Seasonal relationships between Australian rainfall and the Southern Oscillation. *Monthly Weather Review*, **111**, 1998–2004.

Miller, R.G., Jr. (1977). Developments in multiple comparisons 1966–1976. *Journal of the American Statistical Association*, **72**, 779–88.

Miller, R.G., Jr. (1981). *Simultaneous Statistical Inference* (second edition). New York: Springer-Verlag.

Montgomery, R.B. (1940a). Report on the work of G.T. Walker. *Monthly Weather Review, Supplement No. 39*, 1–22.

Montgomery, R.B. (1940b). Verification of three of Walker's seasonal forecasting formulae for India monsoon rain. *Monthly Weather Review, Supplement No. 39*, 23–4.

Mood, A.M., Graybill, F.A. & Boes, D.C. (1974). *Introduction to the Theory of Statistics*. New York: McGraw-Hill Book Company.

Mossman, R.C. (1913). Southern hemisphere seasonal correlations. *Symons's Meteorological Magazine*, **48**, 2–6, 44–7, 82–3, 104–5, 119–24, 160–3, 200–7, 226–9.

Mossman, R.C. (1924). Indian monsoon rainfall in relation to South American weather, 1875–1914. *Memoirs of the India Meteorological Department*, **23**, 157–242.

Neter, J., Wasserman, W. & Kutner, M.H. (1983). *Applied Linear Regression Models*. Homewood, IL: Richard D. Irwin, Inc.

Nicholls, N. (1981). Air–sea interaction and the possibility of long-range weather prediction in the Indonesian Archipelago. *Monthly Weather Review*, **109**, 2435–43.

Niebauer, H.J. (1984). On the effect of El Niño events in Alaskan waters. *Bulletin of the American Meteorological Society*, **65**, 472–3.

Normand, C.W.B. (1932). Some problems of modern meteorology, No. 6. Present position of seasonal weather forecasting. *Quarterly Journal of the Royal Meteorological Society*, **58**, 3–10.

Normand, C.W.B. (1953). Monsoon seasonal forecasting. *Quarterly Journal of the Royal Meteorological Society*, **79**, 463–73.

Palmer, T.N. & Mansfield, D.A. (1984). Response of two atmospheric general circulation models to sea–surface temperature anomalies in the tropical East and West Pacific. *Nature*, **310**, 483–85.

Parthasarathy, B. & Pant, G.B. (1984). The spatial and temporal relationships between the Indian summer monsoon rainfall and the Southern Oscillation. *Tellus*, **36A**, 269–77.

Parthasarathy, B. & Pant, G.B. (1985). Seasonal relationships between Indian summer monsoon rainfall and the Southern Oscillation. *Journal of Climatology*, **5**, 369–78.

Pittock, A.B. (1984). On the reality, stability, and usefulness of southern hemisphere teleconnections. *Australian Meteorological Magazine*, **32**, 75–82.

Preisendorfer, R.W. & Barnett, T.P. (1983). Numerical model-reality intercomparison tests using small-sample statistics. *Journal of the Atmospheric Sciences*, **40**, 1884–96.

Quenouille, M.H. (1952). *Associated Measurements*. London: Butterworths.

Quinn, W.H., Zopf, D.O., Short, K.S. & Kuo Yang, R.T.W. (1978). Historical trends and statistics of the Southern Oscillation, El Niño, and Indonesian droughts. *Fishery Bulletin*, **76**, 663–78.

Ramage, C.S. (1983). Teleconnections and the siege of time. *Journal of Climatology*, **3**, 223–31.

Rasmusson, E.M. & Carpenter, T.H. (1983). The relationship between eastern equatorial Pacific sea surface temperatures and rainfall over India and Sri Lanka. *Monthly Weather Review*, **111**, 517–28.

Rogers, J.C. (1988). Precipitation variability over the Caribbean and tropical Americas associated with the Southern Oscillation. *Journal of Climate*, **1**, 172–82.

Sellick, N.P. (1932). Seasonal foreshadowing by correlation. *Quarterly Journal of the Royal Meteorological Society*, **58**, 226–8.

Shukla, J. & Paolino, D.A. (1983). The Southern Oscillation and long-range forecasting of the summer monsoon rainfall over India. *Monthly Weather Review*, **111**, 1830–37.

Troup, A.J. (1965). The 'southern oscillation'. *Quarterly Journal of the Royal Meteorological Society*, **91**, 490–506.

Tukey, J.W. (1977). Some thoughts on clinical trials, especially problems of multiplicity. *Science*, **198**, 679–84.

Van Heerden, J., Terblanche, D.E. & Schulze, G.C. (1988). The Southern Oscillation and South African summer rainfall. *Journal of Climatology*, **8**, 577–97.

von Storch, H., van Loon, H. & Kiladis, G.N. (1988). The Southern Oscillation Part VIII: model sensitivity to SST anomalies in the tropical and subtropical regions of the South Pacific convergence zone. *Journal of Climate*, **1**, 325–31.

Walker, G.T. (1910). Correlation in seasonal variations of weather, II. On the probable error of a coefficient of correlation with a group of factors. *Memoirs of the India Meteorological Department*, **21**, part II, 22–45.

Walker, G.T. (1914). Correlation in seasonal variations of weather, III. On the criterion for the reality of relationships or periodicities. *Memoirs of the India Meteorological Department*, **21**, part IX, 12–15.

Walker, G.T. (1917). Correlation in seasonal variations of weather. *Quarterly Journal of the Royal Meteorological Society*, **43**, 218–9.

Walker, G.T. (1918). Correlation in seasonal variations of weather. *Quarterly Journal of the Royal Meteorological Society*, **44**, 223–4.

Walker, G.T. (1923). Correlation in seasonal variations of weather, VIII. A preliminary study of world-weather. *Memoirs of the India Meteorological Department*, **24**, part IV, 75–131.

Walker, G.T. (1924a). Correlation in seasonal variations of weather, IX. A further study of world weather. *Memoirs of the India Meteorological Department*, **24**, part IX, 275–332.

Walker, G.T. (1924b). Correlation in seasonal variations of weather, X. Applications to seasonal forecasting in India. *Memoirs of the India Meteorological Department*, **24**, 333–45.

Walker, G.T. (1928a). Ceará (Brazil) famines and the general air movement. *Beitrage zur Physik der Atmosphære*, **14**, 88–93.

Walker, G.T. (1928b). World weather, III. *Memoirs of the Royal Meteorological Society*, **II**, 97–107.

Walker, G.T. (1930). Seasonal foreshadowing. *Quarterly Journal of the Royal Meteorological Society*, **56**, 359–64.

Walker, G.T. (1936). Seasonal weather and its prediction. *Smithsonian Institution Annual Report 1935*, 117–38.

Walker, G.T. & Bliss, E.W. (1930). World weather IV. Some applications to seasonal foreshadowing. *Memoirs of the Royal Meteorological Society*, **III**, 81–95.

Walker, G.T. & Bliss, E.W. (1932). World weather V. *Memoirs of the Royal Meteorological Society*, **IV**, 53–84.

Walker, G.T. & Bliss, E.W. (1937). World weather, VI. *Memoirs of the Royal Meteorological Society*, **IV**, 119–39.

Wishart, J. (1928). On errors in the multiple correlation coefficient due to random sampling. *Memoirs of the Royal Meteorological Society*, **II**, No. 13, 29–37.

World Climate Research Programme (1985). Scientific Plan for the Tropical Ocean and Global Atmosphere Programme. World Meteorological Organization, World Climate Research Programme Publications Series No. 3.

Wright, P.B., Wallace, J.M., Mitchell, T.P. & Deser, C. (1988). Correlation structure of the El Niño/Southern Oscillation phenomenon. *Journal of Climate*, **1**, 609–25.

Yarnal, B. & Diaz, H.F. (1986). Relationships between extremes of the Southern Oscillation and the winter climate of the Anglo–American Pacific coast. *Journal of Climatology*, **6**, 197–219.

13

Impact of ENSO events on the southeastern Pacific region with special reference to the interaction of fishing and climate variability

RÓMULO JORDÁN S.

Scientific Secretary
Comisión Permanente del Pacífico Sur (CPPS)
Santiago, Chile

Introduction

There are only a few places in the world where a recurring natural phenomenon of the magnitude of El Niño so severely affects the economies and populations in that region. The biological productivity of the eastern equatorial Pacific, among the highest in the world, occasionally shifts between high and low levels of productivity – with such fluctuations affecting living marine resources. When a major El Niño event occurs, it often has catastrophic impacts on the means of production, on infrastructure, on services, and on human health and well-being, especially along the western coast of South America.

This chapter discusses the impacts of El Niño events on ocean and land-based resources, stressing the biological responses of living marine resources and the impact of those events on fisheries. Attention is given to the fact that the level of El Niño impacts on natural resources and the economy is not only a consequence of the magnitude, duration, and intensity of the El Niño event, but is also affected by such factors as the health of marine populations and the level of development of human settlements.

The long-term changes sustained by living marine resources and fisheries in the southeast Pacific region are investigated from the perspective of interannual oceanic and climatic variability, fishing intensity, and the state of the resources.

To stress the diversity as well as the scope of impacts associated with an El Niño, the consequences of the 1982–83 event, the second most damaging one this century (the most damaging was that of 1925–26), are highlighted.

The author emphasizes the multidimensional impacts of El Niño along the southeast Pacific to draw attention to their importance to society.

The southeast Pacific region and El Niño

General characteristics

A large section of the southeast Pacific region, particularly the coastal areas located off Peru and northern Chile, including some areas of Ecuador, feature a coastal and maritime climate showing temperate and dry conditions. Sea surface temperatures fluctuate between 14° and 21°C, with salinity rates about 35 per thousand; the mixed layer is known to be rather shallow as a result of a shallow thermocline. Intense upwelling processes provide high concentrations of nutrients to the euphotic zone, giving origin to one of the world's most productive regions in terms of organic matter (Ryther, 1969; Guillén and Calienes, 1981). The records of Chavez and Barber (1985) show average concentrations of up to 4.27 $gC/m^2/day$ (i.e., 1559 $gC/m^2/year$) off the Peruvian coast. This figure is the highest ever estimated. As a result of this high level of productivity, various kinds of pelagic species abound in the region.

In the past, there were large populations of sea birds in the region. Their droppings deposited on offshore islands produced nearly 300,000 tons/year of organic fertilizer (guano) for agricultural use in the early 1950s. By the late 1960s, annual anchoveta catches of about 13 million metric tons were landed off the coasts of Peru and Chile. Today there are active multispecies fisheries exporting sardine, jack mackerel, chub mackerel and anchoveta. Unlike these highly productive coastal waters of the southeast Pacific, the lands adjacent to the coastal areas of Peru and northern Chile are extremely arid, bearing little production.

These general characteristics of the southeast Pacific region are occasionally (every two years or so) disturbed by meteorologicoceanic variations, popularly known as El Niño. Years of high biological production in the marine environment are often followed by six- to 14-month periods during which there are significant physical and ecological changes that have significant adverse societal and economic impacts (Fig. 13.1).

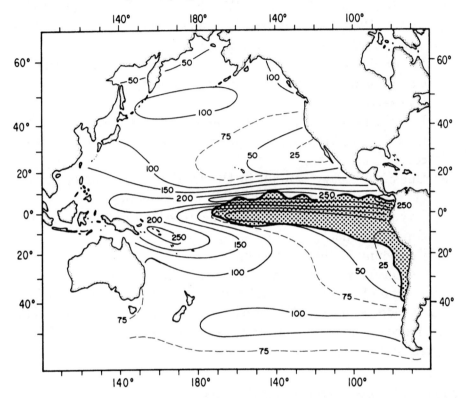

Fig. 13.1 Isolines of annual rainfall distribution in the Pacific (Rasmussen, 1984) and SST anomalies of 2°C or greater off the southeast Pacific, between September 1982 and February 1983 (shaded area) derived from maps of the U.S. Climate Analysis Center.

The term El Niño originated in the last century when Peruvian fishermen and seamen labeled the periodic flow of warm waters appearing around Christmastime, the "El Niño Current." The appearance of warm water was accompanied by heavy rainfall on land and reduced fishing resources along the Peruvian coast. Currently, the term "El Niño/Southern Oscillation" (ENSO) refers to the large oceanic–climatic alterations in the Pacific Basin that have worldwide impacts. The southern oscillation represents the most outstanding indication of interannual climate variability, usually associated with monsoon rains and atmospheric circulation

anomalies in various parts of the world (Rasmussen, 1984). Enfield and Allen (1980) demonstrated the large geographic scope of
oceanographic disturbances that occur during El Niño, affecting
most areas of the world from Alaska to the Antarctic.

Many authors have already referred in detail to the physical
characteristics of El Niño (see Wyrtki, 1975; Zuta et al., 1976;
Rasmussen, 1984). A brief list of the most constant physical characteristics of the phenomenon is given below:

(1) Inflow of northern warm waters towards the coasts of Ecuador
and Peru with 23 to 29°C and low salinity (32.5 to 34.5 per
thousand).

(2) Penetration of oceanic warm waters with high salinity (higher
than 35 per thousand) towards the Peruvian and Chilean
coasts.

(3) Unusually heavy rainfall in the tropical and subtropical regions, particularly when El Niño is very strong.

(4) Low nutrient concentration in the upwelling region within the
euphotic zone.

(5) Deepening of the thermocline.

(6) Sea level rise along the west coast of South America.

Each one of these characteristics shows differing degrees of intensity, depending on the severity of each El Niño event. The
occurrence of each of these characteristics does not seem to follow
a given pattern.

Biological considerations and fisheries

The ocean's biological and ecological responses are rather swift at
the outset of an El Niño, with some organisms reacting from the
very early stages of the phenomenon. These could thereby serve as
early warning biological indicators (Rojas de Mendiola et al., 1985;
Avaria and Alvial, 1985). Productivity, normally exhibiting high
renewal rates in the southeast Pacific, sharply declines, thereby
sparking the collapse of the whole ecological chain and disturb-
ing the distributive patterns of flora and fauna, mainly those of
neritic and pelagic species. Barber et al. (1985) offer a very comprehensive description of upwelling processes and El Niño effects
on productivity and on the trophic levels. These authors stress the
fact that the high yearly mean productivity and the wide-ranging
interannual variability are interrelated. They further demonstrate

that the interannual nutricline variability produced by El Niño is but a causal process of biological variability.

The 1982–83 El Niño, one of the most intense events of the twentieth century, has provided the means to better understand the various aspects related to biological responses. The Workshop on the El Niño Phenomenon, held in Guayaquil (Ecuador) in 1983 (CPPS, 1984), showed chronologically the changes that occurred in the Pacific Ocean from September 1982 onwards; these changes became more prominent in early 1983. A substantial decrease in productivity was observed off the Ecuadorian, Peruvian and Chilean coasts as well as notable changes in phytoplankton quantity and composition, which showed a predominance of dinoflagellates over diatoms. Comparable changes were sustained by zooplankton with an increase of warm water tunicates, coelenterates and quetognates. (See also Avaria, 1984; Santander and Zuzunaga, 1984; Rojas de Mendiola et al., 1985.) Once quality and availability became altered at the lowest trophic levels, deficiencies occurred in the physical conditions of herbivorous fish which then wasted large amounts of energy, resulting in weight loss (Dioses, 1985; Sanchez et al., 1985; Martinez et al., 1985). Simultaneously, reproductive behavior was altered, high mortality rates of juveniles were registered and schools of fish were displaced outside their traditional living grounds; there were, also, significant changes in adult and recruit biomass (see Pauly and Tsukayama, 1987; IFOP, 1985). In this regard, several authors have mentioned the occurrence of changes in components of neritic benthos and demersal fish as well as migration of genus *Penaeus* crustaceans (e.g., shrimp) to the southern areas located more than 1000 miles from their traditional living grounds; other effects included an unusual abundance of lamellibranch molluscans *Argopecten purpuratus* and high mortality rates of a large number of species, such as molluscs, fish, birds, and wolves, as well as the destruction of algae prairies (see CONCYTEC, 1985; IFOP, 1985; Robinson and Del Pino, 1985).

As is shown later, the impact of an El Niño on a particular fishery depends both on the magnitude and duration of the El Niño and on the state of health of the fish population. The severe impacts of the 1982–83 El Niño, more than those of any previously recorded El Niño, showed that in the presence of such severity, maritime and terrestrial ecosystems are likely to suffer major impacts.

The following is a summary of the generic responses to an El Niño event in the southeast Pacific. This summary does not follow a strict temporal progression:

(1) Low primary productivity as a result of poor fertilization of the euphotic zone.

(2) Decrease of food quality and availability in the lowest trophic levels.

(3) Displacement of schools of autochthonous fish species to other regions, high mortality rates of pelagic and benthic organisms, depending on the intensity of the event.

(4) Appearance of tropical species in coastal areas of Peru and Chile.

(5) Change in reproductive patterns; decrease in larvae and egg production of pelagic fish.

(6) Low recruitment rates and decrease in pelagic/neritic resources biomass.

(7) Change in behavior patterns of benthic/pelagic resources.

Although the above-mentioned reactions appear to be somewhat stable for a general (idealized) generic description of the phenomena, each event shows rather distinct characteristics, which are often more varied. Some aspects of the ecological (and societal) impacts of a particular El Niño event may be absent from other events. For example, with respect to fauna considerations, it has been possible to verify that the 1972–73 El Niño event led to the eruption of large numbers of the species *Euphilax dovii* which appeared off the coasts of Ecuador, Peru, and Chile. The 1976 event was characterized by the occurrence of red tides produced by the dinoflagellate species *Gymnodinium splendens*; other past events have witnessed high concentrations of different species such as tuna and swordfish off the coast (Vildoso, 1976).

From the point of view of changes affecting the fauna, the 1982–83 event was characterized by the southward displacement of shrimp (Genus *Penaeus* and *Xiphopenaeus*) as well as of goldfish (*Coryphaena hippurus*) from the high seas, in such large and sustained quantities that a new fishery developed. Likewise, the event produced an unusual concentration of the molluscs *Argopecten purpuratus* (Arntz et al., 1985).

With respect to populations of species which make up fishing resources usually exploited off the western coast of South America, the variability resulting from the most outstanding effects of El Niño is also dependent on several other external components. Scientists investigating El Niño and the Southern Oscillation have produced voluminous bibliographical resources on the physical aspects of the event with particular emphasis on biological fishing characteristics.*

The high population variability related to El Niño has led to a high degree of resilience whereby population contractions are usually followed by population recovery, the influence of other factors notwithstanding, as has been demonstrated in the case of sea birds feeding on anchoveta (Jordán and Fuentes, 1966). In like manner, Barber et al. (1985) have showed how productivity revival is much higher immediately after the event. Arntz (1986) and Tarazona et al. (1988) have also dealt with the issue of the resilience of benthic communities.

Frequency and intensity of the phenomenon

The frequency and intensity of El Niño have spurred an ever growing interest in identifying different criteria to describe it. Rivera (1987) proposed a classification using the surface temperature variability for the 1925–85 period as the main criterion. The author identified four degrees of intensity of El Niño: weak, moderate, strong and extraordinary. Meanwhile, Quinn et al. (1987) made an important contribution on the basis of an historical study. The authors took into account a sequence of 79 El Niño events that occurred over a period of 450 years. Fourteen characteristics constitute the basis on which to measure their intensity, including penetration of warm waters, deviation of navigational routes, storms

---

* Reference can be made to detailed descriptions of the 1982–83 phenomenon given at the Workshop on the 1982–83 El Niño (CPPS, 1984); National Workshop on the 1982–83 El Niño Phenomenon, Investigación Pesquera Special Issue (IFOP, 1985); Ciencia y Tecnología, Agresión Ambiental: El Fenómeno El Niño (CONCYTEC, 1985); Simposio El Niño, su Impacto en la Fauna Marina (Arntz et al., 1985); El Fenómeno El Niño 1982–83 con Particular Referencia a sus Efectos sobre los Recursos Pesqueros y las Pesquerías en el Pacífico Sudeste (CPPS, 1984); El Niño en las Islas Galapagos, El Evento de 1982–83 (Robinson and Del Pino, 1985.)

and rainfall, destruction of buildings, bridges and cultivated fields, significant increases of sea surface temperatures and of sea level, biological reactions, decrease in fishing yields, and so forth. They classified El Niño as *moderate, strong* and *very strong*. Although the authors are aware of the occurrence of a weak El Niño, they have not made a clear distinction of this classification because of a lack of reliable information and parameters. These reasons account for gaps in observations of moderate intensity events in the 1525–1791 period.

Tables 1 and 2 of Quinn et al. (1987) once more confirm the irregularity of the El Niño cycle. Its appearance may be expected over periods of two to seven years, whether or not the phenomenon can be classified as moderate, strong or very strong. Grouping these data into 100-year periods, indicates that in each century there occur two or three very strong El Niño events; six to 10 strong events and 15 to 18 moderate ones (Table 13.1). In other words, it is possible to expect the occurrence of very strong El Niño events every 14 to 63 years; strong ones could be expected to occur every six to 18 years and moderate ones every two to 11 years. From Table 13.1 it follows that the frequency and intensity of El Niño events do not seem to have increased recently.

Table 13.1 Frequency of El Niño events in the southeast Pacific. Summary of Tables 1 and 2 from Quinn et al. (1987).

Period	Moderate (M and M+)	Strong (S and S+)	Very strong (VS)
1525–92	?	6	2
1607–96	?	10	0
1701–91	?	8	2
1803–1900	18	7	3
1902–88	15	6	2

Over the 450-year period mentioned, very strong El Niño events occurred in 1578, 1728, 1791; 1828, 1877–78, 1891; 1925–26 and 1982–83. Even though there are indications of the occurrence of natural disturbances over a 40,000-year period, Wells (1987) and De Vries (1986), through a review of results obtained on the basis of geomorphological, paleontological, sedimentological and archeological methods, reached the conclusion that there is not as yet enough evidence on record to identify an El Niño in prehistoric times. As the 1986 Guayaquil (Ecuador) Chapman Conference on El Niño concluded, it is of the utmost importance to learn how far in the past El Niño events occurred on earth. Such knowledge would allow for a more accurate interpretation of environmental changes and for an improved assessment of biological processes.

This paper uses historical references in accordance with Quinn's classification (Quinn et al., 1987) in order to examine more closely those oceanic–climatic fluctuations affecting natural populations as well as economic activities.

Socioeconomic impact of El Niño

Impact characteristics

Although El Niño events may be clustered together based on certain similarities, each event can be shown to have rather distinctive characteristics. Their intensity, duration, geographical scope and timing vary. From the point of view of their impacts, there are additional factors that make comparisons even more difficult; these include the health conditions of natural populations, the relationship between different populations and fishing pressures exerted on them. Furthermore, the importance of the impacts also depends on the economic value that some living resources have in comparison with others at any given point in time. In the past, large quantities of guano produced by sea birds on coastal islands of the southeast Pacific were very highly valued sources of income. El Niño events not only affected the economy of the guano-exploiting country but also the international community, because El Niño reduced the availability of fertilizer for agricultural use. Since the early 1950s, the main concern shifted from guano production to increased production of anchoveta fish meal (partly due to the collapse of the California sardine fishery). At that time, declining amounts of fish meal were causing major increases in the international prices of poultry and cattle feed. Nowadays, because of changes occurring in species composition beginning in the early 1970s, multispecies fisheries of sardine, jack mackerel and anchoveta have been given greater attention.

The impact of El Niño-related floodings on terrestrial resources often depends on human activities (e.g., the location of towns, cultivated fields and transportation infrastructure, as well as the types of industries). For instance, the economic impacts of the very strong 1578 El Niño, as compared with the 1982–83 event, would obviously be very different in nature, even though the two events may have been equally strong to those who had to suffer their impacts. Nor is the effect of the same event likely to be similar from one place to another, since it depends on such features as geomorphological and soil structures. For example, heavy rains that fall on sandy base soils bring about mud slides more easily than heavy rains in vegetated areas.

The strong 1972–73 El Niño brought Peruvian and Chilean fisheries to a standstill, forcing nearly 1500 fishing boats and 200 fish-processing plants to stop operations; production of fish meal was reduced by some 2 million metric tons per year. More than 100,000 people whose livelihood, directly or indirectly, depended upon fishing activities became unemployed and without income. These countries faced a shortage of foreign exchange which greatly affected their reserves. While workers managed to secure subsistence loans from entrepreneurs, the latter obtained bank loans in the hope that the situation might improve, which it did not. Glantz and Thompson (1981) looked into the socioeconomic implications of financial resources subjected to the risks involved in climatic variability. Glantz (1984) also investigated various national and international levels of decision-making that influenced the exploitation of the anchoveta.

The very strong 1982–83 El Niño dramatically brought to light the fact that the phenomenon not only accounts for ecological disturbances of the marine environment (because its impacts are felt on fisheries and guano production) but also accounts for heavy losses sustained by land-based means of production and services including the agricultural and industrial sectors, social services, housing, schools, hospitals, and communication and transportation networks.*

---

* While scientific research has made inroads into identifying the linkages between El Niño and its worldwide effects, global programs have been set up by such international organizations as the World Meteorological Organization (WMO) and the International Oceanographic Commission (IOC). National programs have also been established to research these linkages and their forecasting potential. Awareness of national governments in the southeast Pacific of the need to pursue joint coordinated research efforts led to the establishment,

The 1925–26 and 1982–83 El Niño were two of the strongest events this century. In the 56-year interval between these events, 11 moderate to strong El Niño events were recorded. During this period, research activities focused on the geophysical causes of El Niño and those biological reactions involved in guano production and fishing activities, while little attention was given to the accompanying socioeconomic impacts of El Niño on land and in the sea. Very strong events, linked to heavy rainfall, are capable of producing more serious damage on land than at sea, as the following section shows.

Socioeconomic impacts of the 1982–83 El Niño

Various regions of the world have faced calamitous conditions as a result of the occurrence of the 1982–83 El Niño event: protracted droughts in some regions and heavy floodings in others adversely affected the economy of many countries (Glantz et al., 1987). The countries in the southeast Pacific were subjected to the extremely severe impacts of the 1982–83 El Niño, both at sea and on land. Three components stand out from the point of view of economic impacts:

(1) The ecological disturbance of the marine environment affecting fisheries along the western coast of South America
(2) The heavy rains over the land resulting in flooding and mudslides
(3) Pacific storms greatly damaging the coastal region

Even though they cannot be directly linked to El Niño, protracted drought conditions simultaneously affected the high Andean regions of Peru, Bolivia and the northeastern section of Chile, and accounted for heavy losses sustained by the agricultural and industrial sector. The northeastern region of Bolivia witnessed

in 1974, of the Regional Program for the Study of the El Niño Phenomenon (ERFEN) within the framework of the Permanent South Pacific Commission (CPPS). CPPS includes Colombia, Chile, Ecuador and Peru. ERFEN is a multidisciplinary effort which addresses meteorological, oceanographic and biological aspects of El Niño and is linked to IOC and WMO programs. The ERFEN Bulletin and the South American East Pacific Climate Analysis Bulletin offer to the international scientific community the results of in situ observations of the El Niño phenomenon (e.g., ERFEN, 1984).

heavy flooding which also extended into Brazil, the intensity of which surpassed, by a factor of three, the average figure so far recorded (CEPAL, 1983).

Damage resulted from the flooding of cultivated fields, and population centers as well as the washing out of roads. Governments were forced to declare a state of emergency in several regions from late 1982 onward (Jordán, 1983a). Moderate rainfall in arid and semiarid areas brought about some changes in terrestrial ecosystems: green crops appeared even in normally arid areas (e.g., Ferreyra, 1985; Luong and Toro, 1985). Changes also occurred in yields of cultivated fields (e.g., Valdivia and Dobrea, 1984; Silva, 1985; Vargas et al., 1985). Altered conditions such as these provided temporary relief to subsistence farmers (Vreeland, 1985).

October 1982 witnessed the beginning of a heavy rainy season along the Colombian and Ecuadorian coasts, progressing southwards towards Peru. Rivers overflowed, leading to severe flooding, landslides and coastal erosion occurred that had very damaging effects on housing and historic monuments, agricultural activities, cattle raising, transportation, health, and education. In Ecuador, seven provinces with a population of about five million, i.e., more than 60 percent of the country's total population, were directly adversely affected. Five provinces on the northern coastal region of Peru, inhabited by some four million people (25 percent of the country's population), also sustained heavy losses. In Bolivia, flooding covered more than 150,000 km^2, affecting some 700,000 people. Drought conditions adversely affected about 380,000 km^2, or about 35 percent of the country's area (CEPAL, 1983).

Peru's Instituto Nacional de Planificación and the U.N. Economic Commission for Latin America carried out many assessments of the impact of the 1982–83 El Niño on Peru (INP, 1983; CEPAL, 1983). Table 13.2, summarized from the two assessments mentioned above, quantifies damages sustained by the various economic sectors in Ecuador, Peru and Bolivia. Table 13.3 provides an estimate of economic losses of nearly $US3 billion sustained by these three countries, clearly a heavy blow to their economies.

The levels of production of consumer goods were affected by heavy losses because of the isolation of production and population centers, causing a serious shortage of supplies. This in turn exacerbated existing market deficiencies and resulted in a sharp increase in the prices for basic commodities, thereby seriously im-

Table 13.2 Account of the principal damages caused in 1982–83 by flooding and drought in Peru, Ecuador, and Bolivia.

	Peru	Ecuador	Bolivia
A. Flooding			
Damaged and/or destroyed			
Housing units	>30,000	>20,000	14,500
Affected population	>800,000	950,000	700,000
Damaged schools	875	>223	w/d
Destroyed bridges	47	25	>10
Highways (kilometers)	2,634	1,800	>100
Lost Agricultural lands			
(hectares)	79,563	30,700	w/d
Loss of cattle	w/d	>1,500	w/d
Damaged drinking water and sewage			
systems (number of towns)	86	29	w/d
Health care centers and			
hospitals	101	>50	w/d
Archaeological remains			
(damages)	32	—	—
B. Drought			
Affected area (hectares)	231,458	0	380,000
Agricultural production			
(metric tons)	525,062	0	s/d
Livestock affected	2,636,900	0	>1,500,000
Affected population	1,200,000	0	1,600,000

pairing the health conditions of the poorest sectors of society. The impacts on foreign trade, shown by the inability of these countries along the western coast of South America to meet their export quotas, resulted in a sharp reduction in foreign exchange income and in their ability to import consumer goods.

Fisheries yields were significantly affected. At the same time, artisanal fishing enjoyed a temporary and comparatively small gain as a result of the growth sustained by some resources off the coast. (Arntz et al., 1985). The displacement of living marine resources from one area to another had an adverse effect on some countries, which experienced a 50 percent decrease in yields, while other regions benefited from increased yields of the same order of magnitude. Due to the multispecies structure of resources as well as to the resilience and adaptation of some of them to wide ranging environmental variability among other adverse circumstances, re-

Table 13.3 Economic losses caused by the 1982–83 El Niño phenomenon in Ecuador, Peru and Bolivia (in millions of US dollars).

	Ecuador*	Peru**	Bolivia*
Agribusiness	233.8	649.0	716.0
Fishing	117.2	105.9	—
Industry	54.6	479.3	—
Electric energy	—	16.1	—
Mining	—	310.4	—
Transportation and communications	209.3	303.1	—
Housing	6.3	70.0	17.8
Health, water and sewage systems	10.7	57.1	4.7
Education	6.6	5.9	—
Archeological remains	—	without appraisal	—
Others	2.1	—	—
TOTAL	640.6	1,996.8	836.5

Source: *CEPAL; **INP and CEPAL.

gional fisheries did not sustain as dramatic a decline in the 1982–83 event as in the 1972–73 collapse (Jordán, 1983b).

The values shown on Tables 13.2 and 13.3 provide some idea of the economic losses. Nevertheless, societal effects are often difficult to quantify: increased malnutrition and morbidity, loss of employment, closing of schools, distress involved in the loss of food supplies and personal belongings and, last but not least, the disappearance of the cultural patrimony that results from the loss of archeological remnants of ancient civilizations.

Six years after the occurrence of the 1982–83 El Niño, restoration activities are still in progress, underscoring the fact that the impacts of a specific El Niño event often extend well beyond the actual physically defined occurrence. This usually results in a diversion of funds earmarked for economic development programs to emergency relief programs. In the midst of the lingering multi-year impacts of the 1982–83 El Niño, yet another weak–to–moderate El Niño occurred in 1986–87, the effects of which were felt in the short-lived alterations of some marine resources and about a 10 percent reduction in fish landings. Apparently, no adverse effects occurred on land.

The interaction of fisheries and climate variability

General framework

The growth and development of natural populations in the marine environment are a direct result of essential elements, generally referred to as biotic components, as well as fluctuations of the physical and chemical environments in which those biotic components thrive. When one deals with populations exploited by society one must remember that fishing is an additional (the third) component having an influence upon the size of fish populations.

Together with the surrounding atmospheric conditions, the marine environment constantly undergoes changes which have an impact on the various processes and biological stages of species through time. The way in which such changes have an effect on population development largely depends, on the one hand, on intraspecific conditions such as the average life span of species, reproductive patterns, the ability to adapt to, and resilience in the face of, environmental changes; and on the other hand the effects of the interacting mechanisms at the population level and the intraspecies relationships such as competition and predation. MacCall (1984) groups the sources of changes affecting fish population abundance into five categories, including fishing. Csirke (1988) suggests that recruitment strength accounts for the most meaningful and proximate reason for fluctuations observed in fish populations. He further states that recruitment is at the same time the outcome of marine and atmospheric conditions and, probably, of the size of spawning adults. Simultaneously, the size of the adult population at a given time depends on the intensity of fishing activities, particularly the activity's pressure on stocks of remaining adults and young recruits.

We have already discussed the way in which climate–ocean variability periodically affects marine populations in general. As an exogenous component of the marine environment, fishing has an impact on native populations by decreasing population size and interfering with such biological processes as reproduction. Mortality rates due to fishing activities may be moderate and therefore have no effect on the population's natural reproductive process; however, fishing levels may become so intense that they could lead to the collapse of the resource. Fishery biology provides a good example of this statement. On the one hand, demersal populations, particularly low fluctuating ones as a result of homogeneous environments, are able to sustain moderate fishing pressures for relatively long periods of time. On the other hand, populations of short-lived, highly fluctuating pelagic fisheries may be more sensitive to intensive fishing (see Saville, 1980; Csirke, 1988). Research on a large number of pelagic fisheries from around the world that underwent collapse-like processes showed the difficulty in making an accurate assessment on how changes affecting the environment and over-fishing alike could have been responsible for such processes. Sharp and Csirke (1983) present several cases of changes in neritic populations that were due both to environmental impacts and fishing. Cañon (1986) stresses the importance of environmental factors in collapses, suggesting that changes caused by fishing did not have a meaningful effect on the disappearance of the anchoveta in northern Chilean waters.

The southeast Pacific region is rich in documented situations in which pelagic populations have been subjected to periodic environmental "shocks" and simultaneously to strong fishing pressures. A typical example are the anchoveta and sardine species as well as the guano-producing birds.

Researchers at the Peruvian Instituto del Mar (IMARPE) have illustrated how the anchoveta population had been reduced by intense fishing pressures and by environmental factors, though it was not always possible to separate the factors responsible for the decline.

El Niño occurrences at intervals of between two and seven years bring about an abrupt and temporary displacement of individuals and populations, introducing changes in reproductive physiology, egg and larvae survival, recruit and adult biomass, and fish school behavior among other changes (CPPS, 1984; IFOP, 1985; Jordán, 1987).

Information about fish population fluctuations worldwide has limited use since it can only be applied to the resource exploitation stage. This makes the process of attributing the causes of population change to fishing activities or to environmental fluctuations more difficult. Scientific literature on the population dynamics of land-based species offers data on fluctuations of native populations showing distinctive periods of low and high population density (Lack, 1954). Likewise, statistical data are available on guano-producing birds covering a 60-year period during which records were kept. This information dates back to the time when the commercial exploitation of anchoveta fisheries was nonexistent. Reference is made therein to strong fluctuations in response to El Niño. Assessing the reaction of a species closely linked to the anchoveta (being the anchoveta's main predator) may be useful in understanding more completely the changes sustained by fish populations resulting from El Niño events as opposed to those resulting from fisheries' impacts. Available information suggests that bird populations were not depressed by environmental change for long periods of time but, rather, that such declines were due to a very different cause, discussed below.

Anchoveta fisheries, sea birds and climate

Anchoveta (*Engraulis ringens*), a short-lived, small-sized fish, has been considered one of the most important resources of the southeast Pacific region since ancient times. In the first place, the species had been able to support some 20 to 30 million sea birds in addition to other types of predators; secondly, it has supported fisheries that have achieved the highest catches (by tonnage) in the world, having reached that figure within a relatively short period of time (Table 13.4). The 1950s and 1960s witnessed the swift growth of the anchoveta fishery, notwithstanding high predation by "guano" birds, and the occurrence of four El Niño events in this period,* i.e., 1953 (m+); 1957–58 (s); 1963(w+); and 1965 (m+).

This rapid rise in the exploitation of fishing resources abruptly ended in 1972. During the seventeen years from that date onward, low yields of anchoveta have been recorded. Two factors account for this situation: the occurrence of the 1972–73 El Niño (as extreme as the 1957–58 event, which was not accompanied by a reduction in anchoveta catches because fishery pressures were relatively low) and high catch levels, considerably above permissible levels, which accounted for a perilously unstable condition

* Year 1963 is not shown as a El Niño year in the Quinn et al. table (1987). The event was recorded by Romero & Garrido (1985), *Invest. Pesq. (Chile)* *32* for the Chilean Coast; IMARPE's internal documents and Jordán (1964) also recorded the event, *Inf. Inst. Recursos Mar. Callao, No. 27*, when the reaction of guano-producing birds was examined.

Table 13.4 Statistics of the anchoveta fishery off the coast of Peru, representing 90 percent of the anchoveta fishery in the southeast Pacific.

Year	Vessels	T.R.B. (tons)	Fishing Days	Catch (ton × 10³)	Processing Plants
1952				16	
1953				37	
1954	126	2,400		43	
1955	176	3,599		59	
1956	220	5,075		119	27
1957	272	7,167		326	39
1958	321	9,273		737	53
1959	414	16,342	294	1,909	68
1960	667	21,949	279	2,944	89
1961	756	43,261	298	4,580	101
1962	1,069	71,991	294	6,275	112
1963	1,655	127,670	269	6,423	
1964	1,744	140,059	297	8,863	
1965	1,623	138,080	265	7,233	
1966	1,650	150,856	190	8,530	171
1967	1,569	154,727	170	9,825	
1968	1,490	138,561	167	10,263	185
1969	1,455	128,652	162	8,960	188
1970	1,499	179,698	180	12,277	
1971	1,473	199,114	116	10,282	
1972	1,399	194,679	89	4,448	
1973	1,256	188,400	27	1,785	

Source: IMARPE

event took place, and reached the following meaningful conclusions:

(1) The average sustainable maximum yield is 10×10^6 according to the exponential method and 10.3×10^6 according to the logistical method (these figures refer to total yield less 0.74×10^6 accounted for by bird predation) with a confidence coefficient of ± 1 million metric tons.

(2) There exists a large excess industrial (fleet and processing) capacity above that required to capture the maximum sustainable yield.

(3) The mean value for the anchoveta biomass in 1963–70 represents only 38 percent of that of the 1960–61 period; this implies that stock instability is on the increase.

(4) The production of large-sized anchoveta shows sustained decrease. The fisheries are therefore forced to concentrate (as well as depend) on one year-class alone each year with the subsequent risk of total biomass depletion in case the recruitment efforts do not meet expectations. The application of maximum sustainable yield methods to anchoveta fisheries made it possible to issue provisions regulating fishing activities on the basis of a catch quota, provided that conditions in the marine environment are more or less uniform and recruitment figures are, on average, favorable.

It should be pointed out that, due to the success reached by fishing efforts on the one hand and the fact that the guano bird effort, on the other, was also taken into account, MSY values were overestimated with regard to actual stock size and with regard to annual recovery possibilities of anchoveta populations.

The close relationship between the condition of anchoveta populations, fishing effort and guano bird populations led scientists to estimate the MSY, bearing in mind both man's and birds' fishing (predation) efforts. Table 13.5 shows values calculated for the critical period of 1960–71, in which the bird population sustained a fourfold decrease compared with the impacts of the 1965 moderate El Niño event. The anchoveta catch for the previous year amounted to about nine million tons. The decreased bird population did not recover in the 24-year period to the present. At the same time, anchoveta catches by fishermen continued to increase after the 1965 El Niño (see Fig. 13.2). It is not coincidental that another important anchoveta predator, the "bonito"

(*Sarda chilienses*) species, decreased to an extent similar to that of guano birds, a situation that was exacerbated from 1972 onwards (Jordán, 1983b).

Table 13.5 Catch and effort applied to the anchoveta by humans and guano birds during 1960–71.

| Fishing season (Sept–Aug) | Fishing (F) | | Guano Birds (B) | | Total catch F + B | Total effort F + B |
	Catch (millions of tons)	Effort (millions of GRT*)	Size of population (millions)	Catch (millions of tons)		
1960–61	3.9	7.3	12.6	2.0	5.9	11.0
1961–62	5.5	9.1	17.0	2.7	8.2	13.6
1962–63	6.9	13.3	18.1	2.8	9.7	18.7
1963–64	8.0	19.1	14.7	2.3	10.3	24.6
1964–65	8.0	17.9	16.5	2.6	10.6	23.8
1965–66	8.1	19.9	4.4	0.7	8.8	21.6
1966–67	8.2	17.4	4.6	0.7	8.9	18.9
1967–68	10.0	17.9	4.3	0.7	10.7	19.2
1968–69	10.1	20.1	5.2	0.8	10.9	21.7
1969–70	10.9	21.8	4.7	0.7	11.6	23.2
1970–71	10.0	17.3	4.5	0.7	10.7	18.6

*Gross registered tonnage.
Source: IMARPE.

The 1965 El Niño, identified as moderate (Quinn et al., 1987), has had the most damaging effect to date on guano bird populations. Its impact was even stronger than the strong 1957–58 El Niño, after which bird populations gradually recovered as they have done since ancient times (Jordán and Fuentes, 1966; Tovar, 1983). However, the 1965 El Niño cannot be deemed as the most significant cause of the decline and lack of recovery of bird populations (Fig. 13.3); rather, the primary reason was the continued growth and efficiency of the anchoveta fishing fleet.

As mentioned above, the strong 1972–73 El Niño had catastrophic effects on anchoveta fisheries; the bird populations were at low levels since 1965. The very strong 1982–83 El Niño was comparatively less damaging to the anchoveta fisheries, which were already depressed since 1972, and to other pelagic fisheries (see Figs. 13.2 and 13.5).

Research on planktonic fish egg density made it possible to conclude that this new prevalence of pelagic species in catch figures in fact responds to changes in the species composition of populations. Santander (1980) shows that sardine eggs, rare in the 1960s, became more prevalent after 1973 and gradually reached widespread distribution along the Peruvian coast, becoming more· plentiful than anchoveta eggs. Bernal et al. (1983) also identified the occurrence of changes in abundance and distribution of sardine eggs in Chilean waters with important spawnings taking place between 35° and 40° south latitude. Tsukayama and Santander (1987) made a detailed analysis of the evolution of sardine, jack mackerel, and horse mackerel concentrations, as well as of their progressive colonization along the Peruvian coast extending to those coastal areas where anchoveta used to predominate (see also Zuzunaga, 1985).

The gradual process of substitution of the anchoveta by other species has been under way simultaneously to a displacement of the traditional centers of heavy pelagic fisheries production. Previously located in Peru, these centers have gradually moved southwards toward the Chilean coasts where sardine and jack mackerel have seemingly encountered better conditions for expansion. This situation also gives rise to a new energy distribution in the ocean because of the large amount of organic productivity currently unexploited in Peruvian waters and advantageously exploited off the coast of Chile where new species flourished. When referring to changes in production centers, Jordán (1983b) noted that while for Peru it is possible to speak of a "substitution" of the anchoveta species and a partial "restitution" of lost biomass, as far as Chile was concerned, it is only possible to speak of an "activation" of the ecosystem through an increase in productivity at the secondary trophic level.

The disappearance of 10–15 million metric tons of biomass from the second trophic level off the Peruvian coast (which the biomass of new species was unable to replace) is bound to have an impact on pelagic communities and, perhaps, to an even larger degree, on benthic and demersal communities.

Circumstances under which changes in the composition of pelagic species in the southeast Pacific were brought about (environmental changes such as El Niño as well as overfishing) have been identified. However, a clear and precise explanation of the mechanisms giving origin to those circumstances, as well as their implications for future forecasting, is still lacking.

Changes in the composition of pelagic fish species off the southeast Pacific coast, and the decreasing anchoveta population, have prompted the incorporation of significant changes into the fishing industry. Such changes include the redistribution of the fishing fleets' fishing grounds, the change in location of fish meal processing plants, the introduction of new fishing technologies designed for fish of larger size and with a better ability to elude capture.

Sardine, which to a certain degree substituted for anchoveta, in terms of both catch figures and fish meal production, possesses an inherent advantage because of the possibilities it offers to the canning industry, to fish meal processing and to other industrial uses. These possibilities have yet to be extensively exploited. Fishmeal production is not a cumbersome process. On the contrary, it offers great marketing possibilities. With respect to canning, however, attention should focus on the fact that international markets impose restrictions on large imports of canned fish products.

Acknowledgement

I would like to express my thanks to Maria Krenz for her assistance in editing the English version of this paper.

References

Arntz, W.E., Landa, A., & Tarazona, J. (eds) (1985). *El Niño: Su impacto en la fauna marina*. Conferencias del Symposio El Fenómeno El Niño y su Impacto en la Fauna Marina; dentro del Noveno Congreso Latinoamericano de Zoologia, Arequipa, Peru, 9–15 October 1983. Special issue of *Boletín del Instituto del Mar del Perú*.

Arntz, W.E. (1986). The two faces of El Niño 1982–83. *Report on Marine Research* Special issue 31, 1–46 (Scientific Commission for Marine Research, Hamburg, FRG).

Avaria, S. (1984). Cambios en la composición y biomasa del fitoplancton marino del norte de Chile durante el fenómeno El Niño 1982–83. CPPS *Revista Pacífico Sur*, **15**, 303–9.

Avaria, S. & Alvial, S. (1985). La investigación ecológica del fitoplancton marino en Chile. *Revista Biológica Marina* (Valparaiso), **21**(1), 61–106.

Barber, R., Chavez, F. & Kogelschatz, J.E. (1985). Biological effects of El Niño. *CPPS Boletín ERFEN*, **14**, 3–30 (Spanish–English).

Bernal, P.A., Robles, F. & Rojas, O. (1983). Variabilidad física y biológica en la región meridional del sistema de corrientes Chile-Perú. In *Proceedings of the Experts Consultation to Examine Changes in Abundance and Species Composition of Neritic Fish Resources*, eds. G. Sharp and J. Csirke, pp. 683–711. FAO Fisheries Report 291, Vol. 3. Rome, Italy: FAO.

Boerema, L.H., Saetersdal, G., Valdivia, J., Tsukayama, J. & Alegre, B. (1967). Informe sobre los efectos de la pesca en el recurso peruano de anchoveta. *Boletín del Instituto del Mar de Peru*, **1**(4), 133–86.

Cañon, J.R. (1986). Variabilidad ambiental en relación con la pesquería nerítica pelágica de la zona norte de Chile. In *La Pesca en Chile*, ed. P. Arana, 195–205. Valparaiso, Chile: Escuela de Ciencias del Mar, Universidad Central de Valparaiso.

CEPAL (UN Economic Commission for Latin America) (1983). Los desastres naturales de 1982–83 en Bolivia, Ecuador y Perú. E/CEPAL/G 1274. New York: United Nations.

CONCYTEC (Consejo Nacional de Ciencia y Tecnología) (1985). *Ciencia, Tecnología y Agresión Ambiental: El Fenómeno El Niño*. Lima, Peru: CONCYTEC.

CPPS (Comisión Permanente del Pacífico del Sur) (ed.) (1984). Taller sobre el Fenómeno de El Niño 1982–83, Guayaquil, Ecuador, 12–16 diciembre 1983. *Revista Pacífico Sur*, **15** (entire issue).

Csirke, J. (1988). Small shoaling pelagic fish stocks. In *Fish Population Dynamics* (2nd edition), ed. J.A. Gulland, 271–302. Chichester: John Wiley & Sons.

Chavez, F. & Barber, R. (1985). La productividad de las aguas frente a la costa del Perú. *CPPS Boletín ERFEN*, **15**, 9–13.

DeVries, T. (1986). A review of geological evidence for ancient El Niño activity in Peru. *Journal of Geophysical Research*, **92**(C13), 14471–9.

Dioses, T. (1985). Influencia del fenómeno El Niño 1982–83 en el peso total individual de los peces pelágicos; sardina, jurel, caballa. *Instituto del Mar de Perú*, special bulletin.

ERFEN (Estudio Regional del Fenómeno El Niño) (1984). Resultados del análisis de las condiciones de El Niño 1982–83. In Resultados de la IV Reunión del Comité Científico de ERFEN, Callao, 28–30 mayo 1984. *CPPS Boletín ERFEN*, **9** (entire issue).

Enfield, D.B. & Allen, J.S. (1980). On the structure and dynamics of monthly mean sea level anomalies along the Pacific coast of North and South America. *Journal of Physical Oceanography*, **10**, 557–78.

Ferreyra, R. (1985). Efectos del fenómeno El Niño sobre el algarrobal. In *Ciencia, Tecnología y Agresión Ambiental: El Fenómeno El Niño*, ed. CONCYTEC, 571–7. Lima, Peru: CONCYTEC.

Glantz, M.H. (1984). Climate and fisheries: A Peruvian case study. *Proceedings of the Climate Conference for Latin America and the Caribbean*. WMO No. 632, 365–78. Geneva, Switzerland: WMO.

Glantz, M.H. & Thompson, J.D. (eds.) (1981). *Resource Management and Environmental Uncertainty: Lessons from Coastal Upwelling Fisheries*. New York: Wiley Interscience.

Glantz, M.H., Katz, R.W. & Krenz, M.E. (eds.) (1987). *The Societal Impacts Associated with the 1982–83 Worldwide Climate Anomalies.* Boulder, Colorado: National Center for Atmospheric Research.

Guillén, O. & Calienes, R. (1981). Productivity and El Niño. In *Resource Management and Environmental Uncertainty: Lessons from Coastal Upwelling Fisheries,* eds. M.H. Glantz & J.D. Thompson, 255–82. New York: Wiley Interscience.

IFOP (Instituto de Fomento Pesquero) (ed.) (1985). Taller Nacional Fenómeno El Niño 1982–83. *Investigacion Pesquera* (Chile), **32**, special issue.

IMARPE (Instituto del Mar del Perú) (ed.) (1972). Report of the Second Session of the Panel of Experts on the Population Dynamics of Peruvian Anchovy, March 1971. *Boletín del Instituto del Mar del Perú,* **2**(7) (entire issue).

INP (Instituto Nacional de Planificación) (1983). Programa Integral de Rehabilitación y Reconstrucción de las Zonas Afectadas por los Desastres de la Naturaleza. Vol. I: Evaluación de Daños en las Zonas Afectadas. Lima, Peru: INP.

Jordán, R. (1964). Las emigraciones y mortandad de las aves en el invierno y otoño de 1963. Informe 27, 43 pp. Callao, Peru: Instituto de Investigación de los Recursos Marinos.

Jordán, R. & Fuentes, H. (1966). Estudio preliminar sobre las poblaciones de aves guaneras. In *Memoria del Primer Seminario Latinoamericano sobre el Océano Pacífico Oriental.* Lima, Peru: Universidad Nacional Mayor de San Marcos.

Jordán, R. (1983a). Preliminary report of the 1982–83 El Niño effects in Ecuador and Peru. *Tropical Ocean–Atmosphere Newsletter,* **19**, 8–9.

Jordán, R. (1983b). Variabilidad de los recursos pelágicos en el Pacífico Sudeste. In *Proceedings of the Expert Consultation to Examine Changes in Abundance and Species Composition of Neritic Fish Resources,* eds. G. Sharp & J. Csirke, 113–29. FAO Fisheries Report 291, Vol. 2. Rome, Italy: FAO.

Jordán, R. (ed.) (1987). Fenómeno de El Niño 1982–83 con Particular Referencia a sus Efectos sobre los Recursos Pesqueros y la Pesquería en el Pacífico Sudeste. *CPPS Revista Pacífico Sur,* **16** (entire issue).

Lack, D. (1954). *The Natural Regulation of Animal Numbers.* Oxford: Clarendon Press.

Luong, T. & Toro, B. (1985). Cambios en la vegetación de las Islas Galápagos durante El Niño 1982–83. In *El Niño in the Galapagos Islands: The 1982–83 Event,* ed. G. Robinson and E.M. Pino, 331–42. Quito, Ecuador: Charles Darwin.

MacCall, A.D. (1984). Part II. Report of the working group on resources study and monitoring. In *Reports of the Expert Consultation to Examine Changes in Abundance and Species Composition of Neritic Fish Resources,* eds. J. Csirke & G. Sharp, 9–39. FAO Fisheries Report 291, Vol. 1. Rome, Italy: FAO.

Martinez, C., Salazar, C., Bohm, G., Mendie, J.J., & Estrada, C. (1985). Efectos del fenómeno El Niño 1982–83. Sobre los principales recursos pelágicos y su pesquería. *Investigacion Pesquera* (Chile), **32**, 129–39.

on Kiladis and van Loon (1988). Although there are several notable exceptions, it is obvious that, in general, there is a strong relationship between monsoon rainfall over India and the ENSO phenomenon. Moreover, this relationship is stronger over the central peninsula than over the northern India (e.g., Kiladis and Diaz, 1989), with rainfall anomalies over southernmost India and Sri Lanka during the northeast monsoon season (mid-September through November) opposite to those over the rest of India during the preceding summer monsoon (Rasmusson and Carpenter, 1983; Ropelewski and Halpert, 1987, 1989; Kiladis and Diaz, 1989).

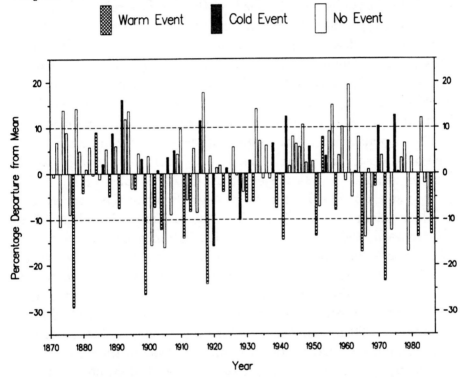

Fig. 14.4 All-India monsoon season (June–September) rainfall from 1871 to 1986, given as a percentage departure from the mean (from Parthasarathy et al., 1988). Warm and cold events in the Southern Oscillation are shown as hatched and solid bars, respectively.

Recent efforts at forecasting the Indian summer monsoon

The above relationships form the basis of much recent work on long-range forecasting of the monsoon over India. For example, Shukla and Paolino (1983) demonstrate a strong relationship between monsoon rainfall and the trend of the surface pressure anomaly at Darwin, Australia, during the months prior to the monsoon. Darwin lies near the center of the western pole of the SO, and pressure tendency there gives some indication of the probability of the development of a warm or cold event.

Other studies have, in the spirit of Walker, developed regression equations based on a combination of SO-related and local parameters to forecast the upcoming monsoon (e.g., Hastenrath, 1987; Parthasarathy et al., 1988). Several observational and modeling investigations have focused on the possible role of SST over the Arabian Sea in determining the strength of the monsoon, all based on the idea that higher than normal SST during the monsoon should enhance monsoon rainfall through increased evaporation. Results from these studies are, however, inconclusive (see Shukla, 1987), and it now appears more likely that the atmosphere is influencing the SST over this region rather than vice versa.

The regional circulation over southern Asia during spring has also been used, with some success, to forecast the upcoming summer monsoon (e.g., Barnerjee et al., 1978; Mooley et al., 1986; Joseph et al., 1981). By far the best predictor in these schemes is the latitudinal position in April of the 500 mb ridge along 75°E (Hastenrath, 1987). The axis of this ridge moves northward during the seasonal transition from spring to summer over southern Asia. If the ridge position is farther south in April, monsoon rainfall tends to be deficient. A related index involves the meridional wind at upper levels over India in May (Joseph et al., 1981). Anomalous 200 mb southerlies during May imply persistence of the pre-monsoon circulation, leading to a delay in the onset of the monsoon, and are statistically associated with deficient monsoon rainfall.

Kanamitsu and Krishnamurti (1978) and, more recently, Joseph (1989) have shown that the frequency of tropical cyclones in the northwestern Pacific is also related to monsoon rainfall, in that more frequent cyclones coincide with deficient monsoon rains.

Yet another large-scale influence on the behavior of the monsoon over India is the so-called 40–50 day wave (or "oscillation"), discovered by Madden and Julian (1971, 1972). The 40–50 day oscillation is associated with an eastward movement of an extensive zone of enhanced cloudiness and rainfall in the tropics. The periodicity and behavior of the oscillation is irregular. In some cases certain features of a particular wave can be traced entirely around the globe, although the largest amplitudes are generally seen over an area extending from the central Indian Ocean to the western Pacific (Murakami et al., 1986). Eastward propagation of the wave in these regions is associated with large shifts in both the tropical and the extratropical circulation (Weickmann et al., 1985; Lau and Chan, 1985).

It appears that there is a relation between the passage of a 40–50 day wave through the Indian Ocean and active monsoon periods (Yasunari, 1980; Krishnamurti and Subrahmanyam, 1982) and monsoon onset over southern India (Webster, 1987). Sikka and Gadgil (1980) have shown a tendency for northward movement of an active rainfall band from southernmost India to the Himalayan foothills, with a period of about 40–50 days. The movement of these bands determined active and break periods within a monsoon season, as well as the monsoon onset. Such a "triggering" mechanism has been noted for the Australian monsoon as well (McBride, 1987).

While many of the regional relationships discussed above have proven useful for extended range prediction of the Indian monsoon, it is not yet clear whether they are independent forcings on the monsoon or are themselves forced by the large-scale environment of the SO. For example, the behavior of the 40–50 day oscillation strongly depends upon the state of the SO, such that during warm events the amplitude of the wave is decreased in the Indian sector and increased in the central Pacific (Lau and Chan, 1985). Clearly the monsoon cannot be viewed in isolation, but must be viewed in the context of the entire global climate system. It is evident that improvements in the extended range forecasts of the large-scale state of the tropical atmosphere will lead to improved forecasts of monsoon rainfall over India, and help to provide better warning of the potentially catastrophic effects of drought and flood.

Impacts of recent droughts in India

Droughts and their impacts in India have had a long, often tragic, history. Many had led to famines for a variety of reasons. Clearly, scientifically sound forecasts of the "health" of the monsoon would be of value to decisionmakers at all levels of society from village to national government. However, there also needs to be a societal component to this approach. In the absence of adequate societal preparation for and response to recurrent droughts of varying intensities, forecasts based on ENSO teleconnections would have greatly reduced value to society.

There was a time when India was identified as a country highly prone to the occurrence of famines. Today, this is no longer valid. The most serious famine of this century occurred in pre-independence India in 1943 in Bengal, during which three million lives were lost. In fact, there was no drought; it was the failure of the administrative system which led to this avoidable famine (Bhatia, 1967; Chattopadhyaya, 1981). In post-independence India, because of two consecutive droughts in 1965 and 1966, food scarcity conditions which led to famine were created in Bihar. Lessons, however, were learned during this period, and subsequent droughts that occurred in 1972, 1979, 1982, and 1987 did not lead to famine. During the droughts of 1979 and 1982 there were no imports of food grains, thereby generating a feeling of achievement in the difficult area of drought-coping mechanisms.

Defining drought and its magnitude

There is a large number of definitions of drought (WMO, 1975; Krishnan, 1979; Wilhite and Glantz, 1985). However, when drought is defined in terms of precipitation or, more generally, agricultural production, the limits of definition, though arbitrary, could be of practical utility to scientists, administrators, planners, and policymakers. While such definitions may not be of direct use to an individual farmer, they are important nevertheless.

A major impact of drought is on agricultural production which influences all other sectors of society directly or indirectly. With the onset of the monsoon, agricultural operations for a *kharif* crop are started. A second crop season, *rabi*, starts with the cessation of monsoon. The major crops of the two seasons are as follows:

Kharif : Rice (irrigated and unirrigated), maize, sorghum, millet, groundnut, pigeonpea, mostly unirrigated cotton.

Rabi : Wheat (mostly irrigated), chickpea, barley, mustard (mostly irrigated), rice, maize and groundnut (irrigated).

More area is planted with crops in *kharif*, but the per hectare average productivity of all the crops is only about one metric ton. The

1965, the economy suffered. As a result, the import of food grains and the provision of economic assistance was difficult for the government to manage.

During 1965–67, almost 13 million metric tons of food grains were imported under the U.S. PL480 program for distribution in the affected areas. This led to the realization that food production should have a major priority in national development planning. Several agriculture and irrigation projects were then begun. Field trials using high-yielding Mexican wheat varieties had shown promise in India. A very important decision was taken by the government on the advice of scientists to import 18,000 metric tons of seed material of wheat from Mexico for cultivation in India. The success of this seed material gave confidence to scientists, administrators and politicians that the country had the potential for producing enough food grains to meet domestic demand. Since the high-yielding Mexican wheat varieties (or their derivatives later produced in India) required irrigation and fertilizers, the plan for developing irrigation resources was strengthened.

Thus, the 1965 and 1966 droughts led to policies toward the modernization of the agricultural sector. The following lessons were among the many learned from the experiences gained during the droughts of 1965 and 1966:

(1) There was a possibility of predicting a forthcoming food shortage. Therefore, the country could prepare itself to combat the impacts of shortage likely to occur somewhere within the country.

(2) Food grains may not be easy to import at the time they are needed.

(3) Efficient transportation is necessary to move food grains from one part of the country to another in order to control price fluctuations in the marketplace.

(4) A well organized system is necessary to monitor drought relief as well as agricultural production operations.

(5) Anticipation, planning and action with respect to drought must be the main objectives of any strategy to mitigate the impacts of drought.

The drought of 1979

The experience of the 1979 drought can be considered a major step in the management of drought in the country. The monsoon onset was late and erratic. M.S. Swaminathan, the Secretary of the Department of Agriculture, warned the state governments to be prepared to face a drought situation if the monsoon proved to be poor. At regular weekly and fortnightly intervals the development of the monsoon was monitored. Prolonged dry spells in August led to severe drought, primarily in the states in northern India.

A Watch Group was set up in the Ministry of Agriculture which consisted of representatives from the Department of Agriculture, the India Meteorological Department, the Indian Council of Agricultural Research and the Ministry of Information and Broadcasting, among others. A two-pronged strategy to prevent food shortages was adopted, one which could be described as curative as well as preventive. A meeting of the chief ministers from the drought-affected states was convened in September and was addressed by the Prime Minister. Watch Groups were established in all state capitals and district headquarters. They were given the responsibility to provide weekly reports of rainfall, agricultural operations, employment, and other aspects.

A 12-point program was created to avert what is called *Trikal* (*Akal, Jalkal* and *Tinkal*), which means the nonavailability of food, water and fodder. Although 240 districts and 220 million people were affected by drought, its severity was not felt because of the buffer stocks in foodgrains (Sinha et al., 1987). This view is substantiated by the fact that the rise in foodgrain prices was much less than in the period between 1965 and 1967 (Table 14.1). The central government allocated 19.6 million tons of foodgrains to the northern states, and incurred an expenditure of 1,569.5 million rupees. As many as 6.2 million people were given employment under special social programs, including the Food for Work Program (Jaiswal and Kolte, 1981).

Relief measures on such a large scale had never been undertaken, but because of them no starvation deaths were reported. If the country had had to import foodgrains to meet the shortage of 22 million tons, the cost would have been US$4.6 billion at the then-prevailing international market price (Jain and Sinha, 1981), an impossible response for the government.

Table 14.1 Average wholesale price of rice in moderately drought-affected and
"normal" states, 1965–67 and 1978–80 (price in rupees per 100 kg).

State	Year					
	1965	1966	1967	1978	1979	1980
West Bengal	68.1	81.3	156.5	176.5	208.0	221.7
Bihar	93.7	133.2	180.8	168.6	190.6	220.4
Andhra Pradesh	65.6	67.8	73.0	162.8	163.1	190.4
Madhya Pradesh	60.9	67.8	71.7	195.0	21.03	234.6
Tamil Nadu	63.1	65.1	68.1	153.7	168.0	192.5

Effect of the 1982–83 drought on production

In 1982 the Indian monsoon arrived late, was erratic, and with-
drew early from key agricultural production areas. The weak 1982
monsoon reduced plantings and yields in the major producing areas during the *kharif* season, resulting in more than a 10 percent
decline in total 1982–83 oilseed production. This monsoon season also resulted in reduced foodgrain production, from 79 million
metric tons in 1981–82 to 69 million tons in 1982–83. However,
the winter production increased from about 54 million metric tons
in 1981–82 to about 59 million metric tons in 1982–83, primarily
because of irrigation. Thus, despite a major drought, the year
ended with a loss of only 5 million tons. In essence, the winter
season cultivation provided a buffering action to the vagaries of
the monsoon season and received considerable additional effort by
farmers as well as by the government.

Tables 14.2 and 14.3 show rice and wheat production in some
states. Rice production was severely affected in Madhaya Pradesh,
Orissa, and Uttar Pradesh in 1979–80, and was reduced by 25 percent in Orissa in 1982–83. In other states the effect was marginal.
In Punjab, where rice is a completely irrigated crop, there was
no reduction of production in 1979–80 and there was an actual
increase in 1982–83. Wheat production was much more stable.
Thus, despite summer drought, there was increased production in
all the five major wheat-producing states, partly as a result of the
availability of irrigation, and partly as a result of favorable winter
rains from January to March 1983.

Table 14.2 Variability in rice production in four states in India (production in million metric tons).

	Year				
State	1978–79	1979–80	1980–81	1981–82	1982–83
Madhya Pradesh	3.56	1.83	4.00	3.83	3.40
Orissa	4.40	2.92	4.32	3.85	2.90
Punjab	3.09	3.04	3.22	3.75	4.15
Uttar Pradesh	5.96	2.55	5.44	5.90	5.53

Table 14.3 Variability in wheat production in five states in India (production in million metric tons).

	Year			
State	1979–80	1980–81	1981–82	1982–83
Madhya Pradesh	2.15	3.06	3.31	3.68
Haryana	3.28	3.60	3.68	4.35
Punjab	7.89	7.70	8.55	9.18
Rajasthan	2.70	2.39	2.93	3.78
Uttar Pradesh	9.89	13.13	12.75	15.29

The effect of the 1982–83 drought could also be seen in the levels of procurement of rice and wheat. There was no increase in the marketable surplus of rice over 1982–83 deliveries, but wheat purchases by the government increased in 1982–83 approximately 15 percent over 1981–82. This shows, once again, that deficit rainfall in 1982 had some effect on production in the monsoon season but had little or no effect on winter season production.

Research stations of the All-India Coordination Project on Dryland Agriculture are distributed around the country. One of these stations in Gujarat (Dantiwada) received somewhat below "normal" rainfall. In addition to starting late, most of the rains came on only two occasions, resulting in a sharp reduction in the yield of pearl millet and a 30–40 percent reduction in cowpea and castor bean. Low-yielding species of green gram (mung bean) were hardly affected, nor were local low-yielding varieties of pearl millet affected, suggesting that low-yielding crops are less sensitive to moisture stress. However, such low-yield but drought-resistant technology has little chance of taking full advantage of favorable conditions in years with better monsoons. This is one area where an accurate forecast of the upcoming monsoon season could help determine the best crops to plant in a particular year.

Since droughts have been a common feature on the Indian sub-continent and have often been followed by famines, most governments have formulated food-related policies. For example, the earliest writings in Arthashastra (Economics) by Chanakya in 321 B.C. noted that "Famine relief was a special care of the state, and half the stores in all the state warehouses were always kept in reserve for times of scarcity and famines." Thus, the concept of buffer stocks maintained by the government is very old in India, and was used by most rulers, until the beginning of the colonial era.

Following the severe droughts of 1965 and 1966, the idea of buffer stocks was again revived but at that time no surplus harvest was available with which to create a buffer. In the 1970s buffer stocks were established through market purchases by the government. In 1982–83, the government had a buffer stock of 18 million metric tons; thus, it was relatively easy to compensate for a 5-million-ton shortage in food grain production. Had the government not intervened and supplied food to the needy at a reasonable price, the situation could have become very difficult for the landless laborer and for nonagricultural communities.

During 1982, the All-India Index of Wholesale Price of all cereals increased by 2 percent between January and August. By this time, the impact of the poor monsoon was apparent. The wholesale price index increased by almost 12 percent between August 1982 and 1983. The maximum increase occurred in the price of rice, followed by pulses and wheat (Fig. 14.6). Thus, relatively speaking, wheat served as a price-stabilizing factor for food grains. A survey of prices in different Indian states showed that, in fact, prices had varied considerably from one location to another. Food price increases were mainly the result of shortages associated with the 1982–83 drought, and with a favorable monsoon and record-setting crop production in 1983, wheat prices, for example, declined.

The drought in 1987

After the experience of the 1965–66 and 1972 droughts, the government of India developed a policy of maintaining buffer food

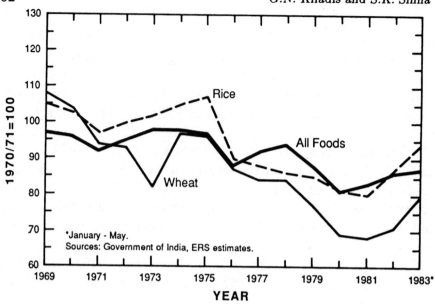

Fig. 14.6 Deflated wholesale price indices for wheat, rice, and foods in India.

stocks consisting mainly of wheat and rice, and small quantities of coarse grains. The desired level of buffer stock is 21.4 million metric tons of food grains. In 1985, the government buffer stock was 28.76 million metric tons, and everyone was concerned about the desirability of such a large buffer, because it required a substantial amount of money to maintain. However, the size of this buffer stock proved to be a great asset when the country faced both the 1986 and 1987 droughts. There was a rapid depletion of the buffer stock, which was reduced to 11.7 million metric tons by 1 July 1988. Formerly, the distribution of food grains through the fair price shops was undertaken only in urban areas. But in 1987 and 1988, public distribution was introduced in some drought affected rural regions as well.

Soon after the 1987 drought, two major efforts were undertaken by the government: (1) to encourage production in *rabi* (winter) of 1987-88, and (2) to encourage production in 1988. As a result of the first effort, there was a marginal increase in *rabi* 1987–88 production over the previous year. Based on the existing rainfall, dependable water availability through ground and surface irrigation in 169 districts was identified in a special drive to increase the production of wheat, rice, maize and pulses. The following tactics were adopted:

(1) An increase in fertilizer use by 20 kg of nutrients per hectare.

(2) The use of improved high-yield variety seeds.
(3) Better management of weeds and timely control of pests and diseases.
(4) Optimal use of groundwater and surface irrigation.
(5) Offer of a bonus/incentive for production/procurement of food grains.
(6) Increase the flow of short- and long-term credit.

Fortunately, 1988 turned out to be one of the best monsoon years in the recent past. A food production target of 170 million metric tons was met. This was an increase of 32 million metric tons over 138 million metric tons produced in the drought year of 1987–88. Thus, the basic underlying strategy is to make the maximum use of a good monsoon year in order to build buffer stocks so that they would be available for use in years with poor monsoons.

Limitations of the policy

Droughts and other natural disasters have occurred in India in the past and will likely occur in the future. It is possible that with a global warming such events could become more frequent in the future. Therefore, there is a strong need for a long-term national policy on the management of natural disasters. Each drought (or flood) receives an immediate response by the government to organize services and various committees. Once the natural calamity (e.g., drought) is over, the events are forgotten. In the government bureaucracy those involved in drought management during one drought episode are often not there during the next one. As a result, planning usually has to start from the beginning when a drought returns, because new bureaucrats often lack experience related to drought. Some limitations on drought management are as follows:
(1) In India one can assess the possibility of a drought soon after the onset of the monsoon in most parts of the country. By the end of the first week of July a reasonable projection about the probability of drought is possible.
(2) When drought does occur, it is often difficult to determine what its ultimate magnitude and intensity might be. The India Meteorological Department's classification for describing deficit rainfall (and drought) is not adequate. Within a block

or district there is often considerable variability in rainfall. Consequently, sometimes those families or villages which had not been adversely affected may gain advantages through aid, whereas those actually affected may receive little or no aid at all. More scientific and socially relevant criteria need to be developed in order to identify those in need in times of drought.

(3) The central government provides major assistance in organizing drought relief and management. However, sometimes this help to states is subjectively based. With more realistic assessment of the impacts, the distribution of aid could become more rational.

(4) Indian agriculture is still based largely on draft animal power. In addition, the cattle population is a major source of income to a large number of people. Even during the last drought, the cattle population suffered extensively. Fodder was not available in drought-affected states, and its price increased considerably. Though cattle camps were established where owners could leave their cattle, this arrangement was not sufficient to minimize the impact of drought on livestock.

(5) A considerable amount of research on the management of drought has been done in India. Techniques for mid-season correction, the efficient use of "harvested water" and fertilizers have been developed. The experimental watersheds in 1987 clearly demonstrated their success. However, their use is not yet widespread.

(6) Often there is not enough preparedness to take advantage of a good monsoon year. While the government has now assured the purchase of excess marketable grains or other agricultural produce, the price of many commodities continues to fall below a remunerative price. This happened to the price of groundnut (and groundnut oil) and mustard in the 1988-89 crop seasons.

(7) There is an ever-increasing need for funds to meet the requirements of drought management. The central government levied a special tax above a taxable income of 50,000 rupees. This was an ad hoc 1-year decision to generate funds for crisis management. There are several programs the government has funded to support industry. Large amounts are spent on irrigated areas where productivity is high. It would be prudent

to consider a long-term objective to develop a "crisis management fund" on an annual basis, which should not be used for any other purpose than to help financial self-sufficiency in crisis management.

(8) The people's representatives must have an important role in monitoring and informing the public, regardless of political party affiliations. The impacts of natural calamities such as drought should be treated as a common concern.

Summary

Recent advances in the monitoring and forecasting of drought (and flood) over India, along with more efficient and timely response programs of the government to deal with climatic anomalies, have greatly reduced the societal impacts of fluctuations in the Indian monsoon. Improved forecasting of the Indian monsoon several months in advance could lead to a further increase in the ability of India to cope with the effects of drought (and flood) on agricultural production. Technical improvements in the observational aspects of ENSO along with deeper understanding of the coupling between the tropical ocean and atmosphere will undoubtedly lead to advances in this direction.

Acknowledgments

We thank Dr. P.V. Joseph and Dr. P.D. Sardeshmukh for helpful comments. Thanks also to J.K. Eischeid for data and graphical support.

References

Angell, J.K. (1981). Comparison of the variations in atmospheric quantities with sea surface temperature variations in the equatorial eastern Pacific. *Monthly Weather Review*, **109**, 230–43.

Barnerjee, A.K., Sen, P.N. & Raman, C.R.V. (1978). On foreshadowing southwest monsoon rainfall over India with mid-tropospheric circulation anomaly of April. *Indian Journal of Meteorology and Geophysics*, **29**, 425–31.

Bhatia, B.M. (1967). *Famines in India*. New Delhi: Asia Publishing House.

Blanford, H.F. (1884). On the connection of the Himalayan snowfall with dry winds and seasons of drought in India. *Proceedings of the Royal Society of London*, **37**, 3.

Chattopadhyaya, B. (1981). Notes towards understanding of the Bengal famine of 1943. *CRESSIDA*, 1, 112–53.

Datta, R.K. (1986). Monsoon dynamics, rainfall and teleconnections. *Pontificia Academy Scientiarvum*, **1988**, 257–392.

Dickson, R.R. (1984). Eurasian snow-cover versus Indian monsoon rainfall–an extension of Hahn-Shukla results. *Journal of Climate and Applied Meteorology*, **23**, 171–3.

Government of India (1898). *Indian Famine Commission Report*, 5–42. New Delhi, India: Government Printing Office.

Hahn, D. & Shukla, J. (1976). An apparent relationship between Eurasian snow cover and Indian monsoon rainfall. *Journal of the Atmospheric Sciences*, **33**, 2461–3.

Hastenrath, S. (1987). On the prediction of India monsoon rainfall anomalies. *Journal of Climate and Applied Meteorology*, **26**, 847–57.

Hildebrandsson, H.H. (1897). Quelques recherches sur les centres d'action de l'atmosphère. *Kon. Svenska Veteus.-Akad. Hundl.*, **9**, 1–36.

Jain, H.K. & Sinha, S.K. (1981). Droughts and the new agricultural technology. Proceedings of the WMO Symposium on Meteorological Aspects of Tropical Droughts., 7–11 December 1981, New Delhi. Geneva: WMO.

Jaiswal, N.K., & N.V. Kolte, 1981: *Development of Drought Prone Areas*. Hyderabad, India: National Institute of Rural Development.

Joseph, P.V. (1978). Subtopical westerlies in relation to large-scale failure of Indian monsoon. *Indian Journal of Meteorology, Hydrology and Geophysics*, **29**, 412–8.

Joseph, P.V., Mukhopadhyaya, R.K., Dixit, W.V. & Vaidya, D.V. (1981). Meridional wind index for long-range forecasting of Indian summer monsoon rainfall. *Mausam*, **32**, 31–4.

Joseph, P.V. (1989). The Indian summer monsoon and tropical cyclones of the western Pacific. *Tropical Ocean–Atmosphere Newsletter*, **4**, 6–8.

Kanamitsu, M. & Krishnamurti, T.N. (1978). Northern summer tropical circulations during drought and normal rainfall months. *Monthly Weather Review*, **106**, 331–47.

Kiladis, G.N. & Diaz, H.F. (1989). Global climatic anomalies associated with extremes in the Southern Oscillation. *Journal of Climate*, **2**, 1069–90.

Kiladis, G.N. & van Loon, H. (1988). The Southern Oscillation. Part VII: Meteorological anomalies over the Indian and Pacific sectors associated with the extremes of the oscillation. *Monthly Weather Review*, **116**, 120–36.

Krishnamurti, T.N. & Subrahmanyam, D. (1982). The 30–50 day mode at 850 mb during MONEX. *Journal of the Atmospheric Sciences*, **39**, 2088–95.

Krishnan, A. (1979). Definitions of drought and factors relevant to specifications of agricultural and hydrological droughts. In *Hydrological Aspects of Droughts*, Proceedings of the International Symposium, 3–7 December 1979, 67–102. New Delhi: Indian National Committee for IHP.CSIR.

Krishnan, A. & Thanvi, K.P. (1971). Occurrence of droughts in Rajasthan during 1941–1960. *Proceedings of the All India Seminar on Dryland Farming*. New Delhi.

Kutzbach, G. (1987). Concepts of monsoon physics in historical perspective: The Indian monsoon (seventeenth to early twentieth century). In *Monsoons*, ed. J.S. Fein & P.L. Stephens, 159–201. New York: John Wiley and Sons.

Lau, K.-M. & Chan, P.H. (1985). Aspects of the 40–50 day oscillation during the northern winter as inferred from outgoing longwave radiation. *Monthly Weather Review*, **113**, 1889–1909.

Lockyer, N. & Lockyer, W.J.S. (1904). The behaviour of the short-period atmospheric variation over the earth's surface. *Proceedings of the Royal Society of London*, **73**, 457–70.

Madden, R. & Julian, P. (1971). Detection of a 40–50 day oscillation in the zonal wind. *Journal of the Atmospheric Sciences*, **28**, 702–8.

Madden, R. & Julian, P. (1972). Description of global scale circulation cells in the tropics with a 40–50 day period. *Journal of the Atmospheric Sciences*, **29**, 1109–23.

McBride, J.L. (1987). The Australian summer monsoon. In *Monsoon Meteorology*, ed. C.-P. Chang & T.N. Krishnamurti, 203–31. New York: Oxford University Press.

Meehl, G.A. (1987). The annual cycle and interannual variability in the tropical Pacific and Indian Ocean regions. *Monthly Weather Review*, **115**, 27–50.

Mooley, D.A. & Parthasarathy, B. (1983). Indian summer monsoon and El Niño. *Pageoph*, **121**, 339–52.

Mooley, D.A. & Parthasarathy, B. (1984). Fluctuations in all-India summer monsoon rainfall during 1871–1978. *Climatic Change*, **6**, 287–301.

Mooley, D.A., Parthasarathy, B. & Pant, G.B. (1986). Relationship between all-India summer monsoon rainfall and location of ridge at 500 mb level along 75°E. *Journal of Climate and Applied Meteorology*, **25**, 633–40.

Murakami, T., Chen, L.-X., Xie, A. & Shrestha, M. (1986). Eastward propagation of 30-60 day perturbations as revealed from outgoing longwave radiation data. *Journal of the Atmospheric Sciences*, **43**, 961–71.

Nehru, J.L. (1947). *The Discovery of India*. New Delhi: Asia Publishing House.

Parthasarathy, B. & Pant, G.B. (1985). Seasonal relationships between Indian summer monsoon rainfall and the Southern Oscillation. *Journal of Climatology*, **5**, 369–78.

Parthasarathy, B., Diaz, H.F. & Eischeid, J.K. (1988). Prediction of all-India summer monsoon rainfall with regional and large-scale parameters. *Journal of Geophysical Research*, **93**, 5341–50.

Rao, Y.P. (1981). The climate of the Indian subcontinent. In *World Survey of Climatology*, vol. 9: Climates of Southern and Western Asia, ed. K. Takahashi & H. Arakawa, 67–182. New York: Elsevier Press.

Rasmusson, E.M. & Carpenter, T.H. (1983). The relationship between eastern equatorial Pacific sea surface temperatures and rainfall over India and Sri Lanka. *Monthly Weather Review*, **110**, 354–84.

Ropelewski, C.F. & Halpert, M.S. (1987). Global and regional scale precipitation patterns associated with the El Niño/Southern Oscillation. *Monthly Weather Review*, **115**, 1606–26.

Ropelewski, C.F. & Halpert, M.S. (1989). Precipitation patterns associated with the high index phase of the Southern Oscillation. *Journal of Climate*, **2**, 268–84.

Ropelewski, C.F., Robock, M. & Matson, A. (1984). Comments on "An apparent relationship between Eurasian spring snow cover and the advance period of the Indian summer monsoon." *Journal of Climate and Applied Meteorology*, **23** 341–2.

Sen, A. (1981). *Poverty and Famines: An Essay on Entitlement and Deprivation*. Oxford: Clarendon Press.

Shukla, J. (1987). Interannual variability of monsoons. In *Monsoons*, ed. J.S. Fein and P.L. Stephens, 399–463. New York: John Wiley and Sons.

Shukla, J. & Paolino, D. (1983). The Southern Oscillation and long-range forecasting of summer monsoon rainfall over India. *Monthly Weather Review*, **111**, 1830–37.

Sinha, S.K., Aggarwal, P.K., & Khanna-Chopra, R. (1985). Irrigation in India: A physiological and phenological approach to water management in grain crops. *Advances in Irrigation*, **3**, 130–213.

Sinha, S.K., Kailashnathan, K. & Vasistha, A.K. (1987). Drought management in India: Steps toward eliminating famines. In *Planning for Drought*, ed. D.A. Wilhite, W.E. Easterling & D.A. Wood, 453–70. Boulder, CO: Westview Press.

Sikka, D.R. & Gadgil, S. (1980). On the maximum cloud zone and the ITCZ over Indian longitudes during the southwest monsoon. *Monthly Weather Review*, **108**, 1840–53.

Spitz, P. (1981). Drought and self-provisioning. *CRESSIDA*, **1**, 18–35.

Subrahmanyam, V.P. (1964). Climate water balance of the Indian Arid Zone. In *Proceedings of the Symposium on Problems of Indian Arid Zone, Jodhpur*.

van Loon, H. & Shea, D.J. (1985). The Southern Oscillation. Part IV: The precursors south of 15°S to the extremes of the oscillation. *Monthly Weather Review*, **113**, 2063–74.

Verghese, B.G. (1967). *Beyond the Famine: An Approach to Regional Planning in Bihar*.

Walker, G.T. & Bliss, E.W. (1932). World weather No. 5, *Memoirs of the Royal Meteorological Society*, **4**, 53–84.

Webster, P.J. (1987). The variable and interactive monsoon. In *Monsoons*, ed. J.S. Fein & P.L. Stephens, 269–330. New York: John Wiley and Sons.

Weickmann, K.M., Lussky G.R. & Kutzbach, J.E. (1985). Intraseasonal (30–60 day) fluctuations of outgoing longwave radiation and 250 mb streamfunction during northern winter. *Monthly Weather Review*, **113**, 941–61.

Wilhite, D.A. & Glantz, M.H. (1985). Understanding the drought phenomenon: The role of definitions. *Water International*, **10**, 111–20.

WMO (World Meteorological Organization) (1975). Drought and Agriculture. Technical Note 138. Geneva: World Meteorological Organization.

Yasunari, T. (1980). Quasi-stationary appearance of 30–40 day period in the cloudiness fluctuations during summer monsoon over India. *Journal of the Meteorological Society of Japan*, **58**, 225–9.

pacts of ENSO events are often characterized by a warming in the equatorial waters off Peru and a subsequent decline in the anchoveta population in this region, along with other species composition changes. The importance of ENSO events, both biologically and physically, point to the need for more research on the relationship between atmospheric phenomena and fisheries.

Research focus

This chapter focuses on the impact of climatic variability and large-scale atmospheric phenomena on the shrimp fishery in the Gulf of Mexico. The shrimp fishery was chosen because of its economic importance to Gulf states and communities. Shrimp not only have a crucial role in the food chain but are also one of the most valuable fisheries in the United States (Thompson, 1986) and in Mexico. For example, in 1985 the shrimp fishery was fourth in landings among all U.S. fisheries but first in value, with the Gulf of Mexico shrimp fishery accounting for approximately 78 percent of the catch (Thompson, 1986). Because of its social, economic and biological importance, the shrimp fishery provides an ideal case study for examining the influences of climate variability, including those attributed to teleconnections.

Despite research indicating the importance of environmental factors in the growth and survival of shrimp, little information exists on whether shrimp throughout the Gulf of Mexico respond similarly to environmental variability, how climatic variability influences the habitat where shrimp live and, ultimately, how changes in environmental parameters affect shrimp landings.

This research focuses specifically on the fluctuations in brown and white shrimp landings in large grid areas in the northwestern Gulf. Regional climatic variability, as measured by temperature, precipitation, wind stress and river runoff, is analyzed in conjunction with shrimp landings. Large-scale atmospheric phenomena which have been shown to affect climate variability in the Gulf are subsequently analyzed, including the Southern Oscillation (SO), the Pacific/North American pattern (PNA), and the North Atlantic Oscillation (NAO).

Background

Shrimp life cycle

Brown and white shrimp, *Penaeus aztecus* and *Penaeus setiferus* respectively, dominate the shrimp catch in the western and northern Gulf of Mexico. The life cycles of these two species are quite similar. Both breed in the deeper waters and lay demersal eggs which hatch into nauplii. After three to five weeks, the post-larval shrimp move onshore, mature into juveniles and approximately four months later emigrate offshore again (Williams, 1960; Kutkuhn, 1966; GMFMC, 1981; Walker and Salia, 1986). Despite these similarities, the life cycles of adult brown and white shrimp are separated both geographically and temporally. Brown shrimp are found predominantly in the moderately saline waters off Texas, while white shrimp more densely populate the less saline Louisiana waters (Gunter and Edwards, 1969). The peak breeding period for brown shrimp is in the winter months; post-larvae immigrate into the estuaries in early to mid-spring, mature, and emigrate offshore in the summer months (Baxter and Renfro, 1967; Gaidry and White, 1973; White and Boudreaux, 1977). White shrimp breed throughout the year with peak production in the fall and spring. Peak immigration occurs in the winter and late spring, followed by maturation and emigration in the spring or late summer and early fall (GMFMC, 1981; Walker and Salia, 1986).

Relatively uncommon among fisheries, Gulf shrimp are dependent on recruit survival rather than adult survival. Regardless of the adult population size, approximately the same level of recruitment will occur (Lindner and Anderson, 1956; Rothschild and Brunenmeister, 1984). Because of the short life cycle of these species, approximately one year, subsequent fishery success directly depends on post-larval and juvenile survival in the estuaries or nurseries (Ford and St. Amant, 1971; GMFMC, 1981; Garcia, 1984; Turner and Boesch, 1988).

Marine environment

Historically, temperature and salinity have been regarded as the environmental variables of primary interest when examining fluctuations in the shrimp fishery. Both have been demonstrated the SO is in its most negative state, bringing about dramatic changes in local and regional fisheries. The anchoveta, normally predominant, becomes scarce, possibly because biological productivity is dramatically reduced (Barber and Chavez, 1983). During some years, shrimp which are typically found farther south on the South American coast, are observed in large numbers off Peru and Ecuador.

As more El Niño–Southern Oscillation (ENSO) events were observed, evidence mounted that the SO not only affected Peruvian waters but also those off the coasts of North America. During the 1982–83 event, temperatures and the structure of the surface layer were altered along the California coast (Fiedler et al., 1986). The impacts from these changes were substantial. Off the California coast, for example, anchovy growth, spawning, larval mortality and size at age were all depressed (Fiedler et al., 1986). Along the Pacific northwest coast, salmon fisheries were depleted in Washington waters but enhanced in Canadian waters, indicating that migratory patterns may have shifted (Hayes and Henry, 1985). Mysak (1986) also reported ENSO-linked changes in migration, recruitment and return time of various fisheries in the Pacific northeast. Such changes in fisheries catch provided further impetus for scientists to understand the effects of the SO at distances far from the eastern equatorial Pacific.

Not all El Niño events occurred simultaneously with a negative state of the SO; however, the relationship was strong enough to prompt further research. The connection between the SO and EN compelled researchers to examine possibilities other than the local marine conditions for explanations of stock alterations. For the first time, large-scale atmospheric phenomena were implicated as a major factor affecting the productivity of fisheries.

Research indicates that large-scale atmospheric phenomena affect climate in the Gulf of Mexico region as well. Rainfall has been shown to be influenced by the Southern Oscillation in this region and winter temperatures have been connected to the PNA and NAO (Meehl, 1978; van Loon and Rogers, 1978; Ropeleweski and Halpert, 1986: Dickson and Namias, 1976).

No research is available on the impact of large-scale atmospheric patterns on biological populations in the Gulf. However, the research described above demonstrates a strong relationship between the success of shrimp in the Gulf and regional climate. It is, therefore, quite possible that large-scale atmospheric phenomena which influence regional-scale climate may affect shrimp populations as well.

Examination of annual rainfall, river discharge, temperature and shrimp catch data in the Gulf further supports the need to examine the response of shrimp populations to large-scale atmospheric variability. From 1955 to 1986, two of the most severe flooding episodes of the Mississippi River occurred in 1957–58 and 1973. These are years when the Southern Oscillation was in its most extreme negative phase. Shrimp catch during these years greatly decreased (Gunter, 1962; Barrett and Gillespie, 1973; White, 1975). White (1975) reported a 40 percent decline in white shrimp landings following record flood levels for the Mississippi and dramatically reduced temperatures.

Work conducted in the Gulf of Carpentaria (Australia), linking the Southern Oscillation to a prawn fishery, further supports the need to examine the response of shrimp to large-scale atmospheric variability. Staples (1983) reported a positive correlation between rainfall, SOI and prawn catch. He further explained that rainfall is a necessary stimulus for prawns to migrate out of the estuaries. Thus, catch increased in years when rainfall was high. Subsequent studies confirmed and expanded on Staples research (Vance et al., 1985; Love, 1987). Vance et al. (1985) demonstrated that not only must sufficiently high levels of rainfall occur but the timing of these rains was crucial as well.

Data

Shrimp landings

Monthly regional shrimp landings data (weight of the catch brought into port) for 1961-1985 were acquired. Only offshore landings data were used. Upon examining the data, regions of low catch and missing data were observed. The regions to be studied, therefore, encompassed only those grid areas with no large gaps in the data. The final study region extended from Alabama (Mobile) to Texas (Rockport) (Fig. 15.1). Both species were caught in abundance in these regions, and there were few years in which no landings were reported in the primary and secondary catch seasons (determined by maximum pounds caught).

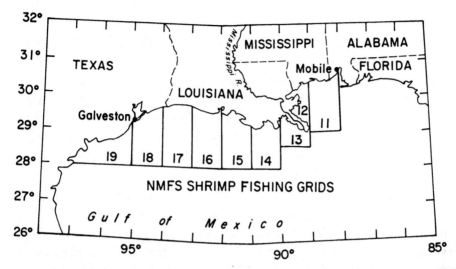

Fig. 15.1 Shrimp fishery grid areas in the Gulf of Mexico, established by the National Marine Fisheries Service. All except grid 12 were used in this study.

Monthly landings were summed to produce three-month seasonal catch data. The seasons, defined as winter (Dec., Jan., Feb.), spring (Mar., Apr., May), summer (June, July, Aug.) and fall (Sept., Oct., Nov.), were based on the biological life cycle of the shrimp (see Christmas and Etzold, 1977; GMFMC, 1981). The primary and secondary catch seasons for brown shrimp were summer and fall, respectively, and for white shrimp were fall and winter, respectively.

In order to analyze the data as accurately as possible, landings data should be adjusted for fishing effort and technological change before environmental impacts on shrimp production are examined (Brunenmeister, 1984). Changes in technology include the number of nets used, the power of the engines and the size of the boats. However, due to inconsistencies in the effort data, it was necessary to use a linear trend to approximate the combined effects of changes in effort and technology. The use of linear trends to account for technological change is a common practice in other climatic impact analyses, such as crop yield/climate modeling (Katz, 1977; Mearns, 1988).

Environmental data

Regional environmental variables were chosen which were expected to affect shrimp abundance. These included river discharge, precipitation, air and sea surface temperature (AT and SST, respectively) and north–south and east–west vector wind stress (WV and WU, respectively). Wind stress was used instead of ordinary wind vectors because it is more closely related to Eckman transport and associated ocean currents. Wind stress vectors toward the north and east are termed WV (+) and WU (+), respectively; wind stress vectors toward the south and west are termed WV (–) and WU (–), respectively.

Rivers chosen included the Mississippi, Atchafalaya, Tombigbee, Alabama, Pascagoula, and Escambia. Selection was based on the level of discharge, therefore possible impact, and on data availability. The last four of these rivers enter the Gulf very close together and their flows are highly correlated; therefore, the sum of their flows, hereafter referred to as "combined Alabama rivers," was used in the modeling.

Coastal precipitation data were available for only 5 of the grids, so precipitation was not included in the initial models but was considered in later stages of the analysis.

Atmospheric indices

Commonly used indices of the three atmospheric phenomena were computed for winter seasons, 1947–85. The Southern Oscillation Index was computed for fall as well, for predictive purposes.

The Southern Oscillation Index (SOI) is the sea-level pressure difference between Tahiti and Darwin (Australia), computed using normalized seasonally averaged sea level pressures. Autumns and winters for which the SOI is < –1 are identified as –SOI seasons and frequently correspond to El Niño (warm events). Autumns and winters for which the SOI is > +1 will be identified as +SOI seasons and correspond to La Niña (cold events).

A commonly used index of the North Atlantic Oscillation (NAO) is the difference in normalized sea level pressures between Ponta Delgada (Azores) and Akureyri (Iceland) (Yarnal and Leathers, 1988). Following Yarnal and Leathers' approach, winters are identified as +NAO when the NAO Index is > +1; −NAO encompasses those winters when the NAO Index is < −1.

An index of the Pacific/North American (PNA) pattern was computed according to the method of Horel and Wallace (1981), using 700 mb height data from grid points in Alberta, Canada, the north Pacific, and northern Florida.

PNA Index =

$$z^*(55N, 115W) - [z^*(45N, 165W) + z^*(30N, 85W)]/2,$$

where z^* is the normalized departure of a particular season from the mean 700 mb height for that particular grid point. Winters having a PNA Index > +1 are called "PNA winters" and those having a PNA Index < −1 are called "Reverse-PNA winters." (Yarnal and Leathers, 1988, use this terminology, although their method of computing the PNA Index is slightly different.)

Methods

The analysis was divided into three phases, with the results of one phase influencing the methods used in the next. This section gives an overview of the resulting methodology.*

Phase I: relating shrimp landings to regional environment

Stepwise multiple regression techniques were used to analyze possible relationships between shrimp landings and environmental variables. For each species, time series were analyzed separately for the two major seasons and the eight grid areas. Use of these multiple data sets provided at least a partial check on the reliability of the models.

The number of potential predictors was very large, primarily because of the life cycle of the shrimp. At different life stages, the shrimp live in different areas, have different tolerances and may

---

* For a more detailed description of both methods and results, the reader is referred to articles currently being prepared for publication by White and Downton. Contact the authors at the National Center for Atmospheric Research, P.O. Box 3000, Boulder, CO 80307. Only information necessary for understanding the results is discussed here.

respond uniquely to environmental stimuli. Hence, environmental conditions two to four seasons prior to the landings had to be considered, requiring that the predictors be examined at several time lags.

The best model was selected for each grid area, based on the criteria that it have the highest adjusted R^2 and that all predictors included in the model be significant at a 90 percent confidence level. Although no two grids had exactly the same predictors, there were often substantial similarities between models in different grids. Comparing the models from the eight grid areas, predictors which were significant and had the same sign in at least three grid areas were considered to be of regional importance in shrimp production.

Phase II: relating regional environment to atmospheric phenomena

Only environmental variables which affect shrimp populations were of interest in this study. Therefore, environmental variables which were found to be related to shrimp landings in Phase I were further examined. Most of the environmental data were available for the years 1949–85. Thus, winter (DJF) values of the environmental variables were compared with the atmospheric indices over that 37 year period.

The atmospheric indices and nearly all of the environmental variables were found to have distributions that were approximately Gaussian. Correlations were first computed between the atmospheric indices and winter values of the environmental variables. Then opposite phases of the atmospheric circulation patterns were contrasted by comparing mean values of the environmental variables in seasons that exhibited the + and – phases of each pattern. Mean values were compared statistically using the two-sample t test. A 95 percent confidence level was used in all statistical tests.

Phase III: relating shrimp landings to atmospheric phenomena

Based on relationships found in Phases I and II, hypotheses were developed about connections between shrimp landings and the SO, PNA, and NAO indices.

These hypotheses required comparison of white shrimp landings in the two phases of each atmospheric phenomenon. The mean detrended landings during or following the two phases of each atmospheric phenomenon were compared using the two-sample *t*-test and the Wilcoxon test (for shrimp data that did not have a Gaussian distribution).

In addition, the ability of the large-scale atmospheric indices to substitute for regional winter weather variables was tested. Comparisons based on high and low values of the atmospheric indices ignore the catch in years with moderate index values. To examine the relationship between white shrimp landings and atmospheric indices in all years, stepwise regressions of white shrimp landings performed earlier were repeated, replacing the winter weather predictors AT, SST, and WV with the SO and PNA indices. The NAO was not used in the regressions because data was missing in three winters; hence, the results would not be comparable with earlier regressions.

Phase I: effects of regional environment on Gulf of Mexico shrimp landings

Results

The physical characteristics of the environment, on the average, are similar across the Gulf of Mexico; however, a gradient from east to west often occurs. Winter temperatures become substantially colder moving westward, with mean SST about 3°C lower and mean AT about 2°C lower in grids 18–19 (Texas) than in grids 11–15 (Mississippi, eastern Louisiana). The temperature gradation is less in spring and fall, and mean temperatures do not differ significantly across the region in summer.

Winds are usually from the south and east in spring and summer, from the north and east in fall and winter. There is a gradation across the region in spring and summer, with stronger winds in grids 18–19 (Texas) than in grids 11–13 (Mississippi).

Examination of the seasonal river flow shows that the largest discharge from all the rivers occurs in the spring of the year. Throughout the year, the Mississippi River dominates the Gulf.

Shrimp landing series for the primary catch seasons are plotted over time in Figs. 15.2 and 15.3. Increasing trends in landings are apparent in about half of the grids for each species. Multiple regression model results are described below for the primary and secondary catch season, first for brown shrimp and then for white shrimp. In some models, predictors are significant in only one grid area or are significant with opposite signs in two or three grid areas. Such results may indicate local rather than regional responses to climate or the presence of other controlling variables such as food availability, predation or currents. It is also possible that some predictors are statistically significant merely as a result of spurious relationships in these particular data samples or that some of the models do not accurately reflect the environment. Thus, only relationships which are significant, and with the same sign, in at least three grids are discussed.

Spring and winter river discharges are the most important variables in summer brown shrimp landings. Summer landings are negatively related to spring river flows in seven of the eight grids. Of all the weather variables, only summer WU (−) is related to catch in half of the grids. Models for fall landings of brown shrimp reflect those observed in the summer landings. River discharge is again dominant. Weather variables show no consistent relationships to landings, never being significant in more than two grids.

White shrimp landings show markedly different relationships compared to brown shrimp. The dominant predictors of fall landings are related to the preceding winter's weather. In all the grids examined, catch is negatively related to either SST, AT or WV in winter. Fall landings are positively related to summer SST and negatively related to summer Mississippi discharge. In the winter, white shrimp landings are related to a greater variety of environmental variables. Landings continue to be negatively related to AT and WV from the preceding winter and spring. However, a greater contribution to the regressions is made by more recent temperature, with winter landings being positively related to the current winter SST and/or the preceding fall AT in six of the grids.

In every grid during the primary catch season of both white and brown shrimp, environmental variables explain at least 50 percent of the additional variance after controlling for trend. In the secondary season, environmental variables explain at least 50

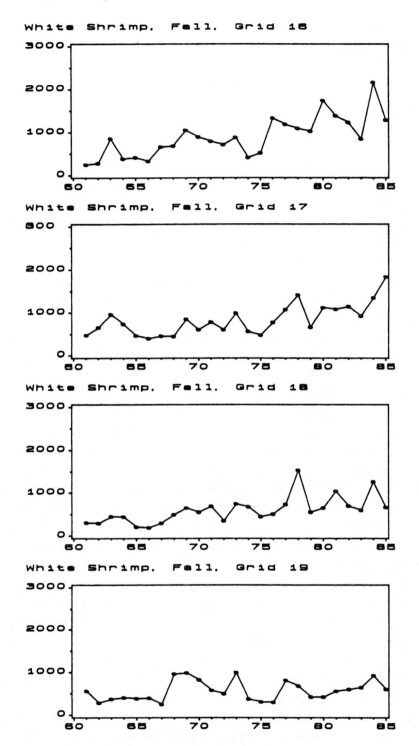

Fig. 15.3 White shrimp, fall landings (in metric tons) by year, 1961–85. Data have not been adjusted for changes in effort or technology.

from the north. Because north winds and cold temperatures often occur simultaneously, it was difficult to distinguish between them as causative agents.

Although it may first appear that cold temperatures and winds from the north would be detrimental to white shrimp, the relationship is a negative one; that is, cold temperatures and northerly winds were related to higher landings. Two explanations are possible. First, northerly winds may hold post-larval shrimp offshore until they reach a larger size and are less likely to be preyed upon when they enter the estuaries (Gary Sharp, National Marine Fisheries Service, Monterey, personal communication). Second, northerly winds which push water out of the estuaries also upwell deeper waters. Upwelling not only renews the water but also suspends bottom sediments, exposing small organisms which are normally protected and providing an additional food source to the bottom-feeding shrimp.

High summer temperatures are related to high landings in some regions. The Gulf of Mexico coastline experiences high temperatures throughout the summer. One possible explanation for this relationship is that metabolic rates, and thus growth, are accelerated at higher temperatures, as long as they are not too extreme. Thus, in the warm summer months shrimp may grow to a larger size before migrating offshore to reproduce. This larger size would be reflected in the subsequent weight measured in landings. Another possibility is that offshore movement in white shrimp is triggered by lower temperatures (Lindner and Anderson, 1956); early cooler temperatures would stimulate smaller shrimp to move offshore earlier, resulting in reduced weight of shrimp landings.

High discharge of the Mississippi River in the summer is associated with low fall and winter landings of white shrimp. It is unlikely that this is directly a result of decreased salinities. The Mississippi discharge is usually substantially lower in summer than in spring, and white shrimp are adapted to much lower salinities which occur in the spring. However, if the discharge was sudden and prolonged, it is possible that the associated decrease in salinity could cause mortality among the juveniles, ultimately reducing landings. In addition, the years with very high summer Mississippi discharge usually also had high spring discharge, thus the duration of lower saline waters may have been extensive. Possibly equally or more important is the hypoxia which results from above normal

Results

<u>Hypothesis 1</u> Fall landings of white shrimp show a strong and consistent relationship with atmospheric teleconnection patterns. –SOI winters lead to higher landings than +SOI winters in seven of the eight grids, +PNA winters lead to higher landings in all eight grids, and –NAO winters lead to higher landings in seven of the eight grids.

Using data for all 25 years, stepwise regressions of shrimp landings in which winter index values of SO and PNA replace the winter predictors AT, SST, and WV show that winter PNA is a significant predictor of fall landings in six grids. By taking the place of either AT or SST in the models, the PNA actually improves the fit in half of the grids. The SOI was not selected as a predictor in any of the best models.

<u>Hypothesis 2</u> Winter landings do not show a strong relationship to the preceding winter's atmospheric patterns. Comparison of landings in the two phases of the indices produced agreement with the hypothesis in the case of the SO, but not the PNA. In stepwise regressions, the preceding winter weather is a significant predictor only in three grids and the SOI offers a reasonable substitute for weather measurements in only two of those grids.

<u>Hypothesis 3</u> When winter landings are compared in opposite phases of the current atmospheric patterns, the SOI and NAO show similar relationships, each apparently having opposite impacts in the eastern and western portions of the study area. During –SOI and –NAO winters, catch appears to be enhanced in the eastern and central grids (11–16) but decreased in the western grids (17–19). The PNA groups, however, do not show the same pattern of differences. These results are consistent with hypothesis 3 in the west, and suggest the opposite affect farther east.

In the regressions, the current SST is a significant predictor in five grids, only one of which is successfully modeled using the SOI. It appears that the SST itself is an important variable in the current winter fishing season, which can not be replaced in a simple statistical model by either the SOI or PNA indices.

Discussion

Regional climate in the Gulf of Mexico appears to be somewhat related to three different atmospheric indices of large-scale circulation. The Southern Oscillation affects precipitation in this region of the Gulf and, in its extreme phases, affects temperatures and winds. The Pacific/North American pattern is associated with cold temperatures and strong north winds in this region. The North Atlantic Oscillation appears to be more weakly related to winter weather in this region, only affecting sea surface temperatures.

Examination of the relationship between white shrimp landings and atmospheric teleconnections shows a consistency in results over all eight shrimp grids; however, differences were statistically significant in only a few of the grids. As a general finding, white shrimp landings in the fall are higher following cold and windy winters. These winter conditions are observed during –SOI, +PNA and/or –NAO winters. In half the grids, the PNA index performed as well in regression models as the weather variables, indicating that this pattern is closely tied to the observed fishery landings. Winter landings are not nearly as strongly related to either the previous or current winter's atmospheric indices.

The analysis of hypothesis 3 suggests that a shift in shrimp populations within the Gulf may occur in response to atmospheric teleconnections. A shift of the shrimp landings from the western grids toward the east during –SOI and –NAO winters or from central Gulf grids toward the west during +SOI and +NAO winters may occur in response to changes in SST. Not only are SSTs colder in –SOI, +PNA, and –NAO winters, but also SSTs in the western grids are colder than those in the eastern grids.

Although not discussed here, brown shrimp landings did not appear to be related to atmospheric teleconnections.* The current

---

* A note on brown shrimp. Regression analyses of brown shrimp landings indicated that brown shrimp are affected primarily by river flows and summer east-west wind stress (see Section I). The dominance of river flow in these regression analyses implies that salinity is an important factor in the production of brown shrimp. Precipitation data, which also may affect salinity, regrettably was unavailable for many of the grids. The relationship between fall and winter SOI and winter precipitation indicates that there may be a relationship between fall or winter SOI and summer catch of brown shrimp. However, investigation

Zimmerman, R.J. & Minello, T.J. (1984). Densities of *Penaeus aztecus, Penaeus setiferus*, and other natant macrofauna in a Texas salt marsh. *Estuaries*, **7**(4A), 421–33.

16

Teleconnections and health

NEVILLE NICHOLLS

Bureau of Meteorology Research Centre
Melbourne, Australia

Introduction

The ENSO phenomenon links climate anomalies across the globe. During an ENSO (or "warm") event a specific spatial pattern of climate fluctuations (droughts in Australia, Indonesia, India and parts of Africa, heavy rain and floods on the Pacific coast of South America) tends to occur. During an "anti-ENSO" (sometimes called a "cold event" or "La Niña"), the pattern of climate anomalies is, typically, the reverse of that experienced during an ENSO event (Ropelewski and Halpert, 1989). So, because of ENSO, climate fluctuations in many locations can appear almost simultaneously. These climate fluctuations are said to be "teleconnected."

Climate fluctuations have the potential to cause suffering. Droughts, floods, storm surges and strong winds from tropical cyclones may directly affect human health or lead to famines or epidemics. Climate-related health problems, including famine, epidemics, death and injury from wildfire, flood or storm surge, in areas affected by ENSO may also be "teleconnected." As with the climate anomalies themselves, the climate-related health impacts in various ENSO-affected areas tend to occur simultaneously. For instance, drought-related food shortages can be a potential problem in several countries bordering the Indian Ocean in the same year; that is, the temporal and spatial distributions of drought-related food shortages are not random. Similarly, epidemics of mosquito-borne diseases, and other diseases associated with widespread flooding, may occur almost simultaneously in several other countries. This was the case during the 1982–83 ENSO event, for instance (e.g., Caviedes, 1985; Cedeno, 1986). During "anti-ENSO" events, the reverse can occur; that is, illnesses associated with flooding become a problem in the countries bordering the Indian Ocean.

Areas affected by ENSO are also teleconnected in a second way. ENSO has a tendency to amplify the interannual climate variability of the regions it affects. These areas, then, are more prone to severe droughts or floods, relative to other areas not affected by ENSO. This factor may provide a partial explanation for the spatial distribution of climate-related health problems.

This chapter starts with a brief description of the climate anomalies and related health problems in three major ENSO events, two from the nineteenth century and the 1982–83 event. Some aspects of major "anti-ENSO" periods, in 1973–75 and 1988, are also described. These cases are used to provide an idea of the human health impacts of ENSO (and "anti-ENSO"), and illustrate the tendency for the same spatial pattern of climate-related health problems to recur. They also indicate the complexity in space and time of the relationship between ENSO-related climate anomalies and health. Similar climate anomalies in two countries do not necessarily lead to identical health problems. Similar climate anomalies in the same country at different times may not lead to identical health outcomes. Other factors can determine whether the potential for climate-related health problems such as famine are realized in a particular area or in any specific year. These other factors provide the reason for the complexity of the health impacts of the climate anomalies.

The relationship between ENSO and climate variability is then discussed. Countries affected by ENSO generally suffer more intense droughts and floods than is the case elsewhere. This tendency for more variable climate is even stronger in tropical countries affected by ENSO. The role of this extra variability in the climate–health relationship is considered.

Finally, some of the implications of these health teleconnections of ENSO are considered, including the implications for the way assistance is distributed, as well as for its timing and magnitude, how we regard countries with severe climate–health problems, and even for long-term solutions to these problems.

In Indonesia a severe drought affected the dry-season crop of 1982 and delayed the planting of the 1982–83 crop (Malingreau, 1987). At the national level, the main effect of the drought was to postpone the realization of self-sufficiency objectives until 1984. At the local level, there were reports of isolated outbreaks of cholera related to the effect of drought on water supplies. Three hundred forty deaths from starvation were reported (Canby, 1984). Once again Australia, despite suffering the worst drought of the century (Gibbs, 1984), was not threatened by food shortages or famine. Nevertheless, human health was affected by the drought. Bushfires in the southeast of the country early in 1983, their severity intensified by the drought, claimed 75 lives (Voice and Gauntlett, 1984). The spatial distribution of tropical cyclones was affected in the Pacific. Six of them struck French Polynesia leaving 25,000 homeless in Tahiti alone (Canby, 1984). In southern Africa drought threatened famine and disease (Canby, 1984).

About 88 percent of the Brazilian Northeast was affected by drought. More than 14 million people were affected; roughly 2.8 million received drought assistance (Gasques and Magalhães, 1987). On the Pacific coast of South America, ENSO events are usually accompanied by very severe flooding. Major flood-related health problems arose in Ecuador, Bolivia and Peru (Caviedes, 1985; Telleria, 1986; Cedeno, 1986; Russac, 1986; Gueri et al, 1986). In Ecuador in early 1983, heavy rains associated with ENSO caused floods and landslides, leaving 600 dead. The rainfall and floods were followed by epidemics of malaria, leptospirosis, respiratory disease and gastroenteritis. Russac notes that "for the inhabitants ... [of some parts of Peru] ... it was a time of anguish, terror, hunger, disease and heartache."

The 1973–75 and 1988 anti-ENSOs

There was a strong ENSO event in 1972/73, accompanied by droughts in Indonesia, Ethiopia, India, Australia, and New Guinea. Elsewhere climate anomalies similar to those experienced in the events described above occurred once again. In some areas, however, the anti-ENSO event that followed caused greater problems.

As was noted earlier, anti-ENSOs are often accompanied by heavy rains in those areas where ENSO events bring drought. In

1973 much of Australia was repeatedly flooded. A warm, wet winter provided ideal conditions for mosquitoes and their faunal hosts and the mosquito population increased dramatically. Early in 1974, this large mosquito population resulted in an epidemic of Murray Valley encephalitis, a severe and often fatal viral illness transmitted to humans by mosquitoes (Forbes, 1978). Such epidemics, in southern Australia, usually occur at the end of major anti-ENSOs (Nicholls, 1986).

Similar mosquito-borne diseases caused problems elsewhere during this anti-ENSO. In southern Africa, a wet summer in 1974 produced an epidemic of West Nile fever, characterized by fever, rash, myalgia and arthralgia (McIntosh et al., 1976). In coastal Brazil, an epidemic of human encephalitis caused 61 deaths (Lopes et al., 1978). The Indian monsoon in 1973 was one of the most intense of the past century (Rasmusson and Carpenter, 1983) and led to an epidemic of Japanese encephalitis (Chatterjee and Banerjee, 1975). *Time* magazine (13 October 1988) reported a very widespread Japanese encephalitis epidemic in Uttar Pradesh in northern India during the strong anti-ENSO of 1988. Unofficial reports put the death toll at 5,000, mainly children.

The 1973–75 period was remarkable from the points of view of both climate and health. An extended anti-ENSO event occurred with heavy rains in many areas that are usually semiarid. These areas, such as Australia, India, southern Africa, and coastal Brazil, tend to be drought-stricken during ENSO events and flooded during anti-ENSOs. Insect-borne diseases were prevalent, as noted above, in many of these areas, and this prevalence is related to rainfall. Heavy rainfall and flooding provides ideal conditions for mosquito populations to explode. The increase in mosquito numbers leads to increased incidents of biting of humans, increasing the likelihood of human infection. Thus the anti-ENSO event, with its simultaneous flooding in many countries, links the periods of occurrence of epidemics of insect-borne disease in these countries. The more "direct" health impacts of floods, e.g., drowning, food shortages due to flood damage to crops, and epidemics due to polluted water supplies, are likely to be linked in the same way. Simultaneous flooding in India and the Sudan (the latter the result of heavy rains in Ethiopia) during the 1988 anti-ENSO event led to these problems in both countries.

The linkage between relative rainfall variability, latitude and ENSO provides another form of climate "teleconnection" between certain areas. Low-latitude areas where the climate is affected by ENSO will usually have much more variable rainfall than elsewhere. So droughts and floods may be much more severe in these locations. There is, therefore, a physical basis for expecting potentially more severe climatic impacts in these areas than, say, in the less variable climates of Europe or the east coast of North America. If droughts and floods are more severe in the tropical countries affected by ENSO, then the potential for more severe impacts on health will be greater.

Many of the areas that have suffered well-publicized health impacts from climate fluctuations in recent times (e.g., India, Ethiopia, Sudan, Ecuador) are in the tropics and are affected by ENSO. It must be recognized that these countries are affected by high relative climate variability, due to their low latitude and the effect of ENSO. Their highly visible health problems result, to some extent at least, from relatively frequent extreme meteorological events.

All countries suffer from droughts and floods. It is sometimes suggested that, since all countries have droughts but not all countries have famines, factors other than climate variability are required to explain why droughts in some areas lead to famine. The higher climate variability (i.e, more severe droughts and floods) in countries affected by ENSO relative to, say, the United Kingdom, may provide a partial explanation of why in these countries droughts can lead to severe food shortages whereas in the United Kingdom they do not.

This is not to say that the high relative climate variability in low-latitude countries affected by ENSO will necessarily lead to highly visible health impacts. Socioeconomic–political factors may exacerbate or mitigate the potentially higher climatic hazard. A highly variable climate can provide the frequent severe droughts and floods which set the scene for a potential food shortages and even famine. But other factors must operate to realize this potential. There are countries with high relative climate variability (e.g., Australia), where severe droughts do not lead to famines. In some other areas affected by ENSO (most notably northern China and India), the famine–drought nexus appears to have been broken in the past few decades, at least at the national level. In other

ENSO-affected countries (e.g., Ethiopia) not every drought leads to famine.

Many observers have noted that drought itself cannot be considered the sole cause of famines. Digby (1969) reports an investigation of the Indian famines of the late nineteenth century by the Rev. J.T. Sunderland, published in 1900, famines which occurred during ENSO-related droughts. Sunderland dismissed the failure of the rains as the cause of the famines. He noted that when drought was severe in some parts, other sections had plenty; that irrigation, which was widespread, provided certainty in cropping; and that transport was good across the country, allowing ready conveyance of food from areas of abundance to areas of scarcity. Sunderland decided that the real cause of the famines was "the extreme, the abject, the awful, poverty of the Indian people." This extreme poverty kept most of the population on the very verge of suffering even in years of plenty and prevented them from storing anything to tide them over years of scarcity. The Indian thus "finds starvation invariably staring him in the face if any disorder overtakes that little crop which is the only thing which stands between him and death."

Sunderland then went on to determine the cause of the poverty. He attributed it to the "enormous foreign tribute," which "drains away her wealth in a steady stream that is all the while enriching the English people, and of course correspondingly impoverishing the helpless people of India." Digby agreed that the nineteenth century famines occurred "not because rains fail and moisture is denied; always, even in the worst of years, there is water enough poured from the skies on Indian soil to germinate and ripen the grain, but because India is steadily and rapidly growing poorer."

Glantz (1987) reviewed recent studies on the famine–drought nexus in Africa today and reached a similar conclusion; namely, that drought cannot be regarded as the "primary or sole source of the agrarian crisis existing in most countries on that continent. Drought is most often a contributing factor to other underlying problems plaguing societies that are dependent on agricultural production." Garcia (1981) in a study of droughts and famines in 1972–73 (another major ENSO event) also concluded that his "case-studies provide confirmatory evidences that droughts ... are not the sole or even primary cause of internal disequilibrium in the society. They merely reveal a pre-existing disequilibrium."

The conclusions reached by Garcia and Glantz (and in several recent studies cited by Glantz) differ substantially from the view of famines prevailing about two decades ago (Glantz, 1989). At that time, many observers regarded the drought itself as responsible for food shortages and famines in various parts of Africa. Late last century, notwithstanding the analysis by Sunderland, drought was widely regarded as the primary cause of Indian famines. Such views led to technological "solutions" to famine. Often such technological solutions tend to favor cash crop production for export which can actually worsen local food availability (Glantz, 1989). The conclusion that social, economic and political problems (what Garcia refers to as a state of "disequilibrium") lead to vulnerability to drought implies that removal of the source of this disequilibrium is necessary to prevent climate-related health problems.

The influence of social, economic and political structures in determining whether climate anomalies in a particular region will lead to famine or other health problems can explain why some countries (e.g., Australia), with a climate strongly affected by ENSO, do not suffer food shortages that lead to famines and why in others (e.g., northern China and India), the inexorability with which famine followed drought during the nineteenth and early twentieth centuries has now been broken. The conclusion that non-physical factors may determine whether the potential health problems caused by climate anomalies will be realised in a specific country or at a specific time does not, however, invalidate the fact that increased climate variability heightens the potential for health problems. Determination of the causes of climate-related health problems, such as famines, requires the examination of both physical and non-physical factors in a case-by-case approach.

Summary and implications

During 1982 and 1983 a sharp increase in climate-related health problems emerged in a number of places around the globe. This "clustering" in time of real and potential health hazards related to severe climate anomalies was not an isolated case. A similar clustering appears whenever a major ENSO event occurs. "Anti-ENSOs" are associated with their own separate set of health outcomes.

The ENSO phenomenon causes these realized or potential climate-related health hazards to be teleconnected in two ways. The obvious teleconnection is that similar potentially life-threatening climate anomalies tend to occur simultaneously in several areas (e.g., droughts in northern China, Ethiopia, India, Indonesia, southern Africa, and Australia). Secondly, climate anomalies in the countries affected by ENSO are usually relatively more severe, and thus pose a greater potential threat to health, than is the case elsewhere.

The first form of teleconnection, the simultaneity of ENSO-related climate anomalies such as droughts, means that requests for assistance can be expected from several areas simultaneously. At the same time, some possible providers of assistance may be suffering economic problems from drought themselves and may be less able to provide aid. Further, other countries such as Indonesia, normally self sufficient, may as a result of simultaneous drought need to import food, leading to increases in price internationally. All these "teleconnected" effects of the ENSO event, or of teleconnected floods associated with an anti-ENSO, complicate the effective provision of assistance. The assumption that climate anomalies and their potential health impacts will be randomly distributed in space and time is incorrect and may lead to problems in providing assistance to countries affected by ENSO events.

The second form of teleconnection, through the higher relative climate variability experienced by those areas affected by ENSO, needs to be considered in seeking long-term solutions to climate-related health problems. This applies whether the solutions sought are social, economic, political, or technological. Whatever forms the solution takes, it may not be sufficient to transplant them from areas such as Europe or North America which have less variable climates. The solutions, whatever their form, will need to be more robust in the face of climate variability than is the case in many developed nations.

The teleconnection of real and potential health hazards brought about by the ENSO phenomenon does not lead just to difficulties, however. It also suggests that these increases in health problems are, to some extent at least, predictable. ENSO events, and "anti-ENSOs," have characteristic life-cycles. Once an event is underway this life-cycle can be predicted with some confidence. This means that any climate-related health problem (e.g., food

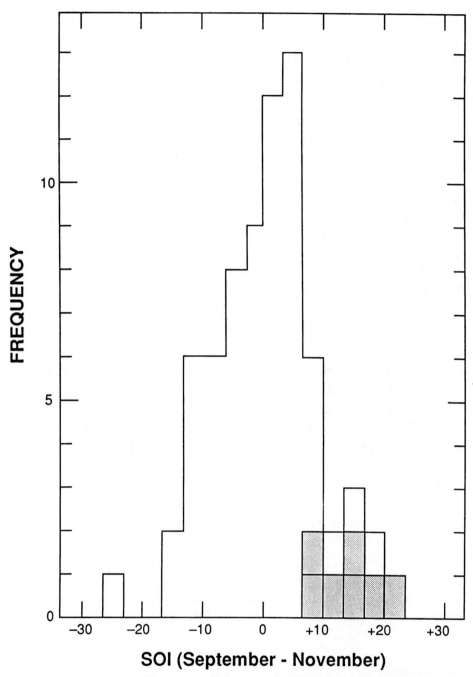

Fig. 16.1 Frequency histogram of the spring (September to November average) values of the SOI from 1916 to 1987. Springs preceding Murray Valley encephalitis (MVE) epidemics in southeastern Australia are stippled. Epidemics occur in the summer after large positive SOI values in spring, i.e., during major anti-ENSO events. The SOI during spring can be used to predict the probability of MVE in the subsequent summer.

shortages, arbovirus outbreaks, increased malnutrition) regularly associated with ENSO events may also be predictable. Nicholls (1988b) suggested that Murray Valley encephalitis (MVE) epidemics in southeast Australia (which occur towards the end of "anti-ENSO" events) can be predicted just by monitoring indices of ENSO. The basis for this suggestion is illustrated in Fig. 16.1, which is a frequency histogram of the Southern Hemisphere spring values of the Southern Oscillation Index (SOI). The SOI is the difference in pressure between Tahiti and Darwin, normalized to a mean of zero and a standard deviation of 10. Values of the SOI in the spring preceding the years with MVE epidemics are stippled. All these epidemics were preceded by very high values of the SOI (i.e., Darwin pressures low, Tahiti pressures high). Such a situation (an anti-ENSO) is accompanied by widespread rainfall and flooding throughout much of eastern Australia. The spring values of the SOI can be used to predict the likelihood of MVE.

Similar approaches might be possible with the other, teleconnected, health problems associated with ENSO. For instance, the Japanese encephalitis epidemics in northern India, such as those of 1973 and 1988, might be predictable through monitoring the SOI. Malaria and other health problems in southern America might be predictable in a similar way; that is, by monitoring (as distinct from predicting) the behavior of the ENSO phenomenon (Nicholls, 1988b). Droughts (with their accompanying potential to lead to food shortages and increased malnutrition) might be predictable in this way also. Of course, prediction of such potential health problems does not guarantee that action can or will be taken to avert them. Glantz (1986) discusses some of the problems involved in using forecasts to avert environmental hazards. Further work is also needed to determine whether the apparent relationships between ENSO and various health hazards, such as Japanese encephalitis in India, are as reliable as that with MVE in southeastern Australia.

References

Canby, T.Y. (1984). El Niño's ill wind. *National Geographic*, **165**, 144–83.
Caviedes, C.N. (1985). Emergency and institutional crisis in Peru during El Niño 1982–1983. *Disasters*, 9, 70–4.
Cedeno, J.E.M. (1986). Rainfall and flooding in the Guayas river basin and its effects on the incidence of malaria 1982-1985. *Disasters*, **10**, 107–11.

Chatterjee, A.K. & Banerjee, K. (1975). Epidemiological studies on the encephalitis epidemic in Bankura. *Indian Journal of Medical Research*, **63**, 1164–79.

Conrad. V. (1941). The variability of precipitation. *Monthly Weather Review*, **69**, 5–11.

Dando, W. (1980). *The Geography of Famine*. London: Edward Arnold.

Digby, W. (1969). *'Prosperous' British India: A Revelation from Official Records*. New Delhi: Sagar Publications (reprint).

Forbes, J.A. 1978. *Murray Valley Encephalitis 1974*. Glebe, Australia: Australian Medical Publishing Co.

Garcia, R.V. 1981. *Drought and Man. Volume 1: Nature Pleads Not Guilty*. Oxford, U.K.: Pergamon.

Gasques, J.G. & Magalhães, A.R. (1987). Climate anomalies and their impacts in Brazil during the 1982–83 ENSO event. In *Climate Crisis*, ed. M. Glantz, R. Katz & M. Krenz, 30–6. New York: United Nations Publications.

Gibbs, W.J. (1984). The great Australian drought. *Disasters*, **8**, 89–104.

Glantz, M.H. (1986). Politics, forecasts, and forecasting: Forecasts are the answer, but what was the question? In *Policy Aspects of Climate Forecasting*, ed. R. Krasnow, 81–96. Washington, DC: Resources for the Future.

Glantz, M.H. (1987). Drought and economic development in sub-Saharan Africa. In *Drought and Hunger in Africa*, ed. M.H. Glantz, 37–58. Cambridge, U.K.: Cambridge University Press.

Glantz, M.H. (1989). Drought, famine and the seasons in sub-Saharan Africa. In *African Food Systems in Crisis*. ed. R. Huss-Ashmore & S.H. Katz, 45–71. New York: Gordon and Breach Science Publishers.

Gueri, M., Gonzalez, C. & Morin, V. (1986). The effect of the floods caused by 'El Niño' on health. *Disasters*, **10**, 118–24.

Kiladis, G.N. & Diaz, H.F. (1986). An analysis of the 1877–78 ENSO episode and comparison with 1982–83. *Monthly Weather Review*, **114**, 1035–47.

Lopes, O.S., Coimba, T.L.M., Sacchetta, L & Calisher, C.H. (1978). Emergence of a new arbovirus disease in Brazil. *American Journal of Epidemiology*, **107**, 444–9.

Magalhães, A.R., Filho, H.C., Garagorry, F.L., Gasques, J.G., Molion, L.C.B., Neto, M., Nobre, C.A., Porto, E.R. & Reboucas, O.E. (1988). Drought as a policy and planning issue in Northeast Brazil. In *The Impact of Climatic Variations on Agriculture. Volume 2: Assessments in Semi-Arid Regions*, ed. M.L. Parry, T.R. Carter & N.T. Konjin, 273–380. Dordrecht, The Netherlands: Kluwer.

Malingreau, J-P. (1987). The 1982–83 drought in Indonesia: Assessment and monitoring. In *Climate Crisis*, ed. M. Glantz, R. Katz & M. Krenz, 11–8. New York: United Nations Publications.

Mallory, W.H. (1926). *China: Land of Famine*, Washington, DC: American Geographical Society, Special Publication No. 6.

McIntosh, B.M., Jupp, P.G., Dos Santos I. & Meenehan, G.M. (1976). Epidemics of West Nile and Sindbis Viruses in South Africa with *Culex (Culex) univittatus Theobold* as vector. *South African Journal of Science*, **72**, 295–300.

Moura, A.D. & Shukla, J. (1981). On the dynamics of droughts in Northeast Brazil: Observations, theory and numerical experiment with a general circulation model. *Journal of Atmospheric Science*, **38**, 2653–75.

Nicholls, N. (1986). A method for predicting Murray Valley Encephalitis in southeast Australia using the Southern Oscillation. *The Australian Journal of Experimental Biology and Medical Science*, **64**, 587–94.

Nicholls, N. (1988a). El Niño – Southern Oscillation and rainfall variability. *Journal of Climate*, **1**, 418–21.

Nicholls, N. (1988b). El Niño – Southern Oscillation impact prediction. *Bulletin of the American Meteorological Society*, **69**, 173–6.

Nicholls, N. (1990). The centennial drought. In *Windows on Australian Meteorology*, ed. E. Webb, Melbourne: Australian Meteorological and Oceanographic Society (in press).

Nicholls, N. & Wong, K. (1990). Dependence of rainfall variability on mean rainfall, latitude, and the Southern Oscillation. *Journal of Climate*, **3**, 162–9.

Pankhurst, R. (1966). The great Ethiopian famine of 1888-1892. *Journal of the History of Medicine and Allied Sciences*, **21**, 95–124 & 271–94.

Porter, A. (1889). *Disasters of the Madras Famine*. Madras, India: Government Press.

Quinn, W.M., Neal, V.T. & Antunez de Mayclo, S.E. (1987). El Niño occurrences over the past four centuries. *Journal of Geophysical Research*, **92**, 14449–61.

Rasmusson, E.M. & Carpenter, T.H. (1983). The relationship between eastern equatorial Pacific sea surface temperatures and rainfall over India and Sri Lanka. *Monthly Weather Review*, **111**, 517–28.

Ropelewski, C.F. & Halpert, M.S. (1989). Precipitation patterns associated with the high index phase of the Southern Oscillation. *Monthly Weather Review*, **117**, 268–84.

Russac, P.A. (1986). Epidemiological surveillance: malaria epidemic following the Niño phenomenon. *Disasters*, **10**, 112–7.

Sinha, S.K. 1987. The 1982–83 drought in India: magnitude and impact. In *Climate Crisis*, ed. M. Glantz, R. Katz & M. Krenz, 37–42. New York: United Nations Publications.

Telleria, A.V. (1986). Health consequences of the floods in Bolivia. *Disasters*, **10**, 88–106.

Voice, M.E. and Gauntlett, F.J. (1984). The 1983 Ash Wednesday fires in Australia. *Monthly Weather Review*, **112**, 584–90.

Walford, C. (1878). The famines of the world: past and present. *Journal of the Statistical Society*, **16**, 433–526.

Wang, S-w. & Mearns, L.O. (1987). The impact of the 1982-83 El Niño event on crop yields in China. In *Climate Crisis*, ed. M. Glantz, R. Katz & M. Krenz, 43–9. New York: United Nations Publications.

17

Teleconnections and their implications for long-range forecasts

NEVILLE NICHOLLS

Bureau of Meteorology Research Centre
Melbourne, Australia

and

RICHARD W. KATZ

National Center for Atmospheric Research*
Boulder, CO 80307

Although seasonal foreshadowing is still very imperfect it has come to stay.

—Sir Gilbert T. Walker
(Walker, 1936, p. 137)

Operational use of teleconnections in forecasting

Teleconnections, especially those related to the Southern Oscillation, provide the scientific basis of operational long-range weather forecast systems run by several national meteorological services. Four systems are described here to demonstrate the potential that the identification of reliable teleconnections offers for providing skillful seasonal forecasts, and the methods that can be used to prepare such forecasts. The first system described is based solely upon the Southern Oscillation and uses values of the Southern Oscillation Index (SOI), the standardized difference in atmospheric pressure between Tahiti and Darwin, as the only predictor. The second example uses the SOI as one of several predictors, while the third example uses a different predictor, the interhemispheric difference in sea surface temperatures. The fourth system described is currently in use on an experimental basis in an attempt to make seasonal forecasts of precipitation in Ethiopia.

* The National Center for Atmospheric Research is sponsored by the National Science Foundation.

Australia

Documented teleconnections, many of which are discussed in earlier chapters, account for much of the variability of climate in many
areas. The Southern Oscillation is the most important of these, at
least for many areas. "The Southern Oscillation is the dominant
pattern of short-term climatic variation over the globe. It accounts for a greater proportion of variance of climatic and oceanic
fields on time scales from a season to ten years than any other single phenomenon, excepting only the annual cycle" (Wright, 1985,
p. 398). The strength of this teleconnection is such that useful
seasonal forecasts for many areas will *not* be possible unless they
involve the Southern Oscillation. This is the situation that applies
to much of Australia. Luckily, the phenomenon is, to some extent,
predictable.

Two features of the Southern Oscillation provide the basis for
prediction: its long life cycle and the "phase-locking" of this cycle
to the annual cycle (Nicholls, 1988). The phase-locking is reflected
in the persistence of the Southern Oscillation being highest between July and February and lowest around April (when the sign
of the SOI often changes).

The strong persistence in the second half of the calendar year
has led many investigators to examine the prospects for using
this persistence to predict climate fluctuations related to the
Southern Oscillation occurring near the end of the calendar year.
Walker (1910) suggested the use of the value of an index of the
Southern Oscillation in October–November to predict Australian
December–April rainfall. Nicholls (1984a) used a similar approach
to predict the date of onset of the north Australian wet season,
which usually commences in December.

The Australian Bureau of Meteorology produces a "Seasonal
Climate Outlook" based on such simple empirical relationships
between the SOI and Australian rainfall. These relationships were
first noticed early this century, by Gilbert Walker and others, and
were confirmed on later, independent, data by Nicholls and Woodcock (1981). Further work (e.g., McBride and Nicholls, 1983)
provided a clearer picture of the seasonal and geographical dependence of these relationships. Predictions are only issued for
seasons and areas where the lag correlation between the SOI and
subsequent rainfall exceeds an arbitrary threshold of 0.4. Linear

regression on the SOI value of the preceding two months is used to predict the rainfall decile range for the ensuing two or three months. In practice, outlooks are prepared monthly for much of eastern and northern Australia, starting in June. The last prediction for the year is made in November.

The simple system used here for prediction could be easily applied to other countries where the Southern Oscillation affects climate in the second half of the calendar year. One such example is Indonesia where the Southern Oscillation can be used to forecast the start of the wet season (Nicholls, 1981). The forecast system could also be changed to provide probabilistic predictions, perhaps with the linear discriminant analysis technique used by the British Meteorological Office in their forecasts of Sahel rainfall (see example 3, below).

U.S.A.

The U.S. National Weather Service's Climate Analysis Center produces forecasts of three temperature and precipitation classes (above, near, and below normal) and of their probabilities at 100 U.S. stations every month for the following 90-day period. A range of tools is used to produce these forecasts.

Their newest forecast tool is the operational multifield analog/anti-analog system described by Livezey and Barnston (1988) and Barnston and Livezey (1989). The use of historical analogs to formulate a forecast assumes that interseasonal changes in the climate system occur similarly from one instance to another, so that when the system is in the same state it was for the same season in a past year, a sequence of events similar to those which occurred in the past instance (the analog) may be expected. Anti-analogs, i.e., past seasons where the climate system anomaly patterns are opposite to those in the current season, can also be used if the following climate anomaly patterns are sign-reversed.

The data sets used to determine the analogs and anti-analogs include Northern Hemisphere 700 mb thickness fields, sea surface temperatures (SSTs) in the Pacific and Caribbean, station temperatures from the U.S.A., and the SOI. Several analogs are chosen and the resulting forecasts use a weighted average of the analogs. The closest analogs are given greater weights. The skill of the forecasts was found to be greater when several analogs and anti-analogs were used, rather than relying on just one "best" analog (Barnston and Livezey, 1989). The skill of the forecast system, overall, is fairly modest but considerable seasonal and area variations occur (Livezey, 1990).

The analog method employed in the U.S.A. could also be adapted for use in other areas where teleconnections affect climate. Such a system can easily provide probabilistic forecasts which should be potentially more useful than categorical predictions. Even areas where the Southern Oscillation is the only teleconnection known to affect climate could benefit from the use of an analog system. As well as allowing probabilistic predictions to be made, such a system may allow information about the temporal changes in the Southern Oscillation to affect the prediction, rather than just relying on, for instance, the value of the SOI for the previous month.

British forecasts of Sahel rainfall

For the past few years the British Meteorological Office has been issuing forecasts of June–October Sahel rainfall (Owen and Ward, 1989). A set of statistical forecasting models based on the technique of linear discriminant analysis are used to forecast the probabilities of five climatologically equiprobable rainfall categories. The predictors are coefficients of SST eigenvectors measured in the spring prior to the summer rainfall season. The downward trend in Sahel rainfall in the past two decades has been accompanied by a warming of the Southern Hemisphere SST relative to the Northern Hemisphere. The main predictor in the British scheme is essentially the difference in hemispheric SSTs. Global general circulation models of the atmosphere with observed SSTs have supported the hypothesis that Sahel rainfall is related to this interhemispheric SST difference. These models are also run with observed SSTs to provide an independent prediction.

The performance of the forecasts in the few years they have been made has been mixed. In 1987 a forecast of very dry conditions was made. Rainfall in that year was the fourth driest this century. The forecast in 1988, which was also for dry conditions, was unsuccessful. Substantial and unanticipated changes in the global

SST anomaly patterns occurred around the middle of that year.

Use of global patterns of SST in prediction, rather than relying on relationships with equatorial Pacific SST related to the Southern Oscillation, may provide better predictions even in areas where the Southern Oscillation is the major determinant of climate fluctuations. The Southern Oscillation strongly affects Australian rainfall, for instance, but recent research (Nicholls, 1989) suggests that inclusion of SSTs from areas not closely associated with this phenomenon, e.g., the south Indian Ocean, could improve the Australian rainfall outlooks.

Ethiopian drought forecasting

The Ethiopian National Meteorological Service Agency (NMSA) has become involved in an attempt to use forecasting analog methods based on ENSO events in order to improve its drought forecasting capabilities. As in several developing countries, resources available to meteorological services are quite limited. Analogs hold out some hope as a relatively inexpensive approach to long-range forecasting. According to the NMSA, ENSO events appear to have significant implications for seasonal precipitation in Ethiopia.

According to Haile (1988, p. 6), "a selection of representative weather charts and analysis of monthly and seasonal rainfall are based on the current state of the ENSO phenomena and the anomalous behaviour of the SST over the Southern Indian and Atlantic Oceans." He goes on to note that "After carefully and analogically assessing the rain-producing components for the selected years in comparison with current ones and by envisaging the rainfall distribution in those chosen seasons, the seasonal weather outlook is disseminated to decisionmakers and other relevant users."

The NMSA forecasters are aware of the problems associated with a reliance on forecasts using ENSO and anti-ENSO analogs. Yet, it is one of the few long-range forecasting tools currently available to them. They have claimed some successes for the past several seasons during which such forecasts were issued. They have also cited examples where the government decision makers used their recommendations to alter agricultural practices on relatively short notice in order to maximize the value of forecasted rains and minimize the impacts of forecasted droughts.

Nowcasting

Although the potential and operational use of ENSO teleconnections for long-range weather/climate forecasting has been stressed so far in this chapter, "nowcasting" based on teleconnections should not be ignored. Nowcasting consists of relying only on the contemporaneous relationships between ENSO and climate variables across the globe. Even without a leading relationship or lacking a way to anticipate the occurrence of an ENSO event, value would potentially accrue by simply possessing the knowledge that certain climate anomalies tend to occur simultaneously (or more anomalies than usual occur) during ENSO events.

This sort of information ought to be employed especially for planning purposes. For instance, international relief agencies could anticipate the competing demands from multiple climate disasters, rather than being "surprised" at their simultaneous occurrence. Moreover, increased vigilance, given the knowledge that an ENSO event is apparently starting, would be helpful for famine early warning systems in regions where drought is teleconnected with ENSO. It would be unfortunate if such opportunities for the utilization of teleconnections were neglected in the rush to incorporate improved long-range weather/climate forecasts into their decisionmaking processes.

Forecast value

Previous case studies

In theory, any real improvements in the skill of forecasts must result in an increase in value to some class of decision makers. This is really just a restatement of the fact that the economic value of any source of information (no matter how imperfect) cannot be negative if the decision maker employs this information in an "optimal" manner and if the cost of purchasing such information is ignored (e.g., Winkler and Murphy, 1985). In addition, applications of decision-analytic techniques to real-world decisionmaking situations have illustrated how current long-range weather/climate forecasts are of much more than negligible economic value to users that are sufficiently sophisticated. For example, Brown et al. (1986) illustrated how seasonal precipitation forecasts in the form currently issued by the U.S. National Weather Service (mentioned in a previous section of this chapter) and with the very limited present skill are still of significant economic value. They considered specific problems faced by wheat producers in the northwestern portion of the U.S. Great Plains of deciding in the spring each year whether to plant a wheat crop or to let the land lie fallow.

On the other hand, some theoretical results concerning the economic value of information from a decision-theoretic point of view indicate that long-range weather/climate forecasts with relatively low skill might well be ignored by many decision makers. In particular, thresholds in skill exist for some simple decisionmaking models below which the forecast information is of no economic value to the decision maker (Katz and Murphy, 1987). Further, the relationship between the skill and economic value of imperfect forecasts can be highly nonlinear (Katz and Murphy, 1990). For instance, some simple decisionmaking models imply a relationship between economic value and skill that is a "convex" curve; meaning relatively large increases in value are associated with improvements in skill for nearly perfect forecasts, whereas relatively small increases in value are associated with improvements in skill for nearly random forecasts. Perhaps these theoretical results suffice to explain the somewhat negative response of potential users when questioned about whether they currently employ long-range weather/climate forecasts (Easterling, 1986).

Optimistic point of view

The potential value to users of seasonal forecasts such as those described in the above section, will depend on the skill of the forecasts. Livezey (1990) examined the skill of the U.S. seasonal forecasts and found that they were only 6 percent better than random "forecasts." Stratifying the forecasts by area and season, however, revealed strong variations in skill. Winter temperature forecasts for the east and southwest of the country, for instance, showed considerable amounts of skill. Livezey concludes that potential users in such areas should be able to "profit considerably from intelligent exploitation of the forecasts."

Overall, however, many teleconnections that might lead to forecast systems involve correlations of only about 0.5. Such relation-

ships account for only one quarter of the variance and this has led
some commentators to suggest that predictions based on teleconnections would not be skillful enough to be useful. The accuracy
of such a relationship can be described more graphically than by
simply quoting the proportion of explained variance by calculating
how frequently a two-category (i.e., above or below the mean) prediction would be correct, for various correlation coefficients. For
a correlation of 0.5 such categorical predictions would be correct
twice as often as they were wrong. Even a correlation of only
0.3 would lead to a correct forecast 50 percent more often than a
wrong forecast. These are quite long odds and seem potentially
useful.

These ratios of correct to incorrect forecasts are a measure of
the accuracy of a forecast system where users take notice of the
forecasts in all cases. At least some users, however, would only use
the forecasts when they predicted a large deviation from the climatological mean. Such users would find that the forecasts they
actually used were correct, when assessed as two-category categorical predictions, more frequently than the above ratios would
suggest. The accuracy and value of forecast systems based on teleconnections may be quite high, especially if only forecasts of large
deviations from the climatological mean are used to affect decision
making, even where the proportion of variance explained by the
predictive relationship seems small.

Using teleconnections to prepare and disseminate probabilistic forecasts, rather than categorical forecasts, may produce even
more useful forecasts. It is well known that probabilistic predictions are potentially more useful than categorical predictions with
the same underlying skill. Madden (1989) calculated, for relationships with different proportions of explained variance (i.e., a correlation of 0.5), the frequency with which one can expect to predict
various two-category probabilities. A relationship that explained
25 percent of the variance would allow predictions of two-category
probabilities of 30/70 percent, or more extreme, on about a third
of occasions. About 15 percent of the time the predicted probabilities would be 20/80 percent or more extreme, that is one category
would be four times more likely than the other one. Such frequent large deviations from the 50/50 percent probabilities that
would be predicted in the absence of any information other than
climatology would seem to be potentially useful to many users.

So, forecasts based on teleconnections that "only explain about 25 percent of the variance" would appear to be potentially valuable. Nicholls (1988) suggested that some of the lag relationships between climate fluctuations and the Southern Oscillation were strong enough to justify producing forecasts of impact variables directly by relating them to the SOI. One example was of the average Australian sorghum yield which is significantly related to values of the SOI observed in the period preceding planting of the crop. So observed values of the SOI before the crop is planted can provide an estimate of the crop yield about six months before harvesting. The correlations explain about 30 percent of the variance so Madden's study indicates that considerable deviations from the average yield could be predicted in this manner. The potential value of such predictions, which are very simple to prepare, to sorghum marketing bodies, to banks considering loans for farm equipment, and even to individual farmers, would appear to be high.

Pessimistic point of view

As just discussed, ample empirical evidence exists that skillful long-range weather/climate forecasts can be made on the basis of teleconnections alone or in conjunction with other long-range forecasting techniques. Moreover, decision-theoretic results establish that such forecasts must be of potential value to decision makers. It remains to address the issue of whether this value will be realized or whether obstacles will be present that act as hindrances to the use of these forecasts.

Recent history indicates that some major obstacles to the general acceptance of long-range weather/climate forecasts do remain. In particular, one difficulty concerns the communication of when an ENSO event is actually occurring. Confusion over the occurrence of an ENSO event stems partly from the lack of adequate real-time observations and partly from the lack of an objective definition for an ENSO event. In the past, some researchers have even denied that an ENSO event was indeed occurring when in its very presence, because their theory on precursors of ENSO events was not exactly satisfied. In recent years, the U.S. Climate Analysis Center has attempted to alleviate this problem by adopting formal terminology akin to that relied on for short-term weather events such as tornados and hurricanes (i.e., "watches") in its monthly publication, the *Climate Diagnostics Bulletin*.

A more serious problem has been the issuance to the public of "unofficial" long-range forecasts of climate anomalies (or of related crop failures) based on teleconnections whose strength or even existence has not been adequately documented. Previous case studies (e.g., a forecast of streamflow in the Yakima Valley, Glantz, 1982) have indicated that imperfect forecasts issued without any provision being made for the possibility of being wrong are likely to result in a backlash against forecasters and their forecasting systems when an erroneous forecast inevitably occurs and has significant economic repercussions. Whenever a forecast is issued, it must be realized that someone is listening and is ready to act on it. If these lessons from the past are not heeded, obstacles to the reliance on long-range forecasts will unfortunately remain.

As a final comment, an earlier study of the value of long-range forecast for the West African Sahel (Glantz, 1977) strongly suggested that even with a hypothetically perfect forecast a year in advance, many societies would be unable to respond effectively because of inadequate infrastructure (e.g., communication and transportation systems). In order to improve the use of long-range forecasts, many developing societies will require economic development assistance – not just improved forecasts.

Climate change

Effect of climate change on teleconnections

If we are to use teleconnections as a basis for seasonal climate prediction, we need to be confident that they are stable, i.e., that they will not change markedly over a period of decades. Most of the relationships comprising the Southern Oscillation have shown themselves to be remarkably stable over the past century. Many of the relationships first identified in the early decades of this century have been confirmed on later, independent data (for an opposing view, see Ramage, 1983). Will they, however, remain stable through the period of rapid climate change expected as a result of the anthropogenic increase of radiatively active gases (the "greenhouse effect")?

Various approaches can be taken to predict the reaction of the Southern Oscillation to the greenhouse effect. None provides satisfactory answers. Examination of the phenomenon's behavior in previous warm epochs results in conflicting predictions and is a flawed approach anyway since these previous warm periods were probably not due to the greenhouse effect, so the changes in atmospheric and oceanic circulation could be rather different from that due to greenhouse warming. No computer model capable of realistically simulating the present-day behavior of the Southern Oscillation has been produced yet, so model predictions of its future behavior are unreliable. Some simple theoretical arguments suggest that changes in the phenomenon might be expected, but such arguments are not completely convincing since they attempt to consider the effects of global warming on just one component of a coupled system. There is also a danger with paleoclimatic arguments and simple theoretical approaches of inferring that changes in average conditions necessarily mean changes in the behavior of the Southern Oscillation, which operates on short time scales unresolvable with typical paleoclimatic data. Prediction of the Southern Oscillation's response to global warming also requires consideration of the possibility that it, as a nonlinear dynamical system, may react dramatically to small changes in governing conditions such as global temperature (rather than just making small adjustments to its behavior). The possibility of feedback between the Southern Oscillation and the greenhouse effect also complicates the prediction of the phenomenon's future behavior.

Thus little can be said, yet, about exactly how teleconnections such as the Southern Oscillation may change as the world warms. The best approach to this problem is probably to routinely monitor the teleconnections to assess their stability in the light of climate changes. Nicholls (1984b) assessed the stability of some predictive relationships involved in the Southern Oscillation and concluded that routine updating of the prediction equations was needed. This was the case even during a period of fairly stable climate, so the case for routine updating during rapid climate changes is even stronger.

Effect of teleconnections on climate change

The existence of teleconnections can complicate the identification of climate trends. Some of the very recent (post-1975) increase in global temperature reflects a fluctuation in the behavior of the Southern Oscillation (Jones, 1988), with several episodes with strong negative values of the SOI (which are usually accompanied by warm tropical oceans leading to an apparent warming). The recent increase in such episodes also accounts for some recent trends in regional climate sometimes attributed to the greenhouse effect (e.g., drought in Australia). These trends might just reflect a random fluctuation in the behavior of the Southern Oscillation. They should not be attributed to the greenhouse effect, for the reasons outlined in the previous section. In other areas these fluctuations in the Southern Oscillation's behavior might be disguising greenhouse-related trends. Routine removal of the influence of teleconnections such as the Southern Oscillation on global and regional climate (that is, formally treating the SOI as a "covariate") would assist in the early detection of trends, perhaps attributable to the greenhouse effect.

Forecast potential

Chapters in this volume reinforce the belief in the value of identifying and understanding teleconnections between ENSO events and climate anomalies worldwide. Such research studies have established that long-range weather and climate forecasts could be potentially improved (e.g., improved in reliability and extended in temporal range) by the incorporation of teleconnections into forecasting systems. Having stated that, it is also be evident that any long-range forecasting scheme will be far from perfect, a conclusion drawn from the empirical basis of these connections as well as from the theoretical limits of predictability inherent in the atmosphere/ocean system. Moreover, certain limitations imposed on the research performed to date should be kept in mind.

Although a rudimentary theoretical basis for teleconnections exists (see, for example, Chapter 9 by Tribbia), the nature and strength of these linkages has primarily been based on simple empirical measures from observational studies, relying on, for example, such statistical measures as cross correlations, regression equations, and contingency tables. Little apparent attention has as yet been given to quantify directly, using either conventional statistical techniques such as prediction intervals or more elaborate forecast verification procedures such as cross validation, the uncertainty that would be present when forecasts are produced. Usually, actual forecasting skill declines relative to that suggested in research studies. Reasons for such deterioration include the selection bias inherent in identifying teleconnections, and the artificial skill introduced when any statistical procedure is fit to a limited sample of observations (as discussed earlier in Chapter 12 by Brown and Katz).

Much of the potential value to society of teleconnections lies in extending temporally the range of weather/climate forecasts. Care must be taken, however, in identifying clearly the exact nature of any apparent leading (as opposed to contemporaneous) relationship between ENSO and a given climate variable. As discussed in Chapter 12, the maximum lead time at which forecasting skill is actually present may be distorted by the serial correlation inherent in climate/oceanic time series.

None of these issues, however, are formidable obstacles to achieving the potential value of reliance on teleconnections in long-range climate and weather forecasting. Nevertheless, their resolution would require additional effort beyond that typical of research studies. As already noted, it is quite common when these issues are addressed for the actual forecasting skill and lead time to be reduced relative to what was apparently indicated by the initial research results. The euphoria over the potential of teleconnections might well produce a backlash when the operational forecasting performance turns out to be less than expected. As Walker (1936, p. 117) recognized:

> *though the prestige of meteorology may be raised for a few years by the issue of seasonal forecasts, the harm done to the science will eventually outweigh the good if the prophecies are found unreliable.*

Would it not be better in the long run to devote the effort to anticipating such disappointments and to adjusting the forecast skill accordingly?

References

Barnston, A.G. & Livezey, R.E. (1989). An operational multifield analog/anti-analog prediction system for United States seasonal temperatures. Part II: spring, summer, fall and intermediate 3-month period experiments. *Journal of Climate*, **2**, 513–41.

Brown, B.G., Katz, R.W. & Murphy, A.H. (1986). On the economic value of seasonal-precipitation forecasts: The fallowing/planting problem. *Bulletin of the American Meteorological Society*, **67**, 833–41.

Easterling, W.E. (1986). Subscribers to the NOAA Monthly and Seasonal Weather Outlook. *Bulletin of the American Meteorological Society*, **67**, 402–10.

Glantz, M.H. (1977). The value of a long-range weather forecast for the West African Sahel. *Bulletin of the American Meteorological Society*, **58**, 150–8.

Glantz, M.H. (1982). Consequences and responsibilities in drought forecasting: The case of Yakima, 1977. *Water Resources Research*, **18**, 3–13.

Haile, T. (1988). Prospects of seasonal weather forecasting as a tool for drought early warning. Abstracts of the Workshop on Drought Early Warning, Addis Ababa, Ethiopia, November 29 – December 2, 1988. Nairobi: UNEP.

Jones, P.D. (1988). The influence of ENSO on global temperatures. *Climate Monitor*, **17**, 80–9.

Katz, R.W. & Murphy A.H. (1987). Quality/value relationship for imperfect information in the umbrella problem. *American Statistician*, **41**, 187–9.

Katz, R.W. & Murphy A.H. (1990). Quality/value relationships for imperfect weather forecasts in a prototype multistage decision-making model. *Journal of Forecasting*, **9**, 75–86.

Livezey, R.E. (1990). Variability of skill of long range forecasts and implications for their use and value. *Bulletin of the American Meteorological Society*, **71**, 300–9.

Livezey, R.E. & Barnston, A.G. (1988). An operational multifield analog prediction system for United States seasonal temperatures. Part I: system design and winter experiments. *Journal of Geophysical Research*, **93**, 10953–74.

Madden, R.A. (1989). On predicting probability distributions of time-averaged meteorological data. *Journal of Climate*, **2**, 922–5.

McBride, J.L. & Nicholls, N. (1983). Seasonal relationships between Australian rainfall and the Southern Oscillation. *Monthly Weather Review*, **111**, 1998–2004.

Nicholls, N. (1981). Air–sea interaction and the possibility of long-range weather prediction in the Indonesian Archipelago. *Monthly Weather Review*, **109**, 2435–43.

Nicholls, N. (1984a). A system for predicting the onset of the north Australian monsoon. *Journal of Climatology*, **4**, 425–35.

Nicholls, N. (1984b). The stability of empirical long-range forecast techniques: A case study. *Journal of Climate and Applied Meteorology*, **23**, 143–7.

Nicholls, N. (1988). El Niño – Southern Oscillation impact prediction. *Bulletin of the American Meteorological Society*, **69**, 173–6.

Nicholls, N. (1989). Sea surface temperatures and Australian winter rainfall. *Journal of Climate*, **2**, 965–73.

Nicholls, N. & Woodcock, F. (1981). Verification of an empirical long-range weather forecasting technique. *Quarterly Journal of the Royal Meteorological Society*, **107**, 973–6.

Owen, J.A. & Ward, M.N. (1989). Forecasting Sahel rainfall. *Weather*, **44**, 57–64.

Ramage, C. (1983). Teleconnections and the siege of time. *Journal of Climatology*, **3**, 223–31.

Walker, G.T. (1910). Correlation in seasonal variations in weather II. *Memoirs of the India Meteorological Department*, **XXI**, 45 pp.

Walker, G.T. (1936). Seasonal weather and its prediction. *Smithsonian Institution Annual Report*, **1935**, 117–38.

Winkler, R.L. & Murphy, A.H. (1985). Decision analysis. In *Probability, Statistics, and Decision Making in the Atmospheric Sciences*, ed. A.H. Murphy & R.W. Katz, 493–524. Boulder, CO: Westview Press.

Wright, P.B. (1985). The Southern Oscillation: An ocean-atmosphere feedback system? *Bulletin of the American Meteorological Society*, **66**, 398–412.

INDEX

9 780521 106849